Communications
in Computer and Information Science   1980

AF148081

Rationale

The CCIS series is devoted to the publication of proceedings of computer science conferences. Its aim is to efficiently disseminate original research results in informatics in printed and electronic form. While the focus is on publication of peer-reviewed full papers presenting mature work, inclusion of reviewed short papers reporting on work in progress is welcome, too. Besides globally relevant meetings with internationally representative program committees guaranteeing a strict peer-reviewing and paper selection process, conferences run by societies or of high regional or national relevance are also considered for publication.

Topics

The topical scope of CCIS spans the entire spectrum of informatics ranging from foundational topics in the theory of computing to information and communications science and technology and a broad variety of interdisciplinary application fields.

Information for Volume Editors and Authors

Publication in CCIS is free of charge. No royalties are paid, however, we offer registered conference participants temporary free access to the online version of the conference proceedings on SpringerLink (http://link.springer.com) by means of an http referrer from the conference website and/or a number of complimentary printed copies, as specified in the official acceptance email of the event.

CCIS proceedings can be published in time for distribution at conferences or as postproceedings, and delivered in the form of printed books and/or electronically as USBs and/or e-content licenses for accessing proceedings at SpringerLink. Furthermore, CCIS proceedings are included in the CCIS electronic book series hosted in the SpringerLink digital library at http://link.springer.com/bookseries/7899. Conferences publishing in CCIS are allowed to use Online Conference Service (OCS) for managing the whole proceedings lifecycle (from submission and reviewing to preparing for publication) free of charge.

Publication process

The language of publication is exclusively English. Authors publishing in CCIS have to sign the Springer CCIS copyright transfer form, however, they are free to use their material published in CCIS for substantially changed, more elaborate subsequent publications elsewhere. For the preparation of the camera-ready papers/files, authors have to strictly adhere to the Springer CCIS Authors' Instructions and are strongly encouraged to use the CCIS LTeX style files or templates.

Abstracting/Indexing

CCIS is abstracted/indexed in DBLP, Google Scholar, EI-Compendex, Mathematical Reviews, SCImago, Scopus. CCIS volumes are also submitted for the inclusion in ISI Proceedings.

How to start

To start the evaluation of your proposal for inclusion in the CCIS series, please send an e-mail to ccis@springer.com.

Grigoris Antoniou · Vadim Ermolayev ·
Vitaliy Kobets · Vira Liubchenko ·
Heinrich C. Mayr · Aleksander Spivakovsky ·
Vitaliy Yakovyna · Grygoriy Zholtkevych
Editors

Information and Communication Technologies in Education, Research, and Industrial Applications

18th International Conference, ICTERI 2023
Ivano-Frankivsk, Ukraine, September 18–22, 2023
Proceedings

 Springer

Editors
Grigoris Antoniou (iD)
University of Huddersfield
Huddersfield, UK

Vadim Ermolayev (iD)
Ukrainian Catholic University
Lviv, Ukraine

Vitaliy Kobets (iD)
Kherson State University
Kherson, Ukraine

Vira Liubchenko (iD)
Odessa National Polytechnic University
Odesa, Ukraine

Heinrich C. Mayr (iD)
University of Klagenfurt
Klagenfurt, Austria

Aleksander Spivakovsky (iD)
Kherson State University
Kherson, Ukraine

Vitaliy Yakovyna (iD)
University of Warmia and Mazury
in Olsztyn
Olsztyn, Poland

Grygoriy Zholtkevych (iD)
V. N. Karazin Kharkiv National University
Kharkiv, Ukraine

ISSN 1865-0929   ISSN 1865-0937   (electronic)
Communications in Computer and Information Science
ISBN 978-3-031-48324-0   ISBN 978-3-031-48325-7   (eBook)
https://doi.org/10.1007/978-3-031-48325-7

This Springer imprint is published by the registered company Springer Nature Switzerland AG
The registered company address is: Gewerbestrasse 11, 6330 Cham, Switzerland

Paper in this product is recyclable.

Preface

This volume contains the contributions to ICTERI 2023, the 18th International Conference on Information and Communication Technologies (ICT) in Education, Research, and Industrial Applications. The conference was held in Ivano-Frankivs'k, Ukraine, during September 18–22, 2023, with a focus on research advances in ICT, business or academic applications of ICT, and design and deployment of ICT infrastructures. To our regret, the current security situation in Ukraine did not allow us to organize the conference as a face-to-face gathering. Therefore, the conference proceeded in a hybrid mode.

Topically, the Main Conference of ICTERI 2023 collected the contributions to: (i) ICT research advances, (ii) information systems technologies and applications, (iii) academic and industry cooperation in ICT, and (iv) ICT in education. As a complement to the Main Conference tracks, ICTERI 2023 continued the tradition of hosting co-located events this year by offering a PhD Symposium and two workshops: the 6th International Workshop on Methods, Resources and Technologies for Open Learning and Research (MROL 2023) and 2nd Workshop on Digital Transformation of Education (DigiTransfEd 2023). The PhD Symposium provided the opportunity for PhD candidates to present, listen to, and discuss research on topics relevant to the scope of the ICTERI conference series. The workshops addressed the following:

- MROL 2023. This workshop focused on new and emerging technologies application in open systems of higher education. It explored the use of open education and research tools, resources, and methods for forming a creative and ICT competent person in the view of European Research Area development.
- DigiTransfEd 2023. This workshop focused on the theory and practice of Digital Transformation of Education. It covered digital transformation of educational processes of all levels, branches, and directions of training

Overall, ICTERI 2023 attracted 90 paper submissions authored by colleagues from eleven different countries. These submissions included: 53 for the Main Conference; 26 for the PhD Symposium; and 11 poster papers. Each submitted paper was reviewed by at least three PC members in a single-blind peer review process. Out of these submissions, we accepted: 13 papers for the Main Conference; 8 papers for the PhD Symposium; and 5 poster papers. The acceptance rate across all the tracks was 28.9 percent. The workshops were run as autonomous co-located events and published their proceedings separately.

The structure of this proceedings volume reflects the conference program. It contains three parts: (i) Main Conference papers; (ii) PhD Symposium papers; and (ii) Poster papers.

The Main Conference was, traditionally, the backbone track of ICTERI 2023. The aim of the track was to offer a forum for presenting and discussing quality contributions focused on research, design, development, deployment, and usage of advanced information systems and ICT infrastructures in industry and education. When forming the program, particular attention was paid to make sure that the presented ideas and solutions can be or have already been put into practice. This could have been demonstrated by a proof-of-concept implementation, a comprehensive prototype, a comprehensive case study, or the analysis of real use cases. Reports on academic-industrial partnerships for ICT innovation and knowledge transfer were as welcome as technological and methodological contributions. The rationale behind holding our PhD Symposium under the umbrella of ICTERI 2023 was to offer an expert environment for the presentation of tractable ideas and early results of PhD projects or other research aiming at receiving a PhD. The poster part of the proceedings is composed of the best selected papers describing poster presentations given at the conference.

This volume would not have been possible without the support of many people. First, we are very grateful to all the authors for their commitment and intensive work. Second, we would like to thank the members of our Program Committee and additional reviewers for providing timely and thorough assessments. Furthermore, we would like to thank all the people who contributed to the organization of ICTERI 2023. Without their efforts, there would have been no substance for this volume.

September 2023

Grigoris Antoniou
Vadim Ermolayev
Vitaliy Kobets
Vira Liubchenko
Heinrich C. Mayr
Aleksander Spivakovsky
Vitaliy Yakovyna
Grygoriy Zholtkevych

Organization

General Chair

Aleksander Spivakovsky   Kherson State University, Ukraine

Steering Committee

Vadim Ermolayev   Ukrainian Catholic University, Ukraine
Heinrich C. Mayr   Universität Klagenfurt, Austria
Mykola Nikitchenko   Taras Shevchenko National University of Kyiv,
     Ukraine
Aleksander Spivakovsky   Kherson State University, Ukraine
Mikhail Zavileysky   DataArt, USA
Grygoriy Zholtkevych   V.N. Karazin Kharkiv National University, Ukraine

Program Chairs

Grigoris Antoniou   University of Huddersfield, UK
Vadim Ermolayev   Ukrainian Catholic University, Ukraine

Proceedings Chair

Vitaliy Yakovyna   Lviv Polytechnic National University, Ukraine,
     University of Warmia and Mazury in Olsztyn,
     Poland

Presentations Chair

Heinrich C. Mayr   Universität Klagenfurt, Austria

Posters Chair

Grygoriy Zholtkevych   V.N. Karazin Kharkiv National University, Ukraine

PhD Symposium Chairs

Vitaliy Kobets   Kherson State University, Ukraine
Vira Liubchenko   Odesa Polytechnic National University, Ukraine,
     Hamburg University of Applied Sciences, Germany

Local Organization Chair

Valentyna Yakubiv   Vasyl Stefanyk Precarpathian National University,
                             Ukraine

Publicity Chair

Maksim Vinnik   Kherson State University, Ukraine

Web Chair

Alexander Vasileyko   Zaporizhzhia National University, Ukraine

Program Committees

Main Conference, Including Poster Track

Andrii Babii	Kharkiv National University of Radio Electronics, Ukraine
George Baryannis	University of Huddersfield, UK
Sotiris Batsakis	University of Huddersfield, UK, Technical University of Crete, Greece
Antonis Bikakis	University College London, UK
Jon Hael Brenas	Welcome Sanger Institute, UK
Tianhua Chen	University of Huddersfield, UK
Olga Cherednichenko	National Technical University "KhPI", Ukraine
Maxim Davidovsky	Zaporizhzhia Regional Institute of Postgraduate Pedagogical Education, Ukraine
Oleksandr Deineha	V.N. Karazin Kharkiv National University, Ukraine
Hennadii Dobrovolskyi	Zaporizhzhia National University, Ukraine
Volodymyr Donets	V.N. Karazin Kharkiv National University, Ukraine
David Esteban	Techforce, Spain
Rustam Gamzayev	V.N. Karazin Kharkiv National University, Ukraine
Brian Hainey	Glasgow Caledonian University, UK
Sungkook Han	Wonkwang University, South Korea
Andrii Hlybovets	National University "Kyiv-Mohyla Academy", Ukraine
Natalya Keberle	Ontotext AD, Bulgaria
Vyacheslav Kharchenko	National Aerospace University "KhAI", Ukraine
Vitaliy Kobets	Kherson State University, Ukraine
Christian Kop	Universität Klagenfurt, Austria
Artur Korniłowicz	University of Bialystok, Poland
Victoria Kosa	Zaporizhzhia National University, Ukraine
Kalliopi Kravari	Aristotle University of Thessaloniki, Greece
Hennadiy Kravtsov	Kherson State University, Ukraine
Vladimir Kukharenko	Kharkiv National Automobile and Highway University, Ukraine

Vira Liubchenko	Odesa Polytechnic National University, Ukraine, Hamburg University of Applied Sciences, Germany
Ievgen Meniailov	National Aerospace University "Kharkiv Aviation Institute", Ukraine
Oleksii Molchanovskyi	Ukrainian Catholic University, Ukraine
Adam Naumowicz	University of Bialystok, Poland
Mykola Nikitchenko	Taras Shevchenko National University of Kyiv, Ukraine
Yuliia Nosenko	Institute for Digitalisation of Education of NAES of Ukraine, Ukraine
Artem Panchenko	V.N. Karazin Kharkiv National University, Ukraine
Andreas Pester	British University in Egypt, Egypt
Dimitris Plexousakis	Institute of Computer Science, FORTH, Greece
Kirll Rukkas	V.N. Karazin Kharkiv National University, Ukraine
Wolfgang Schreiner	Johannes Kepler University Linz, Austria
Dmitry Shabanov	V.N. Karazin Kharkiv National University, Ukraine
Sergey Shmatkov	V.N. Karazin Kharkiv National University, Ukraine
Claudia Steinberger	Universität Klagenfurt, Austria
Martin Strecker	Université de Toulouse, France
Ilias Tachmazidis	University of Huddersfield, UK
Mykola Tkachuk	V.N. Karazin Kharkiv National University, Ukraine
Juri Vain	Tallinn University of Technology, Estonia
Paul Warren	Open University, UK
Borut Werber	University of Maribor, Slovenia
Andriy Yerokhin	Kharkiv National University of Radio Electronics, Ukraine
Iryna Zaretska	V.N. Karazin Kharkiv National University, Ukraine
Chrysostomos Zeginis	Institute of Computer Science, FORTH, Greece

PhD Symposium

George Baryannis	University of Huddersfield, UK
Justyna Dobroszek	University of Lodz, Poland
Scott Erickson	Ithaca College, USA
Tommaso Federici	University of Tuscia, Italy
Magdalena Graczyk-Kucharska	Poznan University of Technology, Poland
Alexander Hošovský	Technical University of Kosice, Slovakia
Touseef Hussain	Sukkur IBA University, Pakistan
Kęstutis Kapočius	Kaunas University of Technology, Lithuania
Ganna Kharlamova	Taras Shevchenko National University of Kyiv, Ukraine
Jurij Klapkiv	Ternopil National Economic University, Ukraine
Vitaliy Kobets	Kherson State University, Ukraine
Joanna Krasodomska	Cracow University of Economics, Poland

Kamil Krot	Wrocław University of Science and Technology, Poland
Kristina Kundelienė	Kaunas University of Technology, Lithuania
Olena Liashenko	Taras Shevchenko National University of Kyiv, Ukraine
Vira Liubchenko	Hamburg University of Applied Sciences, Germany
Kostas Magoutis	University of Crete, Greece
Frederic Mallet	Université Côte d'Azur, France
Andriy Matviychuk	Kyiv National Economic University named after Vadym Hetman, Ukraine
Jacek Maślankowski	University of Gdańsk, Poland
Valentyna Pleskach	Taras Shevchenko National University of Kyiv, Ukraine
Dimitris Plexousakis	Institute of Computer Science, FORTH, Greece
Boris Popesko	Tomas Bata University in Zlín, Czechia
Grigor Stambolov	Technical University of Sofia, Bulgaria
Ilias Tachmazidis	University of Huddersfield, UK
Dmytro Terletskyi	V. M. Glushkov Institute of Cybernetics of NAS of Ukraine, Ukraine
Mykola Tkachuk	V.N. Karazin Kharkiv National University, Ukraine
Borut Werber	University of Maribor, Slovenia

Additional Reviewers

| Oleksii Tkachenko | Taras Shevchenko National University of Kyiv, Ukraine |
| Ludmila Omelchuk | Taras Shevchenko National University of Kyiv, Ukraine |

ICTERI 2023 Sponsors

 **Kherson State University** (**KSU**, http://www.ksu.ks.ua/) is a versatile **scientific, educational, and cultural centre** in the South of Ukraine. It offers full day, extra-mural, external study programs attended by 8,000 students from 23 regions of Ukraine and CIS countries. Its research and teaching staff comprises 18 members of the Ukrainian and foreign Academies of Science, 68 professors, 234 senior lecturers. About 1,000 students are enrolled to its advanced degree programs at Ph.D and Dr.Sci levels.

 **Oleksandr Spivakovsky's Educational Foundation** (**OSEF**, http://spivakovsky.fund/) aims to support gifted young people, outstanding educators, and those who wish to start up their own business. **OSEF** activity is focused on the support and further development of educational, scientific, cultural, social, and intellectual spheres in the Kherson Region of Ukraine.

 **BWT Group** (http://www.groupbwt.com/) are those who do coding all day and night long. For you. In fact, we can do anything. The work we do, and how we do it, is defined in large part by the culture we have developed. We seek to build an organization that every one of us enjoys being a part of, where people feel valued, inspired to grow, and are willing to do what is necessary to take care of each other and our client.

DataArt (http://dataart.com/) develops industry-defining applications, helping clients optimize time-to-market and minimize software development risks in mission-critical systems. Domain knowledge, offshore cost advantages, and efficiency – that is what makes DataArt a partner of choice for their global clients.

ICTERI 2023 Organizers

 Ministry of Education and Science of Ukraine
http://www.mon.gov.ua/

 Vasyl Stefanyk Precarpathian National University, Ukraine
https://pnu.edu.ua/en/

 Kherson State University, Ukraine
http://www.kspu.edu/

 University of Huddersfield, UK
https://www.hud.ac.uk/

 Ukrainian Catholic University, Ukraine
https://ucu.edu.ua/en/

 V.N. Karazin Kharkiv National University, Ukraine
http://www.univer.kharkov.ua/en

 Taras Shevchenko National University of Kyiv, Ukraine
http://www.univ.kiev.ua/en/

 Universität Klagenfurt, Austria
http://www.uni-klu.ac.at/

 Lviv Polytechnic National University, Ukraine
http://www.lp.edu.ua/en

 University of Warmia and Mazury in Olsztyn, Poland
http://www.uwm.edu.pl/en

 Odesa Polytechnic National University, Ukraine
https://op.edu.ua/en

 Ukrainian Scientific and Educational IT-Society (UNIT), Ukraine
https://usit.eu.org/

 DataArt Solutions Inc.
http://dataart.com/

Automated Machine Learning for Knowledge Discovery (Invited Talk Abstract)

Ioannis Tsamardinos[1,2] (iD)

[1] Computer Science Department, University of Crete, Greece
[2] ADBio, Greece
tsamard@jadbio.com

Abstract. Automated Machine Learning, or AutoML, is a newly emerging field in Machine Learning. It promises to automate predictive modelling, democratize machine learning to non-experts, boost the productivity of experts, ensure the statistical validity of the modelling process, and even surpass human experts in quality. AutoML should not only strive to produce a high-quality model, but all information, explanations, interpretations, and decision support a human expert would. In this talk, we presented the challenges of AutoML and the design choices we made to construct the Just Add Data Bio, or JADBio for short, AutoML platform. We also presented industrial and health applications of JADBio.

Keywords: AuroML · JADBio · Predictive modelling · Healthcare

Contents

Ph.D. Symposium

Poster Papers

Invited Paper

The Power of Good Old-Fashioned AI for Urban Traffic Control

Mauro Vallati$^{(\boxtimes)}$ iD

University of Huddersfield, Huddersfield HD13DH, UK
m.vallati@hud.ac.uk

Abstract. The current worldwide increasing trend in urbanisation is aggravating urban traffic congestion's social, economic, and health burdens. The introduction of new means of transport, such as Connected Autonomous Vehicles, and the rise of Artificial Intelligence, is enabling a paradigm shift in urban traffic management and control from existing reactive to proactive traffic control: proactive control paradigms can preemptively address issues, mitigating the negative impact on mobility.

In this paper we provide an overview of the work done in the area by the Huddersfield AI for Urban Traffic Management and Control research team.

Keywords: Urban Traffic Control · Model-Based AI · AI Planning

1 Introduction

According to a recent report from the United Nations, the proportion of world's population living in cities is expected to increase from the current 55% to 68% by 2045. This growth in urbanisation is going to exacerbate existing problems of traffic congestion in urban areas, that are already taking a significant toll in terms of economical and health costs. In the UK alone, the cost of congestion has reached nearly £8 billion in 2021 in lost time and fuel consumption, and has become a major health threat that goes beyond the cardiac and respiratory systems [2,3].

Current deployed real-time urban traffic control methods are predominantly reactive, exemplified by approaches like SCOOT [28], prevalent in Europe, which rely on basic traffic models for connected traffic light clusters. Although reactive to congestion, these methods lack effectiveness due to the use of pre-computed knowledge and simplified models. Notably, they are able to optimise the traffic conditions of the small controlled cluster of junctions, but have no information about the rest of the network; overall network conditions can therefore be detrimentally affected. Additionally, they are bound by predefined conditions for optimising fixed metrics, lacking adaptability to different conditions or requirements.

The rapid evolution of Artificial Intelligence (AI) has spurred novel AI-based strategies for diverse aspects of traffic management. AI approaches are shifting

G. Antoniou et al. (Eds.): ICTERI 2023, CCIS 1980, pp. 3–10, 2023.
https://doi.org/10.1007/978-3-031-48325-7_1

traffic control paradigms from reactive to proactive, pre-emptively addressing potential traffic issues. Consequently, traffic authorities are increasingly embracing AI techniques [1].

This paper offers an overview of the AI-based approaches researched by the Huddersfield AI for Urban Traffic Management and Control (AI4UTMC) group,[1] focusing on Good Old-Fashioned AI [19], or model-based AI. While the merits of model-based AI, such as interpretability and explainability, are well-established, they are not the focus here but are detailed in [31].

The remainder of this paper is organised as follows. First, we provide a short introduction on Automated Planning, that is the AI approach leveraged on in this paper. Second, we show how this class of AI approaches can be used to perform traffic signal control, and then we discuss how it can be used to support traffic routing in the presence of Connected Autonomous Vehicles (CAVs). Finally, we give conclusions.

2   Automated Planning

In this section, we provide a concise introduction to automated planning, with particular attention given to PDDL+.

Automated planning constitutes a foundational discipline of Artificial Intelligence, and it is concerned with the task of generating plans (sequences of actions) based on descriptions of a system to be controlled and its dynamics, initial states, and specified goals to be achieved [17,18].

Within domain-independent planning, a knowledge model is devised to provide the essential information required for an automated reasoner to resolve the corresponding task. This knowledge model encompasses predictive details about actions and specifically outlines how these actions alter the environment's state upon application [21]. Various formalisms exist to articulate this knowledge, with the widely recognised Planning Domain Description Language (PDDL) being prominent. Here, we will explore an advanced version of PDDL that permits the representation of both discrete and continuous state variables, commonly referred to as numeric planning. Building on this foundation, we delve into its extension that explicitly addresses time via continuous processes and events-referred to as the PDDL+ language [12].

Informally, a PDDL+ planning problem encompasses defining the system's behaviour through actions, processes, and events, along with a description of the initial system state and an end goal to be achieved. Actions represent agents' choices, and are under the control of the planning reasoner, while processes and events detail system evolution and can not be directly controlled. The initial state reflects the starting conditions, and the goal outlines the desired outcome for problem resolution. In essence, a PDDL+ planning problem involves discovering a timed sequence of actions (a plan) that guides the system state, adhering to specified processes and events, to fulfil a designated goal requirement.

---

[1] https://www.ai4utmc.info/.

From a formal perspective, a PDDL+ problem over variables V is the tuple (V, A, P, E, I, G) where: V is a set of variables divided into Boolean (F) and numeric variables (X), A is a set of actions, where each action is a pair $(pre(a), eff(a))$ where $pre(a)$ is a logical formula called the precondition of a, and $eff(a)$ is a set of assignments for some subset of V, called the effect of a; I is an initial state (an initial complete assignments for all variables in V); G is a logical formula over V, and P and E are sets of processes and events, respectively.

PDDL+ problems build on the notion of a state, which is an assignment to all variables in X. I is one such states. As a goal, preconditions are formulae over variables from X; We say that a state satisfies a formula if such a state is a model of the formula, following standard conventions from logic. Effects, in the case of actions and events, are assignments to some variable from X. In the case of processes, effects are used to represent the time derivative of some numeric variables in X. Intuitively, processes describe how the world evolve as time goes by.

Solving a PDDL+ problem corresponds to the task of finding a set of *timed* actions, i.e., pairs (t, a) where $t \in R_0^+$ and $a \in A$, such that the all actions are applicable at the corresponding time. The states in which the actions are applied are the result of a combination of effects of continuously changing variables (given by the active processes through time) and discrete changes happening for effects of actions that get applied or events that get triggered.

More details about the syntax and semantics of PDDL+ can be found in [12]. Note also that PDDL+ problems can be described compactly by means of a *lifted* representation. Lifted representations allow the representation of actions, processes and events with free typed variables, which need to be instantiated with objects that are specific for a particular instance of the problem. The lifted representation supports the knowledge engineering process of automated planning applications, by providing a concise and easy to maintain representation of the knowledge needed by the automated reasoner [21]. The introduction of parametric actions and variables does not change the semantics of the problem in that all lifted instances can be transformed into instances with no parameters through a process called *grounding*; yet lifted representations make the models much lighter and easier to inspect. For more details about automatic grounding, the interested reader is referred to [25].

3   AI for Traffic Signal Control

Traffic signal control is the problem of defining the duration of the sequence of stages for each controlled junction, taking into account existing constraints on the minimum and maximum length of each stage, as well as on the minimum and the maximum length of the complete traffic signal cycle. Cycles also include intergreens, which are non-modifiable periods of time between two subsequent stages. Each traffic signal stage is also defined in terms of traffic movements that it enables.

The problem of traffic signal control was tackled by means of PDDL+ automated planning by Vallati et al. in 2016 [4,32], and subsequently re-engineered by McCluskey and Vallati in 2017 [20] and by Percassi et al. in 2023 [23]. In a nutshell, the idea is that the controlled region is modelled in the PDDL+ language in terms of junctions, traffic light phases and cycles, expected demand and expected traffic intentions; the current traffic conditions are then described in terms of snapshot occupancy (i.e., the number of vehicles in each link). A goal is then set for an automated planning engine to find a solution by means of optimising traffic lights. Goals can be defined in terms of number of vehicles expected to navigate through a corridor, number of vehicles expected to enter (exit) the region as soon as possible, or decongesting one or more links of the network. Experiments performed using both synthetic and real-world data demonstrated that the approach can lead to a significant reduction of traffic congestion as well as pollution in urban areas. Figure 1 gives an intuition of the symbolic representation of a problem in the considered formalism, and provides relevant information for a junction J1 in terms of cycle time, sequence of traffic signal phases, and min and max green time for each phase. Details about the overall structure of the network are omitted for the sake of space, but the goal is specified in terms of vehicles that need to leave a considered link of the region, and an excerpt of a generated solution is shown.

To support the automated planning engine in the task of quickly finding a good quality traffic signal setting, a number of heuristics have been introduced [13,23], together with ad-hoc techniques for grounding and reformulation of the complex models [14,25]. Due to the complex nature of the models, the design of tools of supporting validation and verification of the encoded symbolic knowledge has also been investigated [24]. Further, an overall framework supporting the real-time acquisition of knowledge from deployed sensors has been designed, to foster the on-line use of the proposed techniques [6].

Finally, a major benefit of using model-based AI lies in the fact that the model can play the role of a simulator. The approach can then rely on verifiable accurate knowledge models, encoding how traffic moves in the controlled region, to simulate the likely traffic evolution over a considered period of time. Automated Planning lends itself well thanks to its concise and declarative symbolic representation of the dynamics to model, achieved via the standard PDDL+ language. In a recent paper, we introduced the framework that can support the use of the AI-based models as simulators, and show that the accuracy of simulated scenarios makes it a viable option for traffic authorities [5].

4   AI for Traffic Routing

The advent of Connected Autonomous Vehicles (CAVs) paves the way to the use of AI for supporting in innovative ways urban traffic controllers [29]. Vehicle to Infrastructure (V2I) communication allows traffic controllers to collect real-time traffic information from individual CAVs navigating the area, hence maintaining a real-time accurate overview of traffic conditions, and to directly influence the

```
(:init
 (active j1_stage1)
 (= (cycletime j1) 90)
 (= (maxcycletime j1) 120)
 (= (mincycletime j1) 50)
 (= (defaultgreentime j1_stage1) 45)
 (= (defaultgreentime j1_stage2) 14)
 (contains j1 j1_stage1)
 (contains j1 j1_stage2)
 (contains j1 j1_stage3)
 (contains j1 j1_stage4)
 (= (mingreentime j1_stage1) 10)
 (= (maxgreentime j1_stage1) 120)
 (= (mingreentime j1_stage2) 15)
 (= (maxgreentime j1_stage2) 120)
 ...
(:goal
 (> (pculeft link3) 500)
 ...
*** Generated Strategy
15.0: (Switch j1_stage4 j1_stage1)
...
60.0: (Switch j1_stage1 J1_stage2)
976.0: END
```

Fig. 1. Excerpt of a traffic signal optimisation problem, presenting some elements of a single junction with four stages. The block '':init'' defines the initial state; '':goal'' defines the goal conditions that must be achieved. The lower part of the Figure presents a snippet of a potential solution strategy.

way in which traffic moves by providing instructions to vehicles in the controlled region. Here we focus on urban traffic routing, that aims at reducing congestion by distributing vehicles in the controlled network to maximise the exploitation of the available infrastructure.

The intuition of the approach is that, given a urban network and the traffic demand expressed as approaching vehicles and corresponding destinations, the automated planning system is tasked with the identification of routes for vehicles that allows to minimise the average journey time for all the vehicles in the network. The first work on this research thread was by Chrpa et al. [9], that investigated the optimisation of both journey times and air quality. Subsequent works looked more in depth into the optimisation of journey times by considering different planning paradigms and different urban networks [11,30,33]. An excerpt of the problem is provided in Fig. 2, where the current conditions of the network are described as well as the position of the vehicles to be routed, and a desidered goal position for vehicles.

While preliminary results demonstrate the potential of the approach, it emerges that it faces issues in terms of scalability – particularly when large num-

```
;;initial state
(= (length LinkX) 800)   ;;value in metres
(= (occupancy LinkX) 2) ;; estimated vehicles
(= (density-medium LinkX) 40)
(= (density-heavy LinkX) 70)
(= (timeNeeded-light LinkX) 64)   ;;value in seconds
(= (timeNeeded-medium LinkX) 116) ;;value in seconds
(= (timeNeeded-heavy LinkX) 195) ;;value in seconds
(connected LinkX LinkY)
(at Vehicle1 LinkX)
(next Vehicle1 LinkX LinkY)
...
(at Vehicle2 LinkX)
;;goal state
(at Vehicle2 LinkW)
```

Fig. 2. An excerpt of the PDDL+ code used to encode the relevant information for a link (LinkX) of the controlled region, and for 2 vehicles.

ber of vehicles needs to be rerouted. Work to mitigate this issue are undergoing, with promising results already emerging [26,27].

A different line of work explored the use of Answer Set Programming (ASP) [15,16,22] for performing routing while incorporating the notion of risks into the routes of vehicles [7]. For instance, risks consist of increased likeliness of traffic accidents, potential risks for children, or issues related to air quality.

5   Conclusion

In this paper we presented two problems of urban traffic management and control that we are tackling by using model-based AI approaches, in particular automated planning. On the one hand, it is worth noting that the work in the area is by no means concluded, and there is significant room for improvements. It also worth noting that providing a complete overview of the field is beyond the scope of this paper, that is instead focusing only on the work performed by the Huddersfield AI for Urban Traffic Management and Control (AI4UTMC) research team. On the other hand, it is worth mentioning two other loosely related areas where we successfully applied planning techniques: traffic accident management [10] , where the problem is to optimise the movement and deployment of emergency response vehicles according to the severity of accidents, and in-station train dispatching [8] where automated planning has been used to optimise the movements of trains inside train stations.

Finally, we would like to stress that the use of model-based AI approaches can support explainable and verifiable tools and techniques. As the use of AI in traffic control and management continues to grow, the need for explainable and transparent systems becomes increasingly important, and we believe that automated planning is a promising approach to achieve this goal.

Acknowledgements. Mauro Vallati is supported by a UKRI Future Leaders Fellowship [grant number MR/T041196/1].

References

1. Abduljabbar, R., Dia, H., Liyanage, S., Bagloee, S.A.: Applications of artificial intelligence in transport: an overview. Sustainability **11**(1), 189 (2019)
2. Khan, A., et al.: Environmental pollution is associated with increased risk of psychiatric disorders in the US and Denmark. PLOS Biol. **17**, e03000353 (2019)
3. Chang, K., et al.: Traffic-related air pollutants increase the risk for age-related macular degeneration. J. Invest. Med. **67**(7), 1076–1081 (2019)
4. Antoniou, G., et al.: Enabling the use of a planning agent for urban traffic management via enriched and integrated urban data. Transp. Res. Part C: Emerging Technol. **98**, 284–297 (2019)
5. Bhatnagar, S., Guo, R., McCabe, K., McCluskey, T.L., Scala, E., Vallati, M.: Leveraging artificial intelligence for simulating traffic signal strategies. In: 2022 IEEE 25th International Conference on Intelligent Transportation Systems (ITSC), pp. 607–612. IEEE (2022)
6. Bhatnagar, S., Mund, S., Scala, E., McCabe, K., Vallati, M.: On-the-fly knowledge acquisition for automated planning applications: challenges and lessons learnt. In: Proceedings of ICAART (2022)
7. Cardellini, M., Dodaro, C., Maratea, M., Vallati, M.: A framework for risk-aware routing of connected vehicles via artificial intelligence. In: 2023 IEEE International Conference on Intelligent Transportation Systems (ITSC) (2023)
8. Cardellini, M., Maratea, M., Vallati, M., Boleto, G., Oneto, L.: In-station train dispatching: a PDDL+ planning approach. In: Proceedings of the International Conference on Automated Planning and Scheduling, vol. 31, pp. 450–458 (2021)
9. Chrpa, L., Magazzeni, D., McCabe, K., McCluskey, T.L., Vallati, M.: Automated planning for urban traffic control: strategic vehicle routing to respect air quality limitations. Intelligenza Artificiale **10**(2), 113–128 (2016)
10. Chrpa, L., Vallati, M.: On the exploitation of automated planning for efficient decision making in road traffic accident management. In: 2016 IEEE 55th Conference on Decision and Control (CDC), pp. 6607–6612. IEEE (2016)
11. Chrpa, L., Vallati, M., Parkinson, S.: Exploiting automated planning for efficient centralized vehicle routing and mitigating congestion in urban road networks. In: Proceedings of the 34th ACM/SIGAPP Symposium on Applied Computing, pp. 191–194 (2019)
12. Fox, M., Long, D.: Modelling mixed discrete-continuous domains for planning. J. Artif. Intell. Res. **27**, 235–297 (2006)
13. Franco, S., Lindsay, A., Vallati, M., McCluskey, T.L.: An innovative heuristic for planning-based urban traffic control. In: Shi, Y., et al. (eds.) ICCS 2018. LNCS, vol. 10860, pp. 181–193. Springer, Cham (2018). https://doi.org/10.1007/978-3-319-93698-7_14
14. Franco, S., Vallati, M., Lindsay, A., McCluskey, T.L.: Improving planning performance in PDDL+ domains via automated predicate reformulation. In: Rodrigues, J.M.F., et al. (eds.) ICCS 2019. LNCS, vol. 11540, pp. 491–498. Springer, Cham (2019). https://doi.org/10.1007/978-3-030-22750-0_42
15. Gelfond, M., Lifschitz, V.: The stable model semantics for logic programming. In: Proceedings of the Fifth International Conference and Symposium, Seattle, Washington, August 15–19, 1988, vol. 2, pp. 1070–1080. MIT Press (1988)

16. Gelfond, M., Lifschitz, V.: Classical negation in logic programs and disjunctive databases. New Generation Comput. **9**(3/4), 365–386 (1991)
17. Ghallab, M., Nau, D., Traverso, P.: Automated Planning: Theory and Practice. Elsevier, Amsterdam (2004)
18. Ghallab, M., Nau, D.S., Traverso, P.: Automated Planning and Acting. Cambridge University Press, Cambridge (2016)
19. Haugeland, J.: Artificial Intelligence: The Very Idea. MIT press, Cambridge (1989)
20. McCluskey, T., Vallati, M.: Embedding automated planning within urban traffic management operations. In: Proceedings of the International Conference on Automated Planning and Scheduling, vol. 27, pp. 391–399 (2017)
21. McCluskey, T.L., Vaquero, T.S., Vallati, M.: Engineering knowledge for automated planning: Towards a notion of quality. In: Proceedings of the Knowledge Capture Conference, K-CAP, pp. 14:1–14:8 (2017)
22. Niemelä, I.: Logic programs with stable model semantics as a constraint programming paradigm. Ann. Math. Artif. Intell. **25**(3–4), 241–273 (1999)
23. Percassi, F., Bhatnagar, S., Guo, R., McCabe, K., McCluskey, L., Vallati, M.: An efficient heuristic for AI-based urban traffic control. In: 8th International Conference on Models and Technologies for Intelligent Transportation Systems (MT-ITS) (2023)
24. Scala, E., McCluskey, T.L., Vallati, M.: Verification of numeric planning problems through domain dynamic consistency. In: Dovier, A., Montanari, A., Orlandini, A. (eds.) AIxIA 2022. LNCS, vol. 13796, pp. 171–183. Springer, Cham (2022). https://doi.org/10.1007/978-3-031-27181-6_12
25. Scala, E., Vallati, M.: Effective grounding for hybrid planning problems represented in PDDL+. Knowl. Eng. Rev. **36**, e9 (2021)
26. Švadlenka, M., Chrpa, L.: Towards a framework for intelligent urban traffic routing. In: The International FLAIRS Conference Proceedings, vol. 36 (2023)
27. Švadlenka, M., Chrpa, L., Vallati, M.: Improving the scalability of automated planning-based vehicle routing via smart routes identification (2023)
28. Taale, H., Fransen, W., Dibbits, J.: The second assessment of the SCOOT system in Nijmegen. In: IEEE Road Transport Information and Control. No. 21–23 (1998)
29. Vallati, M., Chrpa, L.: A principled analysis of the interrelation between vehicular communication and reasoning capabilities of autonomous vehicles. In: 2018 21st International Conference on Intelligent Transportation Systems (ITSC), pp. 3761–3766. IEEE (2018)
30. Vallati, M., Chrpa, L.: Effective real-time urban traffic routing: an automated planning approach. In: 2021 7th International Conference on Models and Technologies for Intelligent Transportation Systems (MT-ITS), pp. 1–6. IEEE (2021)
31. Vallati, M., Chrpa, L.: In Defence of good old-fashioned artificial intelligence approaches in intelligent transportation systems. In: 2023 IEEE International Conference on Intelligent Transportation Systems (ITSC) (2023)
32. Vallati, M., Magazzeni, D., De Schutter, B., Chrpa, L., McCluskey, T.: Efficient macroscopic urban traffic models for reducing congestion: A PDDL+ planning approach. In: Proceedings of the AAAI Conference on Artificial Intelligence, vol. 30 (2016)
33. Vallati, M., Scala, E., Chrpa, L.: A hybrid automated planning approach for urban real-time routing of connected vehicles. In: 2021 IEEE International Intelligent Transportation Systems Conference (ITSC), pp. 3821–3826. IEEE (2021)

Main Conference

Automated Design of a Neuroevolution Program Using Algebra-Algorithmic Tools

Anatoliy Doroshenko[1,2] , Illia Achour[2] , and Olena Yatsenko[1(✉)]

[1] Institute of Software Systems of National Academy of Sciences of Ukraine, Glushkov prosp. 40, Kyiv 03187, Ukraine
oayat@ukr.net
[2] National Technical University of Ukraine "Igor Sikorsky Kyiv Polytechnic Institute", 37, Prosp. Peremohy, Kyiv 03056, Ukraine

Abstract. The adjustment of the previously developed algebra-algorithmic tools towards the automated design and generation of programs that use neuroevolutionary algorithms is proposed. Neuroevolution is a set of machine learning techniques that apply evolutionary algorithms to facilitate the solving of complex tasks, such as games, robotics, and simulation of natural processes. The developed program design toolkit provides automated construction of high-level algorithm specifications represented in Glushkov's system of algorithmic algebra and synthesis of corresponding programs based on implementation templates in a target programming language. The adjustment of the toolkit for designing neuroevolutionary algorithms consists in adding the descriptions and software implementations of the relevant elementary operators and predicates to a database of the toolkit. The use of the toolkit is illustrated by an example of designing and generating a program for the single-pole balancing problem, which applies the neuroevolutionary algorithm of the NEAT-Python library. The results of the experiment consisting in the execution of the program generated with the algebra-algorithmic toolkit are given.

Keywords: Automated Design · Algebra of Algorithms · Neuroevolution of Augmenting Topologies · Neural Network · Program Generation

1 Introduction

Today, artificial neural networks are used in various areas of computer intelligence applications. Neural networks with a small number of neurons, due to their limited approximation capabilities, do not allow solving real practical tasks [1], and the selection of an excess number of neurons in the network leads to the problem of retraining and loss of approximation properties, an increase in the number of necessary hardware resources for its implementation, and an increase in time delays in processing information and training neural networks.

Traditionally, the structure of neural networks is determined manually by an expert developer of the computing systems that use these networks. Optimization of the structure of the neural network is carried out by removing some insignificant weighting

G. Antoniou et al. (Eds.): ICTERI 2023, CCIS 1980, pp. 13–24, 2023.
https://doi.org/10.1007/978-3-031-48325-7_2

coefficients. To implement this, algorithms of successive reduction or increase of the structure and methods of evolutionary optimization (neuroevolution) are used. The latter are applied to solve various tasks of neural network synthesis, namely: selection of features, adjustment of weights, selection of optimal network architecture, an adaptation of learning rules, and initialization of weight coefficient values.

The method of neuroevolution has become very common in recent years [2, 3]. Well-known AI laboratories and researchers are experimenting with it, some new successes are fueling enthusiasm, and new opportunities are emerging to influence deep learning, which has been and continues to be popular over the past decade. The neuroevolution method automates the creation of neural networks. Unlike manual methods, it allows for simultaneous changes in the weights of connections between nodes and the topology of nodes during the design process. One of the most famous implementations of neuroevolutionary algorithms is the NeuroEvolution of Augmenting Topologies (NEAT) [4–6]. Incremental topology neuroevolution is a form of reinforcement learning that uses evolutionary (genetic) algorithms to generate artificial neural networks, their parameters, topologies, and decision rules. The main advantage of neuroevolution lies in the possibility of its wider application compared to supervised learning, which requires marked correct pairs of input and output data for training. In contrast, neuroevolution requires only the ability to evaluate the performance of the generated network at any stage of training.

In this paper, facilities for automated design and generation of programs that use neuroevolutionary algorithms are proposed. The tools apply high-level specifications of programs based on Glushkov's systems of algorithmic algebras (SAA) [7, 8]. Based on the schemes represented in SAA, automated code generation is carried out in a target programming language. The approach is demonstrated on a program for single-pole balancing problem. The design and generation of the program using the open-source library for the implementation of neuroevolutionary algorithms NEAT-Python [9] were carried out.

2 Neuroevolution of Augmenting Topologies

Neuroevolution is a family of machine learning techniques that use evolutionary algorithms to facilitate solving complex tasks such as games, robotics, and modeling natural processes [6]. Neuroevolutionary algorithms imitate the process of natural selection. Very simple artificial neural networks can become very complex. The result of neuroevolution is an optimal network topology, which makes the model more energy efficient and convenient for analysis. The NEAT method is designed to reduce the dimensionality of the parameter search space in the form of a gradual development of the neural network structure in the process of evolution. The evolutionary process begins with a population of small, simplest genomes and gradually increases their complexity with each new generation.

Briefly, the essence of the NEAT approach is the following. In the beginning, we can have a very simple topology consisting only of input (x_0, x_1, ...) and output (a) nodes, as well as (optionally) a hidden layer (h) of nodes, which may be omitted. Next, a genetic algorithm works, which, through crossing and mutations, automatically produces other topologies, which are evaluated based on the fitness function, from which the best ones are selected to continue the generation of populations. At the same time, the quality of recognition improves due to increasing the number of hidden layers of the neural network. Quality can be monitored by observing a plot of the fitness function versus the number of populations generated. As experience shows, the best quality neural network is created as a result, which also has the advantages of compactness and speed.

One of the popular test tasks, which is effectively solved using NEAT, is pole balancing [3, 6, 10, 11]. The challenge is to steer a simulated cart that can only move in two directions using a pole attached to its top by a hinge (inverted pendulum). The longer the cart (controlled by a neural network) can hold the pole in the air, the higher its fitness. This task is very similar to trying to balance a pencil in the palm of a hand—it requires strong coordination and quick reaction.

Software implementations of the neuroevolution algorithm are represented by many libraries, in particular, NEAT-Python [9], SharpNEAT [12], PyTorch-NEAT [13], Multi-NEAT [14], and NEAT Java [15]. In this work, an experiment was conducted to automate the development of the neuroevolution algorithm, which was implemented using the NEAT-Python library.

NEAT-Python is an implementation of the NEAT algorithm in the Python programming language. The NEAT-Python library provides an implementation of standard NEAT methods for modeling the genetic evolution of genomes of organisms in a population. It contains utilities for converting an organism's genotype to its phenotype (artificial neural network) and provides convenient methods for loading and saving the genome configuration together with NEAT parameters. In addition, it offers researchers useful routines that help to collect statistics about the progress of the evolutionary process and save/load intermediate control points. Control points allow periodic saving of the state of the evolutionary process and later resume the execution of the process from the saved control points.

The advantages of the NEAT-Python library include:

- stable implementation, comprehensive documentation;
- availability for easy installation using the PIP package manager;
- the presence of built-in statistics collection tools and support for saving execution checkpoints, as well as restoring execution from a given checkpoint;
- availability of several types of activation functions, support of continuous-time recurrent neural network phenotypes;
- easy extensibility to support different NEAT modifications.

The NEAT-Python library uses a set of hyperparameters that affect the performance and accuracy of the NEAT algorithms (the configuration file is stored in a format similar to Windows.ini files). Options include, but are not limited to:

- *fitness_criterion*—a function that computes the termination criterion from the set of fitness values of all genomes in the population;
- *fitness_threshold*—a threshold value that is compared with the fitness computed using the fitness_criterion function to check the need to complete the evolution;
- *pop_size*—the number of individual organisms in each generation;
- the initial configuration of the network by the number of hidden (*num_hidden*), input (*num_inputs*), and output nodes (*num_outputs*).

3   Facilities for Automated Design of Algorithms and Programs

To automate the development of programs, this work uses the SAA/1 language and the algebra-algorithmic Integrated toolkit for Designing and Synthesis of programs (IDS toolkit) [16, 17]. The language is intended for multi-level structural design and documentation of sequential and parallel algorithms and programs and is based on Glushkov's SAA [7, 8].

The main operator constructs of the SAA/1 language used in this work are the following:

- composition (sequential execution) of operators: "*operator*1"; "*operator*2";
- branching: IF '*condition*' THEN "*operator*1" ELSE "*operator*2" END IF;
- for loop: FOR EACH ("*variable*" IN "*expression*") "*operator*" END OF LOOP.

The process of automated development of algorithms and programs in the toolkit is shown in Fig. 1. The toolkit includes the following components:

- designer of high-level schemes of algorithms presented in SAA (SAA schemes);
- a database containing a description of SAA constructs and parameterized templates for implementing these constructs in target programming languages (C++, Java, Python);
- a generator of program texts based on algorithm schemes using templates from the database.

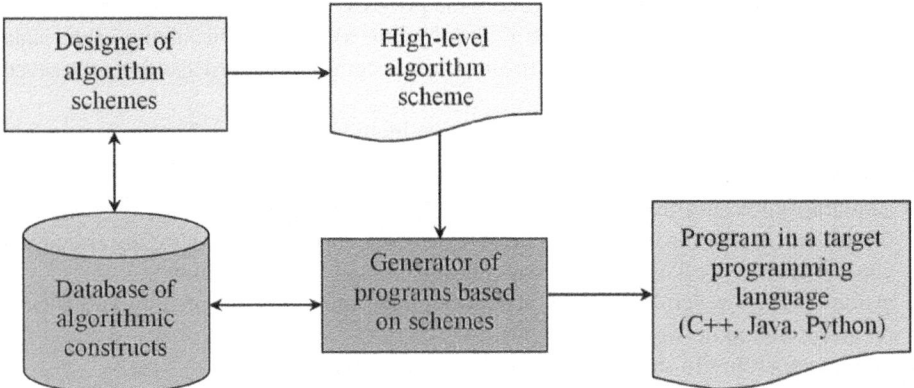

Fig. 1. The development of programs in the IDS toolkit

In the process of working with the designer, the user can edit the description of operator and predicate language constructs, as well as basic operators and conditions stored in the toolkit database.

The description of the database element (operation of SAA or a basic concept) includes its presentation in an analytical and natural-linguistic form, as well as its implementation in a programming language.

Setting up the toolkit for designing neuroevolutionary algorithms, which is the purpose of this work, consists in entering the descriptions of the corresponding basic operators and predicates into the database. Table 1 provides examples of the description of basic elements for neuroevolution algorithms. Identifiers of operators are enclosed in quotation marks, predicates are enclosed in apostrophes. The implementation in programming language uses methods of the NEAT-Python library.

Table 1. Examples of description of basic operators and predicates for neuroevolution tasks in the database of the IDS toolkit

No.	Natural-linguistic form	Implementation in Python language using the NEAT-Python library
1	"Activate the NET (net)(input)"	%1.activate(%2)
2	"Load configuration (config) from file (file)"	%1 = neat.Config(neat.DefaultGenome, neat.DefaultReproduction, neat.DefaultSpeciesSet, neat.DefaultStagnation, %2)
3	"Create the population (p), which is the top-level object for a NEAT run, for configuration (config)"	%1 = neat.Population(%2)
4	"Run neuroevolution for up to (n) generations and display the best genome (best_genome) among generations"	%2 = p.run(eval_genomes, %1 = %1_generations) print('\nBest genome:\n{!s}'.format(%2))
5	"Set (genome) fitness to (value)"	%1.fitness = %2
6	"Adjust (fitness) based on (additional_num_runs) and (success_runs)"	%1 = 1.0 - (%2 - %3)/%2
7	"Create feedforward neural network (net) for (genome) and configuration (config)"	%1 = neat.nn.FeedForwardNetwork create(%2, %3)
8	"Evaluate fitness of the genome that was used to generate net (net)"	fitness = cart.eval_fitness(%1)
9	'Genome (fitness) is greater or equal to fitness threshold (fitness_threshold) for configuration (config)'	%1 >= %3.%2
10	'Genome (fitness) is less than fitness threshold (fitness_threshold) for configuration (config)'	%1 < %3.%2

The basic operators for the implementation of neuroevolutionary algorithms include the following.

1. Activation of the neural network based on the array of input data.
2. Downloading the experiment configuration from a file.
3. Creation of the population p based on the configuration.

4. Starting the process of neuroevolution for n populations and displaying the best genome among populations.
5. Setting the genome fitness.
6. Adjusting the fitness value based on the number of additional simulation runs and successful runs.
7. Creation of a neural network for the specified genome and configuration.
8. Evaluation of the fitness of the genome that was used for network generation.

The predicates (numbers 9 and 10) are intended for checking the fitness value against the threshold value specified in the configuration file.

The natural-linguistic form of the description of basic elements contains the names of the formal parameters indicated in parentheses. The actual parameters specified in the SAA schemes replace the corresponding formal parameters in the text of the implementation of the basic concept in the programming language during program synthesis. Parameters in the implementation of the programming language are marked with the numbers %1, %2, etc.

An example of designing an application using some of the considered basic operators and predicates is given in the next section.

4 Application of the Integrated Toolkit for Designing a Program for Single-Pole Balancing Problem Using the NEAT-Python Library

As an illustration, consider the use of the IDS toolkit for the well-known single-pole balancing problem [6]. This task is a classic example of a reinforcement learning experiment, which is also a commonly accepted benchmark for testing different implementations of control strategies.

In this experiment, the NEAT algorithm is used to implement a controller that regulates a physical device (a cart). The SAA facilities and the IDS toolkit are used to implement a cart device simulator and a neuroevolution-based control algorithm to stabilize the inverted pendulum over a given time interval. The NEAT-Python library [9] is used to implement the neuroevolutionary algorithm.

The statement of the single-pole balancing problem and the mathematical model (motion equations) are described in detail in [6]. A pole balancing controller takes scaled input values and produces an output signal that is a binary value and determines the action to be taken at a given time.

The initial configuration of the neural network of the controller is represented in Fig. 2. It includes five input nodes: for the horizontal position of the cart (x1) and its velocity (x2), for the vertical angle of the pole (x3) and its angular velocity (x4), and an additional input node for bias (x0) (which can be optional depending on the specific NEAT library used). The output node (a) is a binary node that outputs a control signal (0 or 1). The hidden node (h) is optional and can be omitted.

The simulation loop uses the controller's neural network to estimate the current state of the system and select the appropriate action (force to be added to the cart) for the next step. The neural network is created for each genome of the population at a certain generation of evolution, which allows evaluation of the efficiency of all genomes.

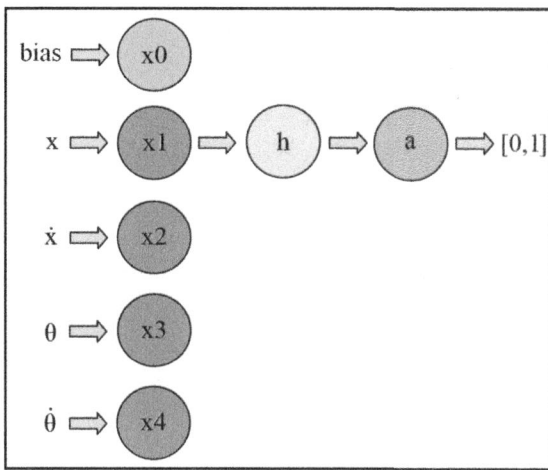

Fig. 2. The initial configuration of a neural network of a single-pole balancing controller

The simulation loop consists of the following steps.

1. Initialization of initial state variables (x, x_dot, theta, theta_dot) with random values within the limits.
2. Simulation loop through a certain number of steps specified by the max_bal_steps parameter.
3. The state variables are scaled to fit the range [0, 1] before loading them as inputs to the neural network controller.
4. The scaled input can be used to activate the neural network, and the output of the neural network is used to obtain a discrete action value.
5. After receiving the action values and the current values of the state variables, a single step of the cart-pole simulation can be run. After the simulation step, the returned state variables are checked against constraints to see if the system state is within acceptable limits.

The SAA scheme implementing the experiment is given below. The scheme begins with defining the libraries and constants used. Next is the compound main statement. The function of evaluating the suitability of all genomes in the population is contained in the compound operator "Evaluate the best net". It receives a list of all genomes in the population and NEAT configuration parameters. For each genome, it creates a neural network phenotype and uses it as a controller to run the simulation.

SCHEME SINGLE POLE EXPERIMENT

"GlobalData"
==== "Import library (*os*)";
    "Import library (*neat*)";
    "Import library (*visualize*)";
    "Import library (*cart_pole* as *cart*)";
    "Import library (*utils*)";
    "Set current working directory (*local_dir*)";
    "Set the directory to store outputs (*out_dir*) (*local_dir*) (*out*)";
    "Set the directory to store outputs (*out_dir*) (*out_dir*) (*single_pole*)";
    (*additional_num_runs* := 100);
    (*additional_steps* := 200)

"main"
==== "Determine path (*config_path*) to configuration file (*local_dir*)
    (*single_pole_config.ini*)";
    "Clean results of the previous run if any or init the output directory (*out_dir*)";
    "Run experiment (*config_path*)";

"Run experiment (*config_file, n_generations* = 100)"
==== "Load configuration (*config*) from file (*config_file*)";
    "Create the population (*p*), which is the top-level object for a NEAT run,
    for configuration (*config*)";
    "Add a stdout reporter to show progress in the terminal for population (*p*)";
    "Run neuroevolution for up to (*n*) generations and display the best
    genome (*best_genome*) among generations";
    "Create feedforward neural network (*net*) for (*best_genome*) and configuration
    (*config*)";
    "Output the message (\n\nEvaluating the best genome in random runs)";
    (*success_runs* := "Evaluate the best net (*net, config, additional_num_runs*)");
    "Output successful (*success_runs*) and expected runs (*additional_num_runs*)
    and check if the best genome is a winner";
    "Visualize the experiment results for configuration (*config*), best genome
    (*best_genome*) from directory (*out_dir*)";

"Evaluate genomes (genomes, config)"
==== FOR EACH (*genome_id, genome* IN *genomes*)
    "Set (*genome*) fitness to (0.0)";
    "Create feedforward neural network (*net*) for (*genome*) and
    configuration (*config*)";
    "Evaluate fitness of the genome that was used to generate net (*net*)";

    IF 'Genome (*fitness*) is greater or equal to the fitness threshold
    (*fitness_threshold*) for configuration (*config*)'

THEN
    (*success_runs* := "Evaluate the best net (*net, config,*
                  additional_num_runs)");
    "Adjust (*fitness*) based on (*additional_num_runs*) and (*success_runs*)"
END IF;
"Set (*genome*) fitness to (*fitness*)"
END OF LOOP;

"Evaluate the best net (*net, config, num_runs*)"
==== FOR EACH (*run* IN range(*num_runs*))
    "Evaluate fitness of the genome that was used to generate net (*net,*
    max_bal_steps = additional_steps)"
    IF 'Genome (*fitness*) is less than fitness threshold (*fitness_threshold*) for
      configuration (*config*)'
    THEN "Return value (*run*)"
    END IF;
END OF LOOP;
"Return value (*num_runs*)"

"Evaluate fitness of the genome that was used to generate net (*net,*
max_bal_steps = 500000)"
==== (steps := "Run cart-pole simulation (*net, max_bal_steps*)");
    IF (*steps = max_bal_steps*)
    THEN "Return value (1.0)"
    ELSE IF (*steps* = 0)
        THEN "Return value (0.0)"
        ELSE
          (*log_steps* := log(*steps*));
          (*log_max_steps* := log(*max_bal_steps*));
          "(*error*) := ((*log_max_steps – log_steps*) / *log_max_steps*)";
          "Return value (1.0 – *error*)"
        END IF
    END IF

END OF SCHEME SINGLE POLE EXPERIMENT

The function of performing the experiment is in the compound statement "Run experiment". It starts by loading hyperparameters from a configuration file and generates the initial population using the loaded configuration. The function then configures reporters to collect statistics related to the execution of the evolutionary process. Output reporters are also added to write the execution results to the console in real time. The evolution process is performed for the specified number of generations, and the results are stored in the source directory. After the best genome has been found during the evolutionary process, a check is made to see if it meets the fitness threshold criteria set in the configuration file. The program returns the genome with the formal best match.

5 Experimental Results

Now, based on the designed SAA scheme, the generation of the program code in the Python language can be automated using the IDS toolkit. Below, there are the results of the experiment on the execution of the generated program. A population of 150 individual organisms was defined in the NEAT-Python configuration file and a fitness threshold of 1.0 was set as the termination criterion. The values of the initial parameters of the neural network are the following: *num_hidden* = 0, *num_inputs* = 4, and *num_outputs* = 1.

When the program is executed, data is displayed for each generation of evolution. The best genome, which is the winner of evolution, encodes a neural network phenotype consisting of only one nonlinear node (output) and three connections from input nodes (size: (1, 3)). The graph of the neural network of the winning controller for the single-pole balancing problem is shown in Fig. 3.

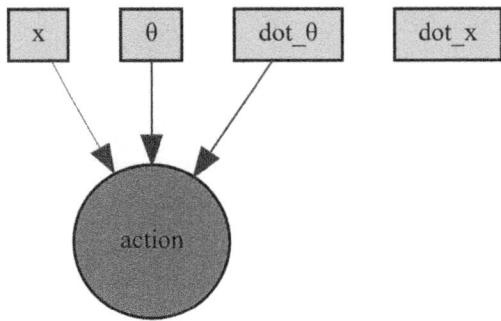

Fig. 3. Graph of the optimal single-pole balancing controller found by the NEAT algorithm

The graph of changes in fitness values over generations of evolution is shown in Fig. 4.

The average fitness of the population in all generations was low, but from the beginning, there was a beneficial mutation that gave rise to a certain line of organisms. From generation to generation, gifted individuals from this line were able not only to preserve their useful traits but also to improve them, which ultimately led to the emergence of an evolutionary winner.

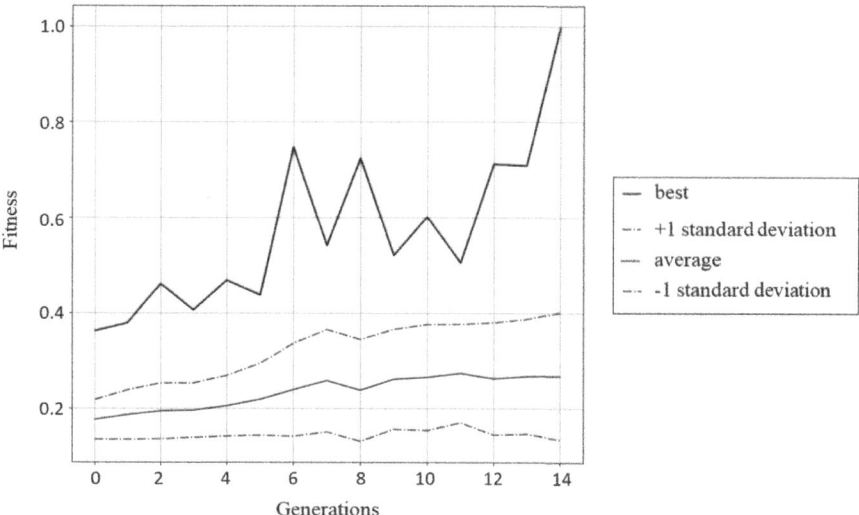

Fig. 4. Average and best values of the fitness function in the single-pole balancing experiment

6 Conclusion

The adjustment of the previously developed algebra-algorithmic toolkit towards the automated design and synthesis of programs using neuroevolutionary algorithms was proposed. The method of neuroevolution of augmenting topologies is intended to reduce the dimensionality of the space for searching for neural network parameters in the form of gradual development of the neural network structure in the process of evolution. The basis of the toolkit is the construction of high-level specifications of algorithms represented in systems of algorithmic Glushkov's algebras, and the generation of corresponding programs based on implementation templates in a target programming language. The approach is illustrated by the example of designing a program for the single-pole balancing problem, which uses the neuroevolutionary algorithm of the NEAT-Python library.

References

1. Subbotin, S.O., Oliinyk, A.O., Oliinyk, O.O.: Non-iterative, evolutionary and multi-agent methods for synthesis of fuzzy and neural network models. ZNTU, Zaporizhzhia (2009). (in Ukrainian)
2. Stanley, K.O., Clune, J., Lehman, J., Miikkulainen, R.: Designing neural networks through neuroevolution. Nat. Mach. Intell. **1**, 24–35 (2019). https://doi.org/10.1038/s42256-018-0006-z
3. Stanley, K.O.: Neuroevolution: a different kind of deep learning. https://www.oreilly.com/radar/neuroevolution-a-different-kind-of-deep-learning. Accessed 05 May 2023
4. Doroshenko, A.Y., Achour, I.Z., Yatsenko, O.A.: Parameter-driven generation of evaluation program for a neuroevolution algorithm on a binary multiplexer example. Radio Electron. Comput. Sci. Control (1), 80–88 (2023). https://doi.org/10.15588/1607-3274-2023-1-8

5. Stanley, K.O., Miikkulainen, R.: Evolving neural networks through augmenting topologies. Evol. Comput. **10**(2), 99–127 (2002). https://doi.org/10.1162/106365602320169811
6. Omelianenko, I.: Hands-on Neuroevolution with Python. Packt, Birmingham (2019)
7. Doroshenko, A., Yatsenko, O.: Formal and Adaptive Methods for Automation of Parallel Programs Construction: Emerging Research and Opportunities. IGI Global, Hershey (2021). https://doi.org/10.4018/978-1-5225-9384-3
8. Andon, P.I., Doroshenko, A.Yu., Zhereb, K.A., Yatsenko, O.A.: Algebra-algorithmic models and methods of parallel programming. Akademperiodyka, Kyiv (2018). https://doi.org/10.15407/akademperiodyka.367.192
9. NEAT-Python. https://github.com/CodeReclaimers/neat-python. Accessed 05 May 2023
10. Chang, O., Kwiatkowski, R., Chen, S., Lipson, H.: Agent embeddings: a latent representation for pole-balancing networks. In: AAMAS 2019. International Foundation for Autonomous Agents and Multiagent Systems, Richland, SC, pp. 656–664 (2019). https://doi.org/10.48550/arXiv.1811.04516
11. Lawrence, W.M.: Solving XOR and Pole-balancing problems using a multi-population NEAT. A thesis presented for the degree of Master of Philosophy. De Montfort University, Leicester (2020). https://dora.dmu.ac.uk/bitstream/handle/2086/20731/William-Lawrence.pdf?sequence=1
12. SharpNEAT. Evolution of Neural Networks. https://sharpneat.sourceforge.io. Accessed 05 May 2023
13. PyTorch-NEAT. https://github.com/uber-research/PyTorch-NEAT. Accessed 05 May 2023
14. MultiNEAT: Portable NeuroEvolution Library. https://github.com/peter-ch/MultiNEAT. Accessed 05 May 2023
15. NEAT Java (JNEAT). https://nn.cs.utexas.edu/soft-view.php?SoftID=5. Accessed 05 May 2023
16. Doroshenko, A., Shymkovych, V., Yatsenko, O., Mamedov, T.: Automated software design for FPGAs on an example of developing a genetic algorithm. In: Burov, O., Ignatenko, O., Kharchenko, V., Kobets, V., et al. (eds.) ICTERI 2021. CCIS, vol. 1635, pp. 74–85. Springer, Cham (2021)
17. Doroshenko, A., Zhereb, K., Yatsenko, O.: Using algebra-algorithmic and term rewriting tools for developing efficient parallel programs. In: Ermolayev, V., Mayr, H.C., Nikitchenko, M., Spivakovsky, A. (eds.) ICTERI 2013. CCIS, vol. 1000, pp. 38–46. Springer, Cham (2013)

On Randomization of Reduction Strategies for Typeless Lambda Calculus

Oleksandr Deineha[1], Volodymyr Donets[2], and Grygoriy Zholtkevych[1(✉)]

[1] School of Mathematics and Computer Science, V.N. Karazin Kharkiv National University, 4, Svobody Sqr, Kharkiv 61002, Ukraine
{oleksandr.deineha,g.zholtkevych}@karazin.ua
[2] School of Computer Science, V.N. Karazin Kharkiv National University, 4, Svobody Sqr, Kharkiv 61002, Ukraine
v.donets@karazin.ua

Abstract. It is well known that the functional programming paradigm provides certain advantages, namely increasing the reliability of software development, improving software verification and validation, providing software artefact reusability, and scaling software solutions. But this paradigm is not free from disadvantages also. The most serious of them are redundant memory usage and low-performance productivity. The ideas concerning the implementation of randomization mechanisms into reduction strategies, which is a significant part of the kernel of functional computing systems, are considered in the paper. The authors present simulation tools for studying such a mechanism and give their grounding. They also present the experimental results of the comparison analysis of variants reduction randomization.

Keywords: lambda calculus · alpha- and eta-conversion · beta-reduction · reduction strategy · randomization · cost of reduction

1 Introduction

The trend in the software industry is to widen the use of ideas related to the functional programming paradigm in software development processes. This trend is in detail analysed in [3]. However, the functional paradigm has not only advantages but also disadvantages (see, for example, [4]). The author of [4] marks as advantages of the functional programming paradigm the simplicity, testability, reusability, and scalability. In contrast, he marks as disadvantages of the functional paradigm

1. the inefficient memory usage and low performance,
2. fewer frameworks and tools compared to the imperative paradigm, and
3. narrower communities of experts and users compared to the imperative paradigm.

G. Antoniou et al. (Eds.): ICTERI 2023, CCIS 1980, pp. 25–38, 2023.
https://doi.org/10.1007/978-3-031-48325-7_3

This paper is focused on the first issue of these disadvantages. More precisely, it is focused on the problem of low computing performance.

It is known that a computational process, which is generated by a functional program, is step-wise applying one of three mechanisms, namely alpha-conversion, beta-reduction, and eta-conversion. So, the computational cost of the process as a whole is the sum of the computational costs of each of these sub-processes and some auxiliary ones, such as redexes recognising. Varying ratios of contributions of the corresponding sub-processes costs to the cost of the process as a whole, one can obtain different models for estimating the performance of functional programs.

Given that the reduction process is not unambiguous, we realise some reduction strategies for the functional engine to eliminate this ambiguity. Of course, the cost of a computational process depends on the reduction strategy used.

Thus, we may pose the following problem.

Problem 1. *For the chosen cost model, find a strategy with the least cost.*

This problem can hardly be solved universally, which Example 1 from Subsect. 3.1 demonstrates. Moreover, the detailed analysis of the problem (see, for example, [1, chapter 13]) led to proving the statement about the absence of an optimal reduction strategy.

The known approaches in similar situations are to choose randomly at each step either of redex or of deterministic strategy for determining this redex for its reduction. We talk about a random strategy in the first of mentioned cases or a mixed one in the second case, respectively.

This article and [5] begin a series of our papers for confirming or refuting the following hypothesis.

Hypothesis. For any combinator F, there exists a random or mixed reduction strategy that optimally reduces FX on average wrt X where X is a combinator representing input data.

This optimal strategy can be found with Reinforcement Machine Learning.

If this hypothesis is true, then we can try to develop a tool for the performance optimisation of executing functional programs.

2   A Brief Survey of Typeless Lambda Calculus

Terms of typeless lambda calculus are constructed from an enumerable set of variables

$$\mathbf{V} = \{v_0, v_1, \ldots, v_n, \ldots\}$$

and four special signs

$$\text{``}\lambda\text{''}, \ \text{``.''}, \ \text{``(''}, \ \text{and} \ \text{``)''}.$$

2.1   Lambda Terms Construction Rules

Members of the set Λ of lambda terms are constructed with the following syntactic rules

– the rule for constructing an atomic lambda term

$$\frac{x \in \mathbf{V}}{x \in \Lambda} \tag{1a}$$

– the rule for constructing a lambda term called an application term

$$\frac{M \in \Lambda \quad N \in \Lambda}{(M\,N) \in \Lambda} \tag{1b}$$

– and the rule for constructing a lambda term called an abstraction term

$$\frac{x \in \mathbf{V} \quad M \in \Lambda}{(\lambda x.M) \in \Lambda} \tag{1c}$$

Usually, terms are represented in the shortened form obtained with these rules

– the outer brackets are dropped;
– the shortened form $M\,N\,K$ refers to the term $((M\,N)\,K)$;
– the shortened form $\lambda x.\lambda y.M$ refers to the term $(\lambda x.(\lambda y.M))$;
– the shortened form $\lambda x.M\,N$ refers to the term $(\lambda x.(M\,N))$.

During the work, authors have developed a modelling tool for the experimental study of computations with lambda terms. The realisation language of this tool is Python. The concept of a variable is represented by the class `Var` of the developed modelling software. This class provides an enumerable set of variable instances having no structure.

```
class Var:
    """is inhabited by instances identified uniquely by
    positive natural numbers.
    """
```

The root class of the developed tool is abstract and responds to the features common to all terms.

```
class Term:
    """is the parent class for classes representing terms of
    specific kinds. It collects properties and methods common
    to all terms indifferently to their kinds
    """

    def __eq__(self, another: Term) -> bool:
        """returns 'True' if the term and term 'another' are
```

```
        equal wrt Leibnitz equality ie they are the same else
        returns 'False'
        """


    @property
    def kind(self) -> str:
        """returns the kind of the term, which may be either
        'atom', 'application', or 'abstraction'
        """

    # A code for other features common to all terms
```

The following three classes inherited from the class **Term** correspond to syntactic rules for constructing terms.

```
class Atom(Term):
    def __init__(self, x: Var):
        """'x' determines the constructed atomic term
        """
        # An atomic term constructor code


class Application(Term):
    def __init__(self, lTerm: Term, rTerm: Term):
        """'lTerm' is the term that applies, and 'rTerm' is
        the term that is applied to
        """
        # An application term constructor code


class Abstraction(Term):
    def __init__(self, x: Var, scope: Term):
        """'x' is variable, and 'scope' is the term which can
        contain (but is not obligatory) a free occurrence of 'x'
        """
        # An abstraction term constructor code
```

2.2   Lambda Calculus Computation Rules

The main computational rule for typeless lambda calculus is the beta-reduction. It means informally the following:

- a term of the form $(\lambda x.M)N$ is called a beta-redex;
- the redex $(\lambda x.M)N$ reduces to replacing all free occurrences of variable x in M with N;

– a reduction process is a sequence (finite or infinite) of replacing a redex with a term obtained as a result of its reduction.

A key restriction of the reduction is that any free occurrences of the variable in N must remain free in the term obtained after the replacement.

Occurrences of a Subterm and a Variable in a Term. To formalize the concept of beta-reduction, we need to consider a "coordinate system" related to a term.

To do this, we consider the set $S = \{\texttt{onleft}, \texttt{inward}, \texttt{onright}\}$. Also, we use the denotation S^* to refer to the set of lists whose elements are in S. Now, we associate the set $S^*(M) \subset S^*$ with each term $M \in \Lambda$ as follows

$$S^*(M) = \{[\,]\} \cup \begin{cases} \{[\texttt{onleft}] + u \mid u \in S^*(L)\} & \text{if } M \equiv L\,R \\ \{[\texttt{onright}] + u \mid u \in S^*(R)\} & \text{if } M \equiv L\,R \\ \{[\texttt{inward}] + u \mid u \in S^*(B)\} & \text{if } M \equiv \lambda x.B \end{cases} \qquad (2)$$

Using the structural induction principle, one can easily prove the following statement.

Proposition 1. *For any $M \in \Lambda$ and $u \in S^*(M)$, the inequality*

$$\text{length}\, u \leq \text{height}\, M$$

is valid and, therefore, $S^*(M)$ is finite where

$$\text{length}\, u = \begin{cases} 0 & \text{if } u = [\,] \\ 1 + \text{length}\, u' & \text{if } u \equiv [s] + u' \end{cases} \qquad (3)$$

and

$$\text{height}\, M = \begin{cases} 0 & \text{if } M \equiv x \\ 1 + \max(\text{height}\, L, \text{height}\, R) & \text{if } M \equiv L\,R \\ 1 + \text{height}\, B & \text{if } M \equiv \lambda x.B \end{cases} \qquad (4)$$

An element $u \in S^*(M)$ can be considered the "location" of subterm M_u in M where M_u is determined as follows.

$$M_u = \begin{cases} M & \text{if } u \equiv [\,] \\ L_{u'} & \text{if } u \equiv [\texttt{onleft}] + u' \text{and } M \equiv L\,R \\ R_{u'} & \text{if } u \equiv [\texttt{onright}] + u' \text{and } M \equiv L\,R \\ B_{u'} & \text{if } u \equiv [\texttt{inward}] + u' \text{and } M \equiv \lambda x.B \end{cases}$$

This definition of M_u is correct due to (2).

The concept of the Brzozowski derivative [2] is fruitful for the below use. Let $X \subset S^*$ and $u \in S^*$, then the Brzozowski derivative of X w.r.t. u is defined as follows

$$u^{-1} \cdot X = \{v \in S^* \mid u + v \in X\}.$$

One can easily check the correctness of the following statement.

Proposition 2. *For any term M and $u \in S^*(M)$, $S^*(M_u) = u^{-1} \cdot S^*(M)$.*

The concept of the Brzozowski derivative allows us to characterise a subset of S^* that is $S^*(M)$ for some term M with the following statement.

Theorem 1. *A subset X of S^* equals $S^*(M)$ for some term M if*

- *X is finite;*
- *X is prefix closed i.e. for any $u \in X$, $u = v + w$ implies $v \in X$;*
- $\mathtt{onleft}^{-1} \cdot X \neq \emptyset$ *implies* $\mathtt{onright}^{-1} \cdot X \neq \emptyset$ *and* $\mathtt{inward}^{-1} \cdot X = \emptyset$;
- $\mathtt{onright}^{-1} \cdot X \neq \emptyset$ *implies* $\mathtt{onleft}^{-1} \cdot X \neq \emptyset$ *and* $\mathtt{inward}^{-1} \cdot X = \emptyset$;
- $\mathtt{inward}^{-1} \cdot X \neq \emptyset$ *implies* $\mathtt{onleft}^{-1} \cdot X = \mathtt{onright}^{-1} \cdot X = \emptyset$.

Moreover, the maximal members of $S^*(M)$, i.e. such that have no extensions lying in $S^*(M)$, refer to atomic subterms of M, which are uniquely identified by variables. Thus, we can consider

$$S^+(M) = \{u \in S^*(M) \mid (u + [s])^{-1} \cdot S^*(M) = \emptyset \text{ for any } s \in S\},$$

which is evidently in one-to-one correspondence with

$$\{u \in S^+(M) \mid M_u = x \text{ for some } x \in \mathbf{V}\}.$$

Thus, one can uniquely associate the variable $x \in \mathbf{V}$ with $u \in S^+(M)$ such that $M_u \equiv x$.

The previous discussion gives us the following method to compute $S^+(M)$

$$S^+(M) = \begin{cases} \{[\,]\} & \text{if } M \equiv x \\ \{[\mathtt{onleft}] + u \mid u \in S^*(L)\} & \text{if } M \equiv L\,R \\ \{[\mathtt{onright}] + u \mid u \in S^*(R)\} & \text{if } M \equiv L\,R \\ \{[\mathtt{inward}] + u \mid u \in S^*(B)\} & \text{if } M \equiv \lambda x.B \end{cases} \tag{5}$$

Classifying Occurrences of a Variable in a Term. We say that

- $u \in S^+(M)$ is a free occurrence of variable x in M if $M_u \equiv x$ and for any prefix v of u (i.e. $u = v + w$ for some $w \in S^*$), M_v coincides no term of the form $\lambda x.B$;
- in contrast, $u \in S^+(M)$ is a bound occurrence of variable x in M if $M_u \equiv x$ and for some prefix v of u, $M_v \equiv \lambda x.B$; if v is the longest such a prefix, then M_v is the scope of x.

We use below the notation $\mathrm{FV}(M)$ to refer to the subset of $S^+(M)$ consisting of free occurrences of variables in M.

One says that a variable x is fresh for a term M if there is no free or bound its occurrence into M and x does not match y in all subterms of M of the form $\lambda y.B$.

Replacing Subterm with Term. Let us consider a technical but utility concept now.

Assume that M is a term, $u \in S^*(M)$, and K is another term. Let us define the term $M[K/u]$ as follows

$$M[K/u] = \begin{cases} K & \text{if } u \equiv [\,] \\ (L[K/u'])\,R & \text{if } u \equiv [\texttt{onleft}] + u' \text{and } M \equiv L\,R \\ L\,(R[K/u']) & \text{if } u \equiv [\texttt{onright}] + u' \text{and } M \equiv L\,R \\ \lambda x.(B[K/u']) & \text{if } u \equiv [\texttt{inward}] + u' \text{and } M \equiv \lambda x.B \end{cases} \tag{6}$$

Renaming a Bound Variable and Alpha-Congruence. Mathematical and programming experience suggests that the specific name of a bound variable is not essential. This motivates introducing the operation of renaming a bound variable in a term.

Let us consider a term $\lambda x.B$, $\text{FV}_x(B) = \{u \in \text{FV}(B) \mid B_u \equiv x\}$, and a fresh variable y for $\lambda x.B$ then the term $(\ldots(\lambda y.B)[y/u_1]\ldots)[y/u_k]$ where $\{u_1, \ldots, u_k\} = \text{FV}_x(B)$ is the result of renaming x with y in $\lambda x.B$.

Definition 1. (alpha-conversion). *Let M be a term and $u \in S^*(M)$ such that $M_u = \lambda x.B$ then a term N is called a result of alpha-conversion of M (symbolically, $M \to_\alpha N$) if $N \equiv M[K/u]$ where K is the result of renaming x in $\lambda x.B$ by some variable that is fresh for M.*

Thus, alpha-conversion is a binary relation on Λ. If terms M and N are related by reflexive-transitive closure of alpha-conversion, then we use the notation $M \to_\alpha^* N$.

Definition 2. (alpha-congruence). *Two terms M and N are called alpha-congruent (symbolically, $M =_\alpha N$) if there is a term K such that $M \to_\alpha^* K$ and $N \to_\alpha^* K$.*

Note 1. The statement $M =_\alpha N$ means actually that $S^*(M) = S^*(N)$ and for any $u \in S^*(M)$ and $v \in S^+(M_u)$, $M_{u+v} \equiv x$ and $N_{u+v} \equiv y$ are both either free or bound and, in the first case, $x = y$.

Similar to mathematics, where we do not distinguish $\int_0^1 f(x)dx$ and $\int_0^1 f(t)dt$, we consider alpha-congruent terms as aliases for referring to the same object.

Beta-Reduction. The key definition in lambda calculus is the following.

Definition 3. *A term is called a beta-redex if it has the form $(\lambda x.M)\,N$.*

To reduce a beta-redex $(\lambda x.M)N$, we take M' such that $M' =_\alpha M$ and free variables of N do not occur bound in M' (one can do it by renaming of bound variables in M by variables are fresh for both N and M); and then we replace free occurrences of x in M with N. Symbolically, the relationship between a beta-redex $(\lambda x.M)N$ and its reduction result K is usually labelled with $(\lambda x.M)N \to_\beta K$.

Now, we can explain what means beta-reduction step. For a term M and $u \in S^*(M)$ such that M_u is a redex and term N is such that $M_u \to_\beta N$, the result of one reduction step is the term $M[N/u]$.

Hence reduction is a binary relation on the set Λ, whose reflexive-transitive closure is denoted by $\to_\beta^*$.

Eta-Conversion. Introducing eta-conversion was motivated by the following. Let us consider terms M and $\lambda x.M\,x$ where x does not occur freely in M then $M\,N$ and $(\lambda x.M\,x)N$ are beta-reduced to $M\,N$ for any term N. Given the principle of extensionality, we are forced to consider these terms equal while they are different in Leibnitz's sense. The eta-conversion

$$\lambda x.M\,x \to_\eta M \qquad \text{whenever } x \text{ does not occur freely into } M$$

eliminates this gap.

Computation Using Typeless Lambda Calculus

Definition 4. *A computation is a step-wise process with steps that are either alpha-conversion, beta-reduction, or eta-conversion only. This process halts if it reaches a term that neither beta-reduction nor eta-conversion can be applied (such a term is said that it is in normal form).*

We label symbolically $M \to_{\beta\eta}^* N$ the statement "there is a computational process transforming M into N".

This definition and Church-Rosser Theorem (see, for example, [1]) ensure that the following defines an equivalence relation on Λ, which is labeled with $M =_{\beta\eta} N$.

Definition 5. *For any terms M and N, $M =_{\beta\eta} N$ means*

$$M \to_{\beta\eta}^* K \text{ and } N \to_{\beta\eta}^* K \text{ for some term } K.$$

Note that normal forms are the results of computation. Church-Rosser Theorem guarantees the uniqueness up to alpha-congruence of the result for any computation.

3   A Terms Reduction Strategy as Choice Procedure

Above, we demonstrated how to model a computation process as stepwise reductions. It means that there exists more than one, as a rule, way to pass from an initial term to its normal form. But for a software realisation, we need to specify some manner for choosing a redex for reducing. Such a choosing manner is called a reduction strategy, more precisely, a one-step reduction strategy (see [1, chapter 13]). The main related result presented in the cited book is that there is no recursive reduction strategy, which is the most productive. In the section, we give examples of the most commonly used recursive one-step reduction strategies and then propose random-based methods to reduce terms.

3.1   Deterministic Terms Reduction Strategies

The most used recursive one-step reduction strategies are the leftmost outer reduction strategy (LO-strategy) and the leftmost inner reduction strategy (LI-strategy). LO-strategy is useful because it is normalizing, i.e. if a term is normalizable, then this strategy provides its normal form.

In contrast, LI-strategy does not achieve the normal form of the term in some cases, even though such a form exists. However, this strategy corresponds to the intuitive idea of how functions should be calculated, namely, function arguments should be computed before computing the function.

Comparing the productivity of LO- and LI-strategies, one can consider the following example, which demonstrates that neither LO-strategy nor LI-strategy is better relative to the number of reduction steps criterion.

Example 1. Let

- ω_n refer to the combinator $\lambda x. \underbrace{x \ldots x}_{n \text{ times}}$ and
- I refer as usually to the combinator $\lambda x.x$.

Then to obtain the normal form of $\omega_n \, (\underbrace{I \ldots I}_{m \text{ times}})$, LO-strategy needs $\Theta(n \cdot m)$ reduction steps, while LI-strategy provides the normal form after $\Theta(n + m)$ reduction steps.

In contrast, consider the term $(\lambda x.\lambda y.y)(\omega_n \, (I \, I))$. LO-reduction strategy provides the corresponding normal form after one reduction step, while LI-strategy needs $\Theta(n)$ reduction steps for normalizing this term.

This example demonstrates that LO- and LI-reduction strategies need different asymptotic numbers of reduction steps for the same combinators. In some cases, the corresponding number for the LO-strategy is greater than the corresponding number for the LI-strategy, but in other cases, we observe the opposite.

3.2   A Deterministic Reduction Strategy as a Choice Procedure

Now, let us reformulate the recursive one-step reduction strategy definition using the concept of "the subterm location". Further, we use this for introducing randomised reduction strategies.

For a term M, $\text{Red}(M)$ is the set formed by $u \in S^*(M)$ such that M_u is either beta-redex or eta-redex.

One can easily understand that the set $\{\text{Red}(M) \mid M \in \Lambda\}$ admits effective enumeration. Thus, one can define a recursive function that associates some member of $\text{Red}(M)$ with M. Such a function is called a recursive one-step reduction strategy.

LO-strategy: let onleft < inward < onright then we choose the lexicographically lower element of $\text{Red}(M)$.

LI-strategy: let onright < inward < onleft then we choose the lexicographically upper element of $\text{Red}(M)$.

3.3   Random Choice and Random Reduction Strategies

The ideas of this and the next subsection were first presented in [5]. These ideas were motivated by estimations of the productivity of random choice in practice for different problem areas.

The random strategy proposed in [5], UR-strategy, is based on the following choosing rule: choose an element of $\mathrm{Red}(M)$ w.r.t. uniform distribution on this set.

Experimental studying shows that UR-strategy is not productive. Therefore, we propose two variants of probability distributions. The first of them determines OR-strategy, which provides reduction behaviour like to LO-strategy, and the second one determines IR-strategy, which provides reduction behaviour like to LI-strategy.

OR-strategy: let $\mathrm{Red}(M) = \{u_1, \ldots, u_k\}$ and

$$\Pr(u_i) = \frac{(\text{length } u_i)^{-1}}{\sum\limits_{j=1}^{k} (\text{length } u_j)^{-1}}$$

then we choose an element of $\mathrm{Red}(M)$ w.r.t. the specified distribution.

This strategy more often chooses redexes located closer to the root of M syntax tree.

In contrast, **IR-strategy:** let $\mathrm{Red}(M) = \{u_1, \ldots, u_k\}$ and

$$\Pr(u_i) = \frac{\text{length } u_i}{\sum\limits_{j=1}^{k} \text{length } u_j}$$

then we choose an element of $\mathrm{Red}(M)$ w.r.t. the specified distribution.

This strategy more often chooses redexes located further from the root of the syntax tree M.

Generalised Random Strategy (GR-Strategy): These strategies and the uniform random strategy from [5] can be joined as follows.

Let $\mathrm{Red}(M) = \{u_1, \ldots, u_k\}$ and $\alpha \in \mathbb{R}$ then

$$\Pr(u_i) = \frac{(\text{length } u_i)^{\alpha}}{\sum\limits_{j=1}^{k} (\text{length } u_j)^{\alpha}}. \tag{7}$$

It is evident that OR-strategy corresponds to the case $\alpha = -1$, IR-strategy corresponds to the case $\alpha = 1$, and the uniform random strategy corresponds to $\alpha = 0$.

3.4   Mixed Reduction Strategies

Another way to randomize the reduction strategy is to mix deterministic ones, as suggested in [5].

More precisely, let us assume that $s_1, \ldots, s_n$ are recursive choice functions related to deterministic one-step reduction strategies, and $0 < w_1, \ldots, w_n < 1$ satisfy the equation $\sum_{i=1}^{n} w_i = 1$.

The corresponding mixed strategy provides the following choice: at the current step, randomly choose a member j of the set $\{1, \ldots, n\}$ using the distribution $\Pr(I) = w_i$, $i = 1, \ldots, n$ and, then, compute s_j for the current situation.

4   Complexity of Terms Reduction Strategies

As was marked above, a computational process corresponding to a functional program is the step-wise application of either alpha-conversion, beta-reduction, or eta-conversion. Hence, the computational cost of the process as a whole is the sum of the computational costs of each of these sub-processes and some auxiliary ones, such as redexes recognising.

Here, we focus on the assumption that the number of beta reductions is the most significant part of the computational cost. In further research, this limitation will be removed.

5   Simulation of Reduction Processes

For studying variants of reduction processes, special instrumental software was developed. The authors plan to present this software solution and evidence of its correctness in a separate paper.

As we stress above, the number of reduction steps is here used as the computational cost of the modelled process.

In this section, we present the results of the experimental comparison of different variants of random reduction strategies with LO- and LI- strategies. For presenting the reduction specificity of terms, we use two typical examples.

Let us assume that

$$\omega_n = \lambda x.\underbrace{x \ldots x}_{n\,\text{times}}, \quad \Omega_n = \omega_n(\underbrace{I \ldots I}_{n}\,\text{times}), \quad I = \lambda x.x, \quad \Phi_n = (\lambda x.\omega_n)\Omega_n I.$$

We consider Ω_n and Φ_n as two terms with quite different behaviour regarding reduction.

5.1   Experimental Study of Random Strategies

Let us consider $n = 5, 10, 15, 20, 25, 30$ and perform normalization of Ω_n and Φ_n using LO- and LI- reduction strategies. Figure 1 shows the number of reduction steps in this case.

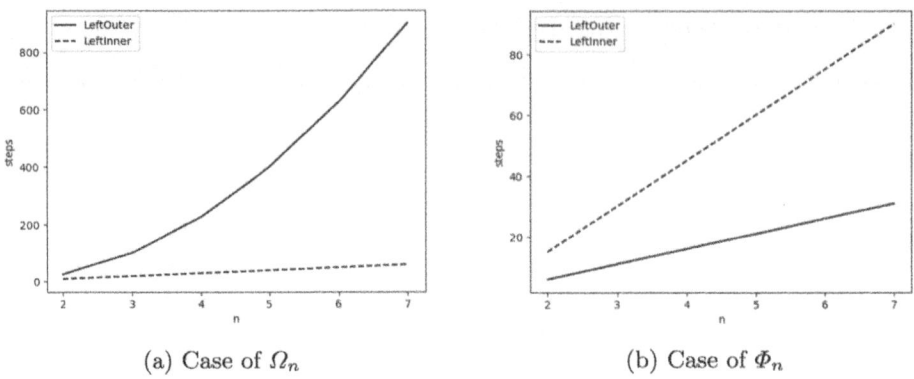

(a) Case of Ω_n (b) Case of Φ_n

Fig. 1. Comparison of the number of reduction steps for LI- and LO-strategies

Figure 2 reduction steps for the generalised random strategy using $\alpha = -2, -1, -0.5, 0, 0.5, 1, 2$.

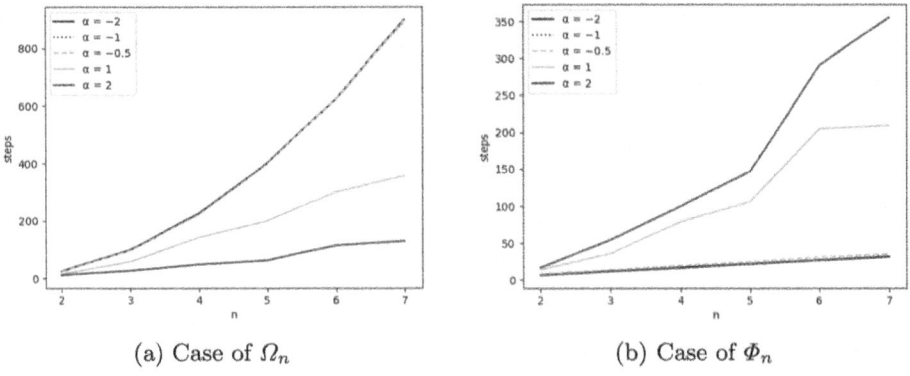

(a) Case of Ω_n (b) Case of Φ_n

Fig. 2. Comparison of the number of reduction steps for the generalised strategy

The presented graphs are the reason for the hypothesis that the condition $\alpha < 0$ provides the behaviour of a random reduction process similar to the one of LI-reduction, and the condition $\alpha > 0$ provides the behaviour of a random reduction process like to LO-strategy.

5.2 Experimental Study of Mixed Strategies

Let us consider $r = 0.25, 0.5, 0.75$ and perform normalization of Ω_n and Φ_n using the strategies obtained by mixing LO- and LI- strategies with weights r and $1-r$.

Figure 3 shows the result of simulating.

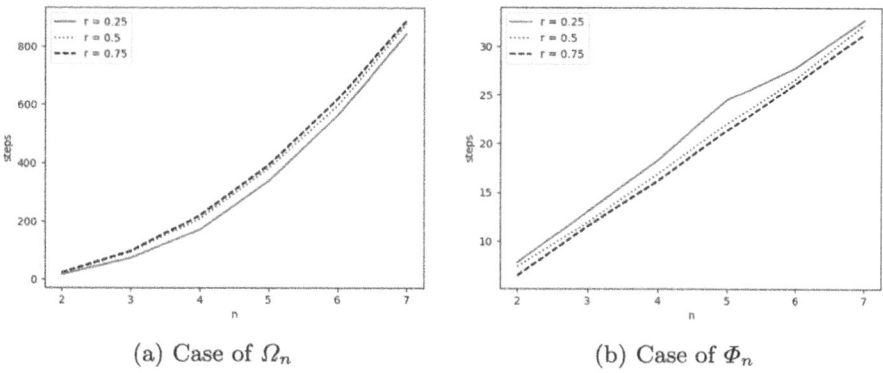

(a) Case of Ω_n (b) Case of Φ_n

Fig. 3. Comparison of the number of reduction steps for the mixed strategy

6   Conclusions

The presented study is motivated by the well-known fact that there is no optimal recursive one-step reduction strategy for the typeless lambda calculus. This fact has a practical consequence: compilers of functional programming languages do not support any mechanism for optimising the reduction strategy.

The existing experience in dealing with optimization problems in other subject areas shows that changing deterministic algorithms by procedures based on random search gives good results. This allows us to suppose that for Problem 1 formulated in Sect. 1. In that section, the research hypothesis is also posed.

Based on this hypothesis, the authors began an experimental study. To ensure the confirmation or refutation of the hypothesis, an environment for simulating the processes of reduction of lambda terms has been developed. For developing the results presented in [5], the authors have proposed a parameterised random strategy that generalises the known random strategies. The possibility to variate the strategy parameter admits discussing the machine learning approach to optimise the computational behaviour of lambda terms in each particular case. Perhaps, further research provides methods for developing the special compiler component, which improves the productivity of running functional programs.

In the paper, another way for reduction strategy randomising has also been considered. The carried-out experimental comparison of this way and the way based on the generalised random strategy has shown that the second way is looked more promising. But an additional study is needed for the ultimate conclusion.

The authors consider reinforcement machine learning as a perspective approach for developing a toolkit providing an optimisation component of compilers. Thus, the possibility of using reinforcement machine learning is the focus of our further research.

References

1. Barendregt, H.P.: The Lambda Calculus, Its Syntax and Semantics. Studies in Logic and the Foundations of Mathematics, vol. 103. North Holland (1985)
2. Brzozowski, J.A.: Derivatives of regular expressions. J. ACM **11**, 481–494 (1964)
3. Hu, Z., Hughes, J., Wang, M.: How functional programming mattered. Natl. Sci. Rev. **2**(3), 349–370 (2015)
4. Neumann, J.: Advantages and disadvantages of functional programming (2022). https://medium.com/twodigits/advantages-and-disadvantages-of-functional-programming-52a81c8bf446. NEW IT Engineering, Medium. Accessed 02 Apr 2023
5. Shramenko, V., Kuznietcova, V., Zholtkevych, G.: Studying mixed normalization strategies of lambda terms. In: Chumachenko, D., et al. (eds.) Proceedings of the 2nd International Workshop of IT-professionals on Artificial Intelligence. ProfIT, vol. 3348, pp. 57–68. CEUR Workshop Proceedings (2022)

Solving Sokoban Game with a Heuristic for Avoiding Dead-End States

Oleksii Ignatenko[2]([☒])[iD] and Ruslan Pravosud[1][iD]

[1] Kyiv Academic University, Kyiv, Ukraine
[2] Ukraine Catholic University, Lviv, Ukraine
o.ignatenko@ucu.edu.ua

Abstract. This paper focuses on applying reinforcement learning methods to solve the game Sokoban. This game is a popular puzzle, relatively easy for humans to solve. However, it poses a significant challenge for computer algorithms due to the irreversible nature of certain moves. To predict which actions will lead to such undesirable states is often difficult for a learning agent – a common problem in tasks requiring planning. We propose using a Monte-Carlo tree search (MCTS) algorithm and a heuristic convolution neural network (CNN) specially trained to separate undesirable, neutral, and desired game states to address this issue. We experimented with different heuristic variations of algorithms and compared them against each other. We have implemented MCTS in two different setups: one with a CNN trained using data obtained during the solving process and one without such training. We also varied the number of rollouts for each move in MCTS and compared the results. The paper's research question was how to improve the performance of learning agents in tasks that require planning to avoid unwanted states.

Keywords: Reinforcement Learning · CNN · Sokoban · MCTS

1 Introduction

The development of machine learning (ML) and reinforcement learning (RL) in rule-based games goes back to the 1960s when researchers began to explore the possibility of creating machines that learn and make decisions autonomously. During this time, Arthur Samuel developed the Samuel Checkers program, the first program based on machine learning techniques, specifically reinforcement learning, to improve its performance with playing games. The program utilized pattern recognition, search algorithms, and self-play to refine its gameplay strategy. Since then, RL has gained even more attention, solving many problems. The recent success of AlphaGo, which used deep learning and RL to learn to play the game of Go at the highest level, attracted worldwide curiosity. The AlphaGo program defeated the world's best human players, demonstrating the potential of the used approach in solving complex problems.

RL is a rapidly developing field of machine learning that has found applications in many areas, including robotics, game playing, and natural language

© The Author(s), under exclusive license to Springer Nature Switzerland AG 2023
G. Antoniou et al. (Eds.): ICTERI 2023, CCIS 1980, pp. 39–52, 2023.
https://doi.org/10.1007/978-3-031-48325-7_4

processing. Its ability to learn from trial and error without explicit instructions makes it a powerful tool for solving complex problems one cant address using traditional programming techniques [3]. Recently, the field of RL has seen significant achievements in several areas. Deep RL, for instance, has enabled the creation of AlphaZero, which learned to play Go, chess, and most board games solely through self-play. AlphaZero is an example of combining deep RL with other approaches like Monte Carlo tree search (MCTS). This approach has also been successful in other games like poker and shogi. In robotics and autonomous systems, model-based RL has enabled the training of agents to make decisions in various environments. For example, in robot control tasks, model-based RL has been used to learn models of the robot and the environment, enabling the robot to plan actions and make decisions with greater accuracy and efficiency. Meta RL has also been applied to robotics, allowing agents to adapt quickly to new environments and tasks. In addition, multi-agent RL has been used in social network analysis (SNA) to model the interactions between individuals and predict their behavior. Finally, offline RL has the potential to improve sample efficiency and learn from previously collected data in situations where online data collection is not feasible. This approach has been applied in recommender systems, where collecting online user feedback data is difficult or expensive. Overall, the field of RL is continuously evolving, and the combination of these approaches leads to breakthroughs in many different areas.

2   Models with Irreversible Actions

Systems with open spaces or irreversible actions are a significant challenge in many real-world applications. Open space of actions leads to models with infinite possible paths, and irreversible actions must be tried many times before the algorithm learns their value. Suppose an agent in such systems performs an incorrect action. In that case, it can lead to unpredictable and undesirable consequences, potentially harming the company or people's lives and health.

One example of such systems is robots that have access to various equipment in production. If such a robot performs an incorrect action, such as turning off critical equipment, it can significantly damage the company. Another example can be energy-efficient building management systems, where some actions can lead to increased electricity consumption or, conversely, to a decrease in the management system's efficiency.

Also, models with open spaces and irreversible actions can be problematic in autonomous systems, such as unmanned aerial vehicles (UAVs). If such a system performs an incorrect action, it can lead to an accident and significant risk to the lives and health of people. These systems require careful design and testing before being put into operation and constant monitoring and support to prevent possible unforeseen consequences.

We can use RL to address this issue in our particular domain. The main idea involves teaching the agents to avoid these undesired actions. The process may involve designing a model that distinguishes between desirable, neutral, and undesirable game states and using the Monte-Carlo tree search algorithm and a heuristic CNN to generate fictitious samples and optimize policies.

2.1 Sokoban Game

Sokoban is a classic puzzle game that originated in Japan in the early 1980s but is based on an even older board game. The word "Sokoban" translates to "warehouse keeper" in Japanese. In this game, the player plays a warehouse worker who must push crates or boxes to their designated storage locations within a maze-like environment.

The game typically uses a grid-based board, where each grid cell represents a specific location within the warehouse. The crates and storage locations are also represented as cells on the grid. The player's character usually starts at a fixed position within the warehouse.

The objective of Sokoban is to move the crates strategically, utilizing the limited space and maneuvering capabilities of the warehouse, to push each crate onto a designated storage location. The puzzle is solved once all the crates are placed in their corresponding storage locations.

The biggest problem in solving the game Sokoban is the irreversibility of some states. One can lose the game because of one wrong move at the beginning. Usually, one can understand this only after extensive calculations of all possible sub-states.

Most reinforcement learning algorithms rely on exploring new states because choosing greedy actions often is a realistic heuristic. However, it also leads to situations when the agent is likely to get stuck in a local maximum from a reward point of view, thus missing out on potentially greater rewards in states it has yet to visit. One solution to this dilemma is the ε-greedy strategy. The agent chooses a greedy action with probability $1 - \varepsilon$, according to the current estimate of possible actions from the current state or estimates of the following states, and chooses a random action with probability ε. Thus, increasing ε increases the level of exploration of new states.

Unfortunately, such a strategy would be fatal in Sokoban, as many moves lead to irreversible states. Therefore, the agent's actions would be highly inefficient compared to humans. Avoiding harmful states is much more important than exploring new ones. Nevertheless, we do not entirely abandon exploration in the algorithm. The agent will randomly choose a state among those with an equal estimate, for example, the unexplored ones.

As an algorithm for solving the game, we propose to use the Monte Carlo tree search with a specific modification: on the third step of the algorithm, instead of simulating, we use an estimate of the state function that takes into account our unwillingness to end up in an irreversible state and, accordingly, assigns a low value to such a node. A similar algorithm was presented in [2]. However, unlike the one presented in this paper, none of the heuristics used in that work considered the drawbacks of ending up in an irreversible state and did not try to avoid them. As a result, the proposed method improved solving rate for 6×6 games with two boxes, which was an adequate problem for our current computational capacity and enough to make conclusions about the potential of this approach.

3 Related Works

Sokoban was first used as a benchmark problem in artificial intelligence in the early 1990s. In particular, the target was to evaluate search algorithms and heuristic techniques for solving problems. Since then, Sokoban has been a benchmark problem for evaluating the performance of various AI planning algorithms. Reinforcement learning methods have been applied to Sokoban in recent years, with some promising results. However, it is difficult to determine precisely when RL methods were first applied to Sokoban, as there has been ongoing research on this topic. Nonetheless, several recent studies have demonstrated that RL-based approaches can achieve state-of-the-art performance on Sokoban, and the development of RL algorithms continues to be an active area of research in this field. The current state-of-the-art (SOTA) methods for Sokoban involve combining deep reinforcement learning with search-based methods. One such approach is using Monte Carlo tree search (MCTS) guided by a deep neural network. Another SOTA approach uses a hybrid method that combines a deep Q-network (DQN) with symbolic planning [8].

We can use several performance metrics to measure these methods' effectiveness. The most commonly used metric is the average number of moves required to solve a set of randomly generated Sokoban levels. Other metrics include the percentage of levels solved within a certain number of moves, the success rate of the algorithm, and the average time taken to solve a level. Researchers may also use additional metrics to evaluate the efficiency or complexity of the algorithm, such as the number of expanded nodes during a search, the number of training episodes required to converge, or the memory usage of the algorithm. Overall, the effectiveness of a Sokoban solver is typically evaluated based on its ability to solve a wide range of levels quickly and efficiently.

In the article [2], a solution to the problem uses a method called Expert Iteration. It involves using two phases of learning, which have the following components:

– the apprentice policy that uses heuristics for the given state to generate an answer;
– the expert policy that models future game trajectories and uses predicted results to make more accurate decisions.

An example of an apprentice policy could be a convolutional neural network that takes the game state image as input and outputs the necessary evaluations. [2] suggests using the A2C architecture, which results in a scalar value representing the evaluation of the current state and four scalars representing the probabilities for each corresponding action. As an expert policy, they suggest using a Monte Carlo tree search.

Paper [6] proposes a slightly similar approach. The method also uses a Monte Carlo tree search and AlphaZero neural network for a heuristic. Using AlphaZero, instead of A2C, leads to much more complex architecture. However, the principle remains the same: the network receives the game state image as input and outputs the evaluation of the state and the probabilities of performing actions.

In the paper [5] authors propose a novel approach to solving the game Sokoban using forward-backward reinforcement learning (FbRL), which consists of two phases. In the forward phase, the agent uses an estimate of the state value function to select actions that lead to high-value states. In the backward phase, the agent updates its estimate of the state value function using the outcomes of those actions.

The authors also introduce a new approach to exploring the state space that considers the irreversible nature of some states. Specifically, the agent uses a mixture of greedy and random actions, with the probability of a random action decreasing when the value function becomes better known. The paper evaluates the proposed approach on a set of Sokoban levels and shows that it outperforms previous RL methods on several metrics, including success rate and efficiency. The authors also conduct experiments to explore the impact of different hyperparameters on the algorithm's performance. Overall, the paper provides a promising new approach to solving Sokoban and demonstrates the effectiveness of FbRL in tackling RL problems with irreversible states and actions.

Also, an overview paper [4] should be mentioned. The paper surveys deep model-based reinforcement learning (RL) algorithms for high-dimensional problems with large state and/or action spaces. The authors focus on model-based RL approaches that learn a model of the environment dynamics and use this model to plan actions rather than directly learning a policy. They note that model-based RL can be more sample-efficient than model-free methods, especially for high-dimensional problems. The paper begins with an introduction to the challenges of high-dimensional RL and the advantages of model-based RL for these problems. It then provides an overview of different models that can be used for model-based RL, including linear and nonlinear models and deep neural networks. The authors also discuss the different components of a model-based RL algorithm, such as data collection, model training, planning, and policy optimization. Next, the paper reviews several deep model-based RL algorithms, including Model-Based Value Expansion (MBVE), Deep Model-Based Reinforcement Learning (DMBRL), and Stochastic Value Gradient (SVG). The paper also discusses some open research questions and challenges for deep model-based RL, such as dealing with uncertainty in the learned models, handling continuous action spaces, and scaling up to large problems. The authors conclude by summarizing the paper's main contributions and providing directions for future research in this area.

While the papers above also used curriculum learning and other modifications of the algorithms, we focus only on MCTS + neural network combination in this paper. We propose replacing the apprentice policy in the algorithm with a heuristic trained to separate irreversible states.

4   The Heuristic

4.1   Evaluation of the Sokoban Game State

Since the game state is quite complex and simple evaluations, such as linear ones, do not fully capture all its properties, the best solution is to use neural networks. A common approach is to use CNN's that take the game's current state as input and output, an evaluation of the game state, probabilities of making certain moves, or evaluations of those moves. Here we use a single scalar value in the output data, which evaluates the input image of the game state.

Each element of the game field is a 16×16 pixel square corresponding to a specific field element. This approach significantly improves the neural network results by adding a convolutional layer with a kernel size of 16×16 and a stride of 16 as the first layer.

To improve the evaluation quality and the training speed, we also normalize the image for each channel to the range $[0, 1]$.

Overall, the network architecture is quite simple and consists, excluding the input and output layers, of two convolutional layers and two inner fully connected layers.

4.2   Preparing Data for Model Training

We aim to train the network separating irreversible losing states from other states. In particular, we can identify winning and intermediate states from the starting state to the winning state. For model training, we define three types: losing, winning, and starting states from which the agent begins the game. The use of intermediate states is not necessary, as the starting states of the game are similar to any intermediate state.

The target value for losing states in the training dataset is -5; for winning states, it is 10; and for starting states, it is 0. The choice of these values is arbitrary but influenced by the properties of the environment. The experiments used game fields of size 6×6 with two boxes. Rewards are the followings: $+10$ for winning, $+1$ for placing a box in the target position, -1 for pushing a box off the target position, and -0.1 for each agent step.

At the same time, there is a limitation on the maximum number of steps the agent can take to solve the game, which is 50. Such a limitation is necessary because the environment does not provide any information about whether the agent is losing. Once the agent reaches this state, the game will continue to infinity because winning is impossible. Therefore, after reaching 50 steps and not winning the game, the agent would receive a reward of -5 – the value we used to label the training dataset. At the beginning of the game, the agent does not have a reward, so the current value of the total winnings is 0, and upon winning, the agent receives a reward of 10, which explains our choice of reward for the other part of the dataset.

4.3 Training of the Network

We conduct experiments with two types of output data. In the first case, the output was a value estimation, and in the second case, the network performed the classification task for states by itself. We prepared the dataset for training using [7] by generating new games and modifying starting positions to the desired ones. All the experiments were conducted with Python programming language using Tensorflow framework for network representation. The code provided in [9], which uses [7]. The generated dataset contains 3431 states of each type. The training was performed in batches of 64 states over 20 epochs.

The metric for the network with state estimation was the mean squared error between the estimate and the true value. The error plot for the training and validation dataset is shown in Fig. 1. The metric value on the test dataset was 1.921.

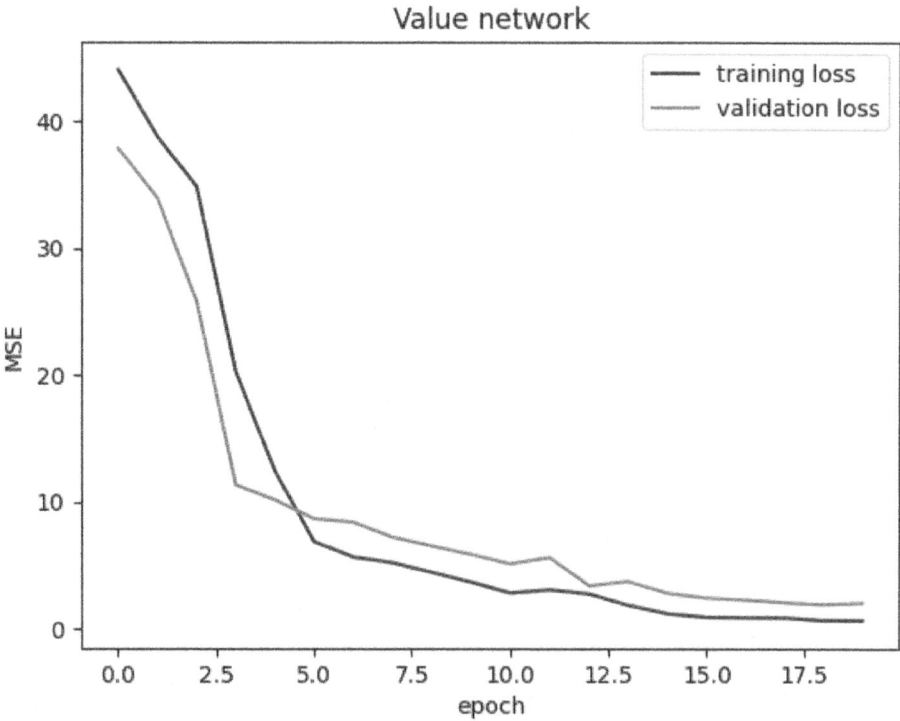

Fig. 1. Value network performance

For the network solving the classification problem, the fraction of correctly classified images (accuracy) was used as the metric. The classification accuracy plot for the training and validation dataset is shown in Fig. 2. The metric value on the test dataset was 0.9396.

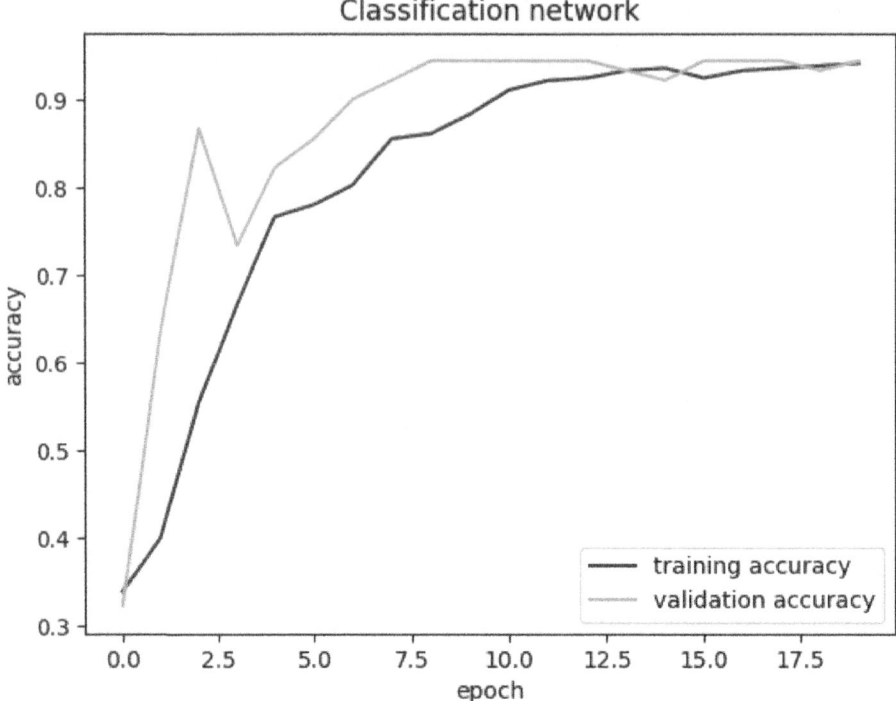

Fig. 2. Classification network performance

Overall, the described model performs well on the given task, achieving a classification accuracy of almost 94% on the test dataset. The mean error value for the network that estimates the state is slightly less than two on the test dataset. Considering the values used (-5, 0, and 10) for different types of states, we can state that the network separates them reasonably well.

5   Experiments

5.1   General Information

We solve the Sokoban game during the experiments using a heuristic similar to the one described in the previous section.

The general technique in solving games using Monte Carlo trees is as follows: the more iterations the algorithm allows for choosing an action from the current state, the more accurate the evaluation of the value of the next states or actions becomes, and accordingly, the ability to win the game increases. However, the number of iterations cannot grow infinitely since as the number of computations increases, the time necessary to find the next step also rises. At some point, it can become impractically large.

In these experiments, the number of solved games measured for 20, 40, 60, 80, and 100 iterations of tree expansion per one move of the agent. If the agent makes 50 moves, the game is considered lost. We calculate the proportion of solved games among 50 generated game fields. We used a fixed set of identical generated starting fields to compare the ratio of solved games with different numbers of iterations of tree expansion.

Since we aim to avoid unwanted states, we do not include searching actions in the algorithm. Nevertheless, this does not mean the agent always moves in the optimal direction. Occasionally, the algorithm faces a situation where choosing between states with the same evaluation is necessary. A simple example is the first iteration of the MCTS algorithm, during which all the states are unexplored. In such cases, the algorithm evenly selects candidates among child vertices. This behavior leads to fluctuations, so sometimes Sokoban game setup is solved for one amount of steps during training experiments but for another during evaluation. This is because of the presence of randomness, which leads to suboptimally and moves in a direction that does not lead to a win.

There is another problem that makes the game challenging – sometimes, quite a large number of actions are required to direct the agent toward the optimal trajectory. A simple case of this can be seen by imagining an agent located in the adjacent position to the left of a box, and the target position is to the left of the agent. To move the agent to the other side of the box, four actions are required, which requires conscious planning that is only sometimes easy for algorithms not designed for such modeling. A similar approach to solving the game was proposed in [1]. Combining the described method with heuristics for avoiding irreversible states may be the subject of further research.

Having obtained the estimate for a leaf node of the tree, we update the estimates of all its ancestors as follows:

$$V(s)^* = \max_i[r_i + V(s_i)^*] \tag{1}$$

$V(s)^*$ – current node value estimation
r_i – reward for performing action i, obtained from interacting with the environment
$V(s_i)^*$ – child nodes value estimation

To choose an action to perform from the current state, we apply the following formula:

$$a^* = arg\max_i[r_i + V(s_i)^*] \tag{2}$$

Here, $V(s_i)^*$ is the estimate of the values of the child nodes of the root node.

5.2 Classification Network vs Value Network as a Heuristic for MCTS

In this section, we train the classification network to separate only two types of states – winning and losing. The underlying idea is to make preliminary labeling using data from MCTS and apply it to improve move evaluation.

The classification network was trained to provide a high probability close to 1 for winning fields and a low probability close to 0 for losing ones.

We propose the following transformations: When receiving an estimate of 0.9 or higher, we consider the position classified as winning and give it a score of 10. For an estimate of 0.1 or lower, we consider the position classified as losing and give it a score of −5. In all other cases, the network is uncertain about the classification of losing/winning, so there is a high probability of observing an intermediate state, to which we give a score of 0.

The value network evaluations are obtained directly from the model.

Results for different numbers of rollouts are illustrated in Fig. 3.

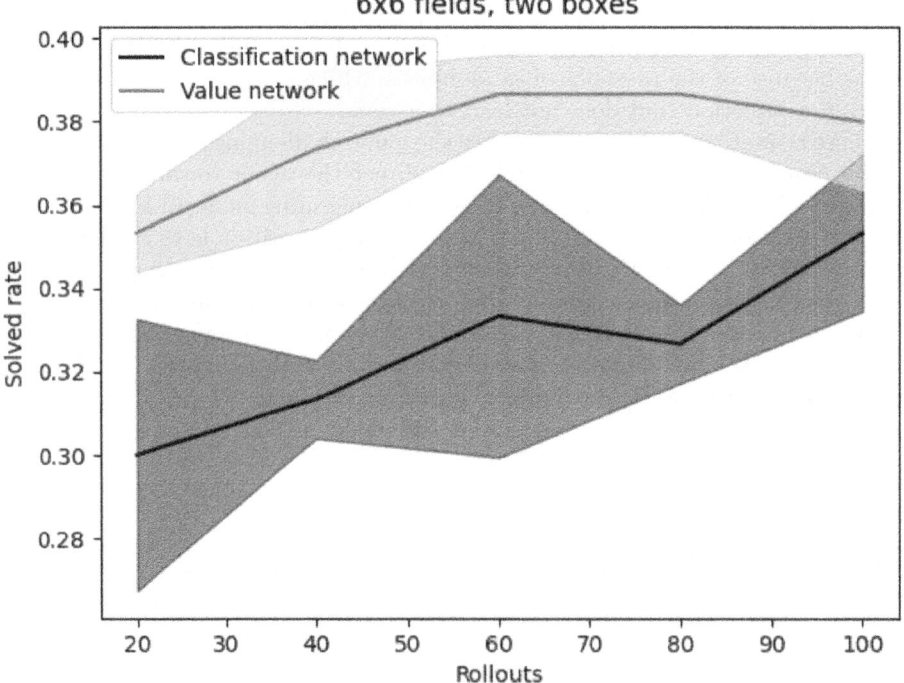

Fig. 3. Classification network vs Value network

As can be seen from the plot, the performance of the value network yields significantly better results for any number of rollouts. From now on, all experiments are conducted using only this type of network.

5.3   Attempt to Fine-Tune the Network During Training

The next step was to attempt to retrain the network using data obtained during the operation of the main Monte Carlo tree search algorithm.

We only used the state value estimation network for these experiments, as previous experiments showed that using the classification network and the main algorithm was ineffective.

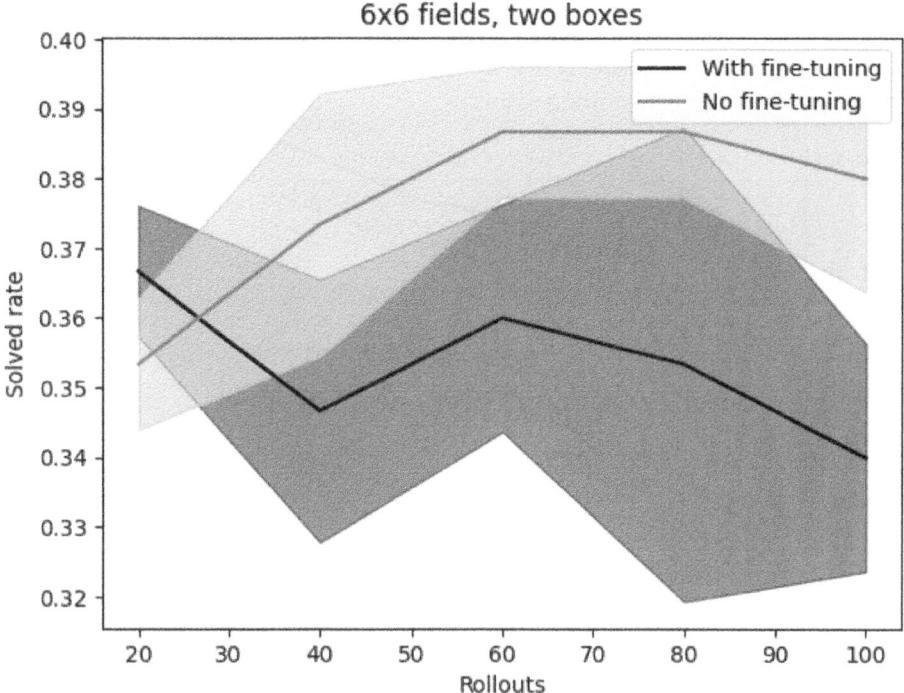

Fig. 4. Fine-tuning vs no fine-tuning

As supplementary, we also used a network trained to separate only winning and losing states.

To retrain the network, we must collect data in a specific way. For each play, after the algorithm's operation, we get the final game state, all intermediate states between the final and initial states, and the result – whether it was a win or a loss. The loss did not always mean reaching a dead-end state, but more often, it was due to reaching the limit on the number of moves allowed to the agent during the game. If the agent ended up in the final state due to winning, we assign a value of 10 to that state. Otherwise, we assign it a value of 0. For all intermediate states, we move from the final to the initial state and assign them a value using the following formula:

$$prev = current - -0.1 \qquad (3)$$

The state preceding the current one receives a value equal to the current value decreased by 0.1, corresponding to a penalty for making an empty move. Thus, we try to label the obtained game states according to the real value function. When the accumulated dataset contains 64 states, we launch network fine-tuning on the corresponding batch. The results are shown in Fig. 4.

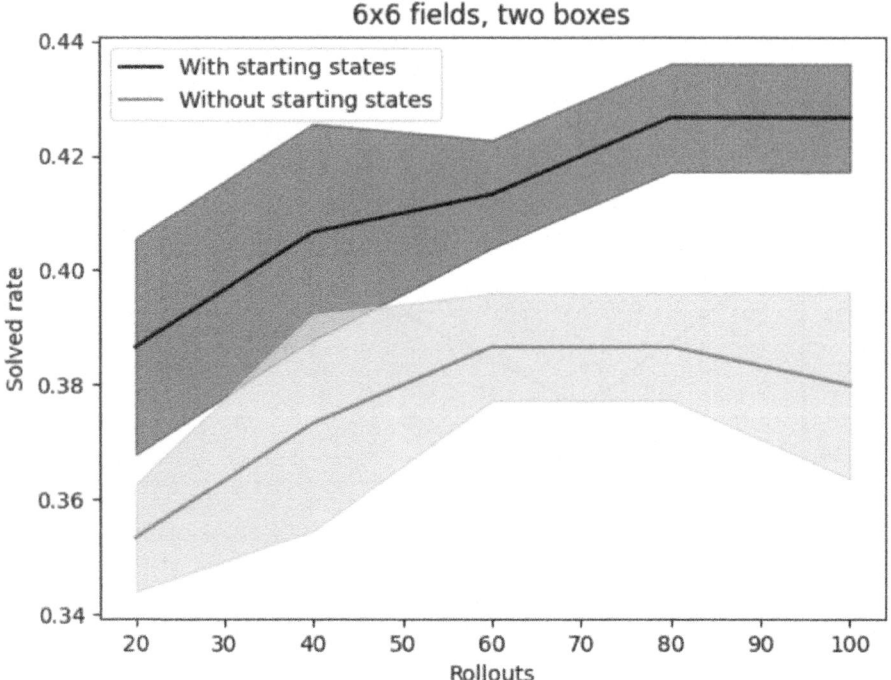

Fig. 5. With starting positions vs No starting positions

Unfortunately, this approach gives worse results than using the network as it is after training on a prepared dataset. A possible reason can be that, in this way, the network begins to lose information about dead-end states and separates them worse.

5.4   Comparison of Heuristics Trained on Different Datasets

In this experiment, the quality of the network's performance is compared when using the main algorithm in two cases:

1. The network is trained only on winning and dead-end states.
2. The network is trained on winning, dead-end, and starting states.

The results are shown in Fig. 5.

As we can see, the plotted trend with simple states in the dataset looks more promising, having a higher value for each number of rollouts.

Table 1. Highest solved games percentage

Method/Rollouts	20	40	60	80	100
Classification network	34	32	38	34	38
Value network	36	40	40	40	40
Value network with fine-tuning	38	36	38	40	36
Value network trained on 3 types states	40	42	42	**44**	**44**

6   Conclusion

All experiments results are summarized at Table 1. From which we conclude that the value network heuristic trained on three types of states without fine-tuning gave the best results using 80 or 100 MCTS rollouts: 44% solved games.

Compared to previous works, the heuristic provides decent results, although somewhat worse. Nevertheless, its simplicity and ability to combine with other algorithms are noteworthy. One possible future usage is to calculate initial weights for more complex networks.

In future work, we plan to improve the algorithm's performance by modifying standard methods by adding new learning stages. For example, the use of the Expert Iteration method described in [2], the Curriculum Strategy [6], or modifications of trajectory encoding methods in neural networks described in [1].

Acknowledgments. We would like to thank the Armed Forces of Ukraine for providing security, that made this work possible.

References

1. Racanière, S., et al.: Imagination-augmented agents for deep reinforcement learning. In: Advances in Neural Information Processing Systems, vol. 30 (2017). https://arxiv.org/abs/1707.06203
2. Ge, V.: Solving planning problems with deep reinforcement learning and tree search (2018). https://hdl.handle.net/2142/101086
3. Sutton, R.S., Barto, A.G.: Reinforcement Learning: An Introduction. MIT Press, Cambridge (2018)
4. Plaat, A., Kosters, W., Preuss, M.: Deep model-based reinforcement learning for high-dimensional problems, a survey. arXiv preprint arXiv:2008.05598 (2020)
5. Shoham, Y., Elidan, G.: Solving Sokoban with forward-backward reinforcement learning. In: Proceedings of the International Symposium on Combinatorial Search, vol. 12, no. 1 (2021)
6. Feng, D., Gomes, C.P., Selman, B.: A novel automated curriculum strategy to solve hard Sokoban planning instances. In: Advances in Neural Information Processing Systems, vol. 33, pp. 3141–3152 (2020). https://arxiv.org/abs/2110.00898
7. Gym Sokoban. https://github.com/mpSchrader/gym-sokoban. Accessed 30 Apr 2023

8. Kissmann, P., Edelkamp, S.: Improving cost-optimal domain-independent symbolic planning. In: Proceedings of the AAAI Conference on Artificial Intelligence, vol. 25, no. 1 (2011)
9. https://github.com/Pronod/Sokoban

Information System for Calculating the Shortest Route for a Mobile Robot in a Multilevel Environment Based on Unity

Oleksandr S. Gerasin[1]($\boxtimes$) (iD), Andriy M. Topalov[1] (iD), Valeriy V. Zaytsev[1] (iD), Dmytro V. Zaytsev[1] (iD), Oleksandr M. Susak[2] (iD), and Oleg V. Savchenko[1]

[1] Admiral Makarov National University of Shipbuilding, 9 Heroes of Ukraine Av., Mykolaiv 54025, Ukraine
oleksandr.gerasin@nuos.edu.ua, topalov_ua@ukr.net
[2] DESIGN BUREAU IMT LLC, 5/1, Buznika str., Mykolaiv 54038, Ukraine

Abstract. An integral part of the global task of increasing the efficiency of movement processes and performing specified technological operations by a mobile robot is the choice of a movement path from the initial (starting) to the final (target) point. At the same time, a preliminary assessment of the proposed route of movement is an important stage in the development of a program for its movement. In this paper, we consider the problem of finding the shortest path for moving the mobile robot in a multilevel environment, taking into account the features of transitions between levels, restrictions, the presence or absence of obstacles along the way. An information system for calculating the shortest route from the starting point to the target point is proposed based on the NavMesh navigation system in the Unity virtual environment. The results of computer simulation of robot movement in a multilevel environment show that the system successfully finds the shortest route for various scenarios.

Keywords: Information System · Shortest Route · Mobile Robot · Multilevel Environment · Navigation System

1 Introduction

A characteristic feature of modern scientific and technological progress is the wide implementation of the robots in the field of production and scientific research. The robots are the universal machines for reproducing human motor and intellectual functions. The practical purpose of creating the robots is to transfer to them those types of activities that are time-consuming, difficult, monotonous, harmful to health and life for humans. The robots are able to replace people during emergency rescue operations after man-made disasters in the nuclear power industry, in the chemical, oil and gas, and mining industries, during the elimination of the consequences of natural disasters [1], and in operations to combat terrorism [2]. They can be used to protect objects and patrol territories [3].

The mobile robots are used mainly in extreme conditions, where people cannot be inside the vehicle, or if their stay there is associated with a risk to life. In addition, in

G. Antoniou et al. (Eds.): ICTERI 2023, CCIS 1980, pp. 53–64, 2023.
https://doi.org/10.1007/978-3-031-48325-7_5

conditions of increased danger, a person may begin to make mistakes, his work capacity and effectiveness of actions decrease. The remote use of the mobile robots is justified in normal conditions for performing heavy or long monotonous work, as well as intra-workshop transport, in automated warehouses, when carrying out earthworks in quarries, etc. [4].

Productive, correct and economical functioning of the mobile robot is ensured by a built-in or remote control system, which processes the given task signals of moving the robot and forms the appropriate control influences on the robot's actuators (motors, valves, distribution mechanisms). Usually, such task signals come from the robot's motion planning information system [5], which provides calculation and selection of an effective motion trajectory in various conditions based on a cartographic base, taking into account information from technical senses and the navigation system [6]. In a broad sense, traffic planning information systems solve the task of laying out a route on a digital map, displaying the location of the robot on this map, displaying a three-dimensional model of the observed scene on the screen and combining it with a video image, warning of a possible occurrence abnormal situations in case of erroneous actions of the operator, recording of all his actions in the electronic log, etc.

The method of generating task signals is largely responsible for the efficiency of further movement and performance of assigned tasks by a mobile robot in various operating environments [7]. Most often, it is based on finding the optimal trajectory of the robot's movement according to the criteria of the minimum path, minimum energy consumption and the possibilities of overcoming it by a specific type of robot. Let's consider in more detail the existing information technologies for finding the optimal distance routes for moving the robot from the initial to the final position, taking into account the features of the surrounding environment of its functioning.

2 Review of Existing Information Systems for Calculating Optimal Routes

Planning and calculating the trajectories of robots, transport and other moving objects are key components of many modern information systems aimed at ensuring the most efficient and safe routes. These systems typically rely on a combination of algorithms, mathematical and simulation models to determine the best paths for moving objects. Generally, in addition to calculating trajectories, these systems also need to take into account a number of other factors, such as collisions with other objects, environmental obstacles, changes in terrain or weather conditions. The creation of such systems requires the development of complex models that can take into account a number of variables and provide accurate predictions of how objects will move over time, so consider the latest developments of researchers in this direction.

Methods for planning the trajectory of the mobile robot in an unknown environment with obstacles are proposed in [8], while the optimization of ant colonies to plan a research path in dynamic environments is used in [9]. There are cases of development of the mobile robots trajectories in construction, for example [10] presents a fully integrated system of measurement and control of the mobile robots for high-precision construction of structures. A method for assessing the patency and planning the trajectory of unmanned ground vehicles with suspension systems on rough terrain is proposed for difficult areas in [11]. The algorithm developed in [12] evaluates the accuracy of tracking the path of the three-wheeled omnidirectional mobile robot created as a personal assistant. Trajectory planning problems for applications in the agricultural sector are also often discussed, for example [13] presents a reconfigurable unmanned ground vehicle for simulating the agricultural tasks.

The development of maritime transport can shift the transportation of goods and people to waterways to reduce car traffic, so [14, 15] describe autonomous land vehicles Roboat and Roboat II for urban waterways. The authors of [16] propose a decentralized approach to traffic planning for several autonomous marine vehicles using a sampling-based algorithm. Web mapping projects such as Google Maps, OpenStreetMap, Bing Maps, etc. can be used to obtain data for autonomous robot navigation.

Paper [17] presents an autonomous navigation system based on OpenStreetMap for a four-wheeled robot that uses a 3D lidar and a CCD camera. The system generates a global path plan based on the input map and localizes the robot using a 3D lidar and visual odometry system. The CCD camera is used to detect and classify objects in the robot's environment. To ensure the safety of road traffic, attention is paid to the planning of the lane change trajectory for automated vehicles, which takes into account different types of roads [18]. The proposed method generates a smooth collision-free trajectory by combining the minimal jerk and spline techniques. The work [19] shows the early stage of development of a robotic system for studies of polluted environments, construction of maps to assist in task planning, and also for research of waste sorting manipulations.

Based on the results of the review it is possible to conclude that researchers have paid considerable attention and a high level of scientific knowledge to the determination of optimal routes for the movement of the mobile robots, taking into account restrictions and obstacles on level surfaces, such as: roads between individual destinations [11, 12, 18] or industrial buildings [8], floors of work and warehouse spaces [9], water surfaces of river and sea water areas [14–16].

However, in existing publications almost no attention is paid to multilevel environments, which are found in many applications of the mobile robots [10, 13, 17, 19]. Sometimes transitions between levels in such environments can occupy a significant part of the robot's travel path, which directly affects the calculation and selection of the shortest route from the starting point to the end (finish) point, which are located on different levels. So, the aim of the work is the development the information system for calculating the shortest route for the mobile robot taking into account transitions between levels in the multilevel environment. To achieve the stated aim it is necessary to propose: a) an algorithm for the process of selecting and forming the shortest path, taking into account transitions between levels in a multilevel space, and b) computer automation tools (an information system) to speed up this process and analyze the resulting path.

3   Algorithm for Finding the Shortest Path in a Multilevel Environment

A characteristic feature of multilevel environments is the presence of heterogeneous transitions between separate, relatively homogeneous or non-homogeneous levels. As an understandable general example, consider the design of an ordinary residential multi-story building (Fig. 1). In the building (see Fig. 1) it is possible to single out the uniform floors of the stories, which are located horizontally, and the stairs, which have a broken profile and are located between the levels obliquely at a certain angle to the horizontal. Traditionally, for a one-level representation the task of finding the optimal path can be solved by searching for the route "start point on level A"-"lift on level A"-"lift on level B"-"end point on level B" (of course, taking into account the restrictions and obstacles on the levels A and B). Similar situations arise when using elevators, escalators or winches.

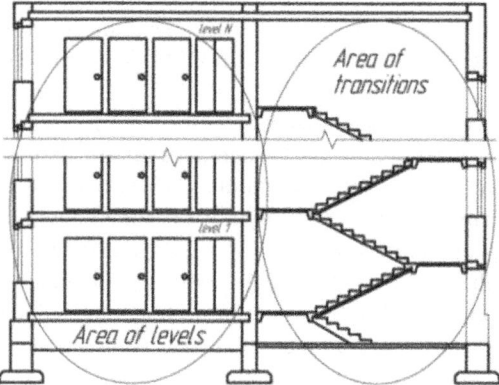

Fig. 1. Scheme of a multi-storey building.

However, not all buildings are equipped with such lifting systems. In addition, there are cases of partial destruction of communications or transport shafts, which makes their use impossible. Then, in this case, it is not necessary to calculate the route of the robot's movement taking into account the physically existing transitions between individual levels. For example, such a problem arises, during demining from unexploded shells or aerial bombs, which are located randomly inside large constructions of apartment buildings, warehouses, industrial buildings. In such cases the search for the shortest path from the starting point to the end point can be further complicated by the simultaneous presence of different types of destruction, for example in the case of partial destruction of large buildings: in different parts of the building and on different floors (something exploded, and something did not). Weights or scores should be used for each elementary path accounting safety risks and feasibility.

Usually, when compiling algorithms for solving complex problems, they are divided into separate, smaller parts, which together form the final solution [7]. The task of finding the shortest path in a multilevel environment can be formulated more broadly by distinguishing the following main separate subtasks (see Fig. 2):

1) assignment of numbers of the starting level i and target level n; 2) input of vectors of coordinates of entry points to the current level $\mathbf{V}_{en}(i)$, transition to another level $\mathbf{V}_{ten}(i)$ or finish $V_f(i)$, obstacles $\mathbf{V}_{ol}(i)$ and restrictions $\mathbf{V}_{ll}(i)$ on the corresponding level; 3) checking the possibility of moving the robot from the entry points of the current level to the points of transition to the next level or the finish point, selection of possible routes; 4) search and calculation of optimal length routes $\mathbf{L}_l(i)$ within the current level among all possible ones considering shortest path target function (TF) f; 5) input of coordinates vectors of exit points from transitions $\mathbf{V}_{tex}(i)$ (in the future, they will be points of entry to the next level), obstacles $\mathbf{V}_{ot}(i)$ and restrictions $\mathbf{V}_{lt}(i)$ at the corresponding transitions; 6) checking the possibility of moving the robot along transitions from entry points to the transition to exit points from the transition, selection of possible transitions; 7) search and calculation of optimal length routes $\mathbf{L}_t(i)$ within transitions from current level to the next among all possible ones considering shortest path TF f; 8) repeating stages 2–7 to determine optimal routes within levels and transitions from the starting level to the target level; 9) determining the shortest possible route L_{opt} or concluding that there is no possible route.

It should be noted that the algorithm in Fig. 2 is made according to the following assumptions: the map of the environment is known; and the environment is static (the availability of routes does not change over time); within the framework of each subtask, the route from the entry point to the transition point is calculated taking into account restrictions and obstacles, and the physical possibility of overcoming the transition is tested [20]. Generally, the different algorithms may be used and then investigated for finding the trajectory with function f in Fig. 2, e.g. calculus-based, graph-based or numerical methods [21].

There are cases when it is impossible to make a path from a certain level L_k to the neighboring level L_{k+1} due to the impossibility of physically overcoming such a transition with the available means of mobile robotics (features of the design of a particular robot). Therefore, in the calculation algorithm, it is advisable to include the possibility of "returning" to the previous level L_{k-1} and finding a way to go to the L_{k+1} level, bypassing the L_k level and immediately making the transition from the L_{k-1} level to the L_{k+1} level in the area of passage between the levels (for example, on the stairwells).

In general, the following two types of transitions can be distinguished: with continuous and discrete description of movement route. Examples of the first are inclined flat surfaces (ramp, sloping building board) and curved surfaces with a constant or variable radius of curvature. The second type is characterized by blockage and, as its partial case, stairs. It should be noted that, for the most part the possibility of a robot transition from one level to another is assessed by experts based on data from visual monitoring of the state of such transitions and the type of the robot used. Experts can conduct an inspection of transitions "on the spot", remotely use means of objective control (for example, using a drone), and sometimes, in difficult cases, third-party experts can be involved using Internet of Things technologies [22].

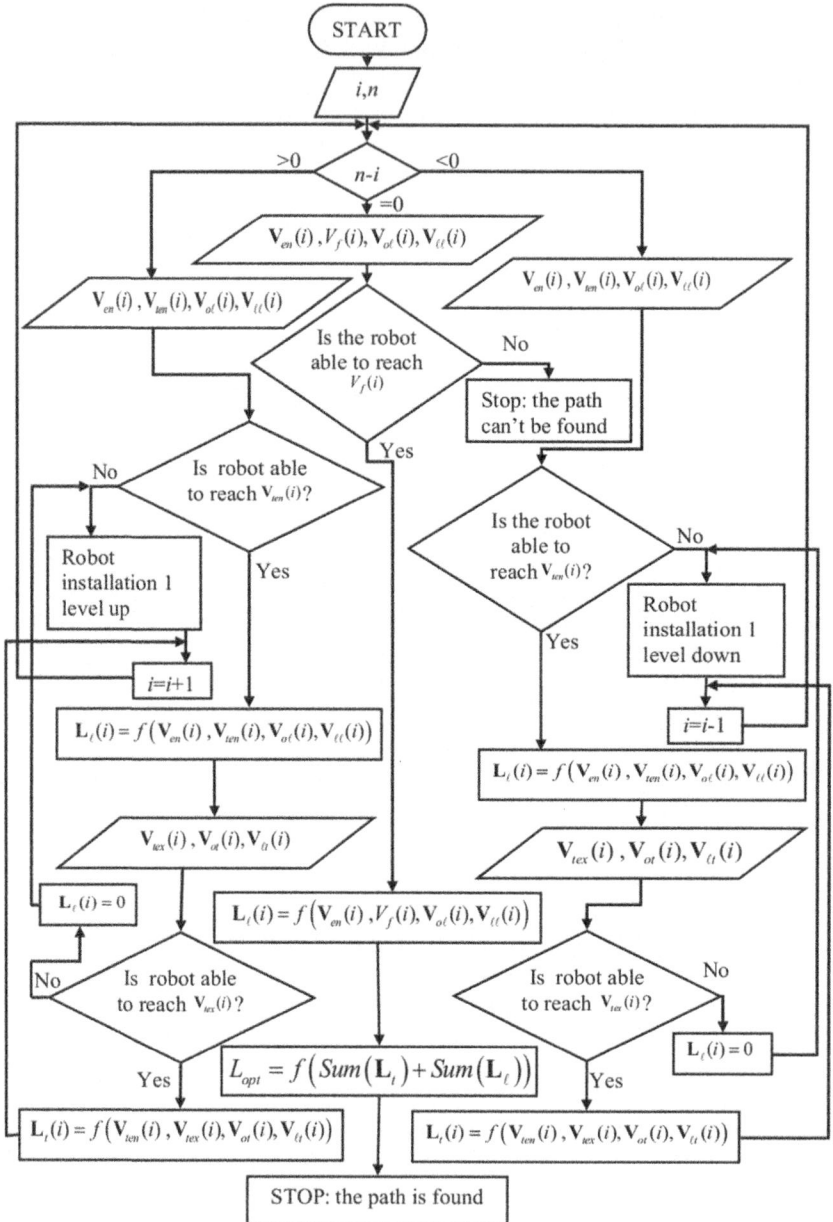

Fig. 2. Algorithm for finding the shortest path in the multilevel environment.

4   Information System for Calculating the Shortest Path in a Multilevel Space Based on Unity

It is known that the use of modeling tools allows to significantly speed up the development or solution of a certain problem, which is related to a physical experiment [23]. In general, the process of developing an information system for finding the shortest path can be described in the form of three stages:

1) formation of a virtual map of a multilevel environment, which contains existing restrictions and obstacles;
2) provision of means of assignment of the initial and final point of movement (initial information);
3) programming the algorithm for finding the minimum path by given points.

For the initial task of starting and target points, as well as finding and calculating the length of the robot movement routes within individual levels and between levels, it is recommended to use the building plan of the structure (house) that acts as the environment for the robot's operation (for example, as in Fig. 1). Such a plan contains drawings of walls and floor slabs with the main dimensions of the building during construction, which can be used to clearly and accurately determine the position in the space of individual rooms, the presence of doors, windows, stairs, as well as the necessary distances for laying the route. Reproduction of such a drawing or its parts using an information system is usually the initial stage for creating a virtual map of a deterministic multilevel working environment of the mobile robot.

In cases where there is no building plan or there is a significant re-planning of the building, which is not marked on it, it is advisable to apply experimental methods of orientation in the building. In the absence of coordinates of placement of obstacles and limitations of the mobile robot, when it functions in unstructured or weakly structured environments, local navigation methods can be used [24]. Such tasks arise when performing reconnaissance or territory research in an a priori unknown environment. The local map of the mobile robot's environment describes the placement of obstacles at distances determined by the range of sensors and can be represented in various ways. The mobile robot searches for the optimal path to move (closest to the target) for navigation to the finish based on the local map of the environment. At the same time, as a rule, the map of the environment presents information about areas free for movement.

To implement an information system for calculating the optimal movement of a the mobile robot to a given goal, this work uses the Unity application development environment. Systems created using Unity support DirectX and OpenGL have a high level of visibility and wide opportunities for creating and configuring elements of virtual or augmented reality [25, 26]. The information system for finding the minimum path is based on the use of the NavMesh component, which is a data structure that describes restrictions and static obstacles in the virtual environment, and allows forming the geometry of the environment in the form of passing surfaces of levels and transitions.

Figure 3 shows the created map of a multilevel virtual environment, which consists of 4 horizontal levels and 9 transitions between levels, such as: stairs and inclined surfaces. Combining levels with each other using different types of transitions allows to more deeply explore the obtained solutions to the problem of finding the minimum path.

In particular, the system has the ability to adjust (based on expert information on the physical ability of a specific robot to overcome): the distance to a wall, a ledge, a turn, an obstacle; height of passages; the maximum slope of the surface; step height (how high obstacles the robot can step on).

Fig. 3. 3D-map visualization (side view).

The virtual model of the multilevel environment was baked after the model formation. Then the abstract mobile robot has the ability to move along the traversable surfaces. For this, the procedures for collecting coordinates from the starting point and the end point, as well as finding the optimal trajectory of movement according to the algorithm in Fig. 2. In addition, movement routes in the proposed information system can be specified manually by clicking the mouse.

5   Analysis of the Results of Simulation the Operation of the Information System

In this paragraph, we will model the process of moving an abstract robot for a virtual multilevel environment in the described information system. Let's consider the cases of finding the optimal route in the absence and presence of obstacles, as well as taking into account the physical possibility of the robot overcoming transitions. The time required for the robot to overcome the determined optimal trajectory is determined for each of the testing cases.

Figure 4 shows the shortest path for moving the robot without obstacles from the starting position to the final position, which the robot covers in 24.25 s. The information system analyzed the entire map and built the optimal trajectory without any obstacles on the map.

Figure 5 illustrates the optimal route according to the presence of an obstacle at the beginning of the last level (almost at the end of the previous route). In this case, the system rebuilt the optimal trajectory taking into account this obstacle. The time taken by the robot to overcome the received route was 26.31 s, which is longer than the previous one, but it is the shortest of all others for this configuration of the multilevel environment.

The result of the system operation when the two entry points to the last level are blocked by obstacles is shown in Fig. 6. When laying the optimal path, the possibility of physically moving the robot along the inclined surface immediately from the lower

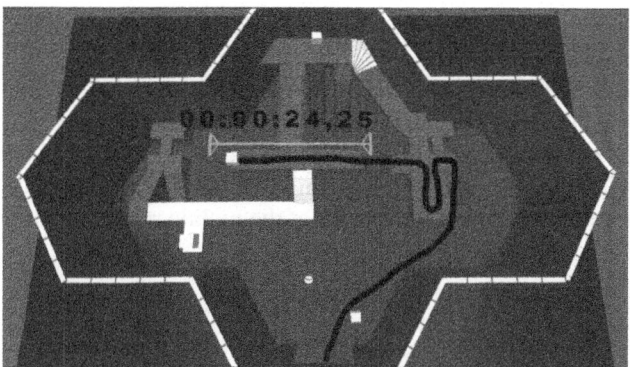

Fig. 4. Shortest route and time of movement from the starting point to the target point without obstacles.

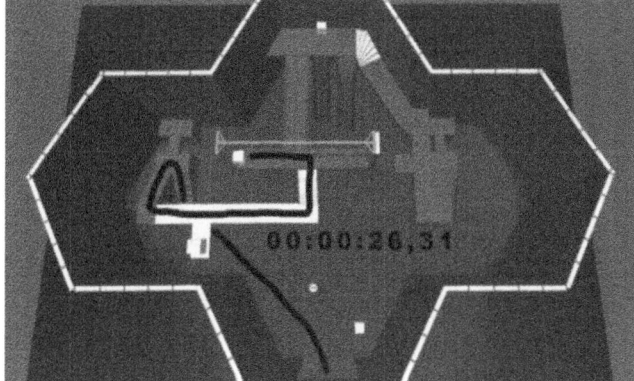

Fig. 5. Shortest route and time of movement from the starting point to the target point with one obstacle.

to the upper level is not taken into account (in terms of length, this would be the shortest path). As can be seen in this case not the shortest route is chosen, the chosen route is the only one of all possible ones.

The results of simulation of the optimal route determination by the information system are summarized in the Table 1, where the number and range of turns of the robot, as well as the number and % of straight sections from the entire path are indicated. It should also be noted that in the considered cases, transitions between levels are carried out sequentially: from the initial (first) to the final (fourth) with the same number of transitions involved (3 each). In a broader case other options are possible, in particular for the case with two obstacles (Fig. 6), a shorter path could be found through the transition from level 1 to level 4 (if the robot can overcome it).

In turn, presented in Table 1 parameters further allow to estimate the energy costs for overcoming levels and transitions, that is especially important for the autonomous mobile robots to predict the sufficiency or insufficiency of their embedded power sources to overcome a particular route.

The obtained results show the possibility of determining the optimal route for the movement of a mobile robot in a multilevel environment by the proposed system using the example of a multilevel composition. The practical significance of the system based on multilevel space concept lies in the preliminary modeling and evaluation of the most suitable route for the subsequent movement of the mobile robot along this route.

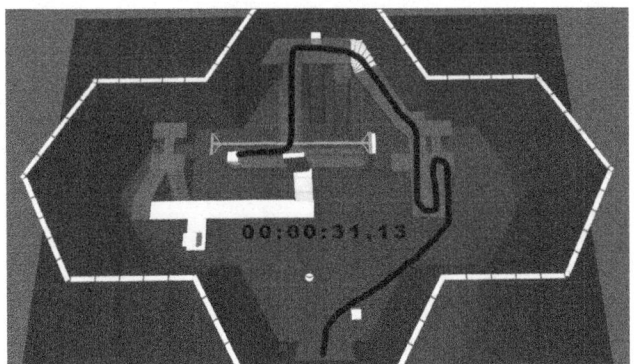

Fig. 6. Shortest route and time of movement from the starting point to the target point with two obstacles.

Table 1. Analysis of the results of simulation optimal robot movement routes in the multilevel environment.

Route with	Time, s	Number of		Range of turning angles, °	Straight areas of the entire path, %
		turns	straight areas		
0 obstacles	24,25	5	5	60–180	47
1 obstacle	26,31	6	6	30–180	57
2 obstacles	31,13	8	9	60–180	53

6   Conclusions

The overview of modern information systems for determining optimal routes shows the problem of considering transitions at moving a mobile robot in a multilevel environment. The proposed shortest path algorithm takes into account the peculiarities of transitions between the surfaces of levels, iteratively calculates the optimal lengths of paths both within individual levels and within transitions between levels. An information system for calculating the shortest route from the starting point to the target point is proposed based on the algorithm and NavMesh navigation system in the Unity virtual environment. The results of computer simulation of robot movement in a multilevel environment show that the system successfully finds the shortest route for various scenarios taking into account presence or absence of obstacles, restrictions on movement and the robot features. Thus, the system makes it possible to visualize a complicated multilevel environment and correctly determine the shortest route, which can then be overcome by an experimentally real robot. Further research should be directed towards combining the proposed system with augmented reality tools (for example, from ARCore [27]) in order to expand the possibilities for orienting the real robots in unfamiliar areas with estimation of energy costs for moving.

References

1. Trevelyan, J., Hamel, W.R., Kang, S.-C.: Robotics in hazardous applications. In: Siciliano, B., Khatib, O. (eds.) Springer Handbook of Robotics, pp. 1521–1548. Springer, Cham (2016). https://doi.org/10.1007/978-3-319-32552-1_58
2. Wang, Y., Xing, J.-P., Xu, Y., Hu, B.: Structural design of driving system for anti-terrorism robot. In: Mengda, X., Kunyu, Z. (eds.) 2nd International Conference on Advances in Mechanical Engineering and Industrial Informatics, AMEII, vol. 73, pp. 1430–1435 (2016)
3. Lee, M.-F.R., Shih, Z.-S.: Autonomous surveillance for an indoor security robot. Processes **10**(11), 2175 (2022)
4. Gerasin, O.S., Kozlov, O.V., Kondratenko, G.V., Rudolph, J., Kondratenko, Y.P.: Neural controller for mobile multipurpose caterpillar robot. In: 10th IEEE International Conference on Intelligent Data Acquisition and Advanced Computing Systems: Technology and Applications, IDAACS, vol. 1, pp. 222–227. IEEE, Metz (2019)
5. Mao, W., Liu, H., Hao, W., Yang, F., Liu, Z.: Development of a combined orchard harvesting robot navigation system. Remote Sens. **14**, 675 (2022)
6. Pivarčiová, E., et al.: Analysis of control and correction options of mobile robot trajectory by an inertial navigation system. Int. J. Adv. Robot. Syst. 1–15 (2018)
7. Kozlov, O.V., Kondratenko, Y.P., Skakodub, O.S., Gerasin, O.S., Topalov, A.M.: Swarm optimization of fuzzy systems for mobile robots with remote control. J. Mob. Multimed. **19**(3), 839–876 (2023)
8. Tebueva, F., Pavlov, A., Satybaltina, D.: A method for planning the trajectory of a mobile robot in an unknown environment with obstacles. In: ES3 Web of Conferences, WFCES, vol. 270, p. 01035 (2021)
9. Santos, V.D.C., Otero, F.E.B., Johnson, C., Osorio, F.S., Toledo, C.F.M.: Exploratory path planning for mobile robots in dynamic environments with ant colony optimization. In: Genetic and Evolutionary Computation Conference, GECCO, pp. 40–48 (2020)
10. Gawel, A., et al.: A fully-integrated sensing and control system for high-accuracy mobile robotic building construction. In: IEEE/RSJ International Conference on Intelligent Robots and Systems, IROS, Macau, China, pp. 2300–2307 (2019)
11. Zhang, K., Yang, Y., Fu, M., Wang, M.: Traversability assessment and trajectory planning of unmanned ground vehicles with suspension systems on rough terrain. Sensors **19**(20), 4372 (2019)
12. Palacín, J., Rubies, E., Clotet, E., Martínez, D.: Evaluation of the path-tracking accuracy of a three-wheeled omnidirectional mobile robot designed as a personal assistant. Sensors **21**(21), 7216 (2021)
13. Saeed, R.A., Tomasi, G., Carabin, G., Vidoni, R., von Ellenrieder, K.D.: Conceptualization and implementation of a reconfigurable unmanned ground vehicle for emulated agricultural tasks. Machines **10**(9), 817 (2022)

14. Wang, W., Gheneti, B., Mateos, L.A., Duarte, F., Ratti, C., Rus, D.: Roboat: an autonomous surface vehicle for urban waterways. In: IEEE International Conference on Intelligent Robots and Systems, IROS, pp. 6340–6347. IEEE, Macau (2019)

15. Wang, W., et al.: Roboat II: a novel autonomous surface vessel for urban environments. In: IEEE/RSJ International Conference on Intelligent Robots and Systems, IROS, pp. 1740–1747. IEEE, Las Vegas (2020)

16. Volpi, N.C., et al.: Decoupled sampling-based motion planning for multiple autonomous marine vehicles. In: OCEANS 2018 MTS/IEEE Charleston, pp. 1–7. IEEE, Charleston (2018)

17. Li, J., Qin, H., Wang, J., Li, J.: OpenStreetMap-based autonomous navigation for the four wheel-legged robot via 3D-lidar and CCD camera. In: Levi, E. (eds.) IEEE Transactions on Industrial Electronics, vol. 69, no. 3, pp. 2708–2717. IEEE, Liverpool (2021)

18. Wang, Y., Wei, C., Liu, E., Li, S.: Lane-changing trajectory planning method for automated vehicles under various road line-types. IET Smart Cities **2**(1), 14–23 (2020)

19. West, C., Arvin, F., Cheah, W., West, A., Watson, S., Giuliani, M., Lennox, B.: A debris clearance robot for extreme environments. In: Althoefer, K., Konstantinova, J., Zhang, K. (eds.) TAROS 2019. LNCS (LNAI), vol. 11649, pp. 148–159. Springer, Cham (2019). https://doi.org/10.1007/978-3-030-23807-0_13

20. Gerasin, O., Kondratenko, Y., Topalov, A.: Dependable robot's slip displacement sensors based on capacitive registration elements. In: IEEE 9th International Conference on Dependable Systems, Services and Technologies, DESSERT, pp. 358–363. IEEE, Kyiv (2018)

21. Wang, J., Meng, M.Q.-H., Khatib, O.: EB-RRT: optimal motion planning for mobile robots. IEEE Trans. Autom. Sci. Eng. **17**(4), 2063–2073 (2020)

22. Kondratenko, Y., Gerasin, O., Kozlov, O., Topalov, A., Kilimanov, B.: Inspection mobile robot's control system with remote IoT-based data transmission. J. Mob. Multimed. **17**(4), 499–522 (2021)

23. Topalov, A., Kondratenko, G., Gerasin, O., Kozlov, O., Zivenko, O.: Information system for automatic planning of liquid ballast distribution. In: 2nd International Workshop on Information-Communication Technologies & Embedded Systems, pp. 191–200. CEUR Workshop Proceedings, Mykolaiv (2020)

24. Gul, F., Rahiman, W., Alhady, S.S.A.: A comprehensive study for robot navigation techniques. Cogent Eng. **6**(1), 1–25 (2019)

25. Axak, N., Korablyov, M., Ushakov, M.: The development of a multi-agent system for controlling an autonomous robot. In: 17th International Conference on ICT in Education, Research and Industrial Applications. Integration, Harmonization and Knowledge Transfer, pp. 96–105. CEUR Workshop Proceedings, Kherson (2021)

26. Unity Manual: Building a NavMesh. https://docs.unity3d.com/560/Documentation/Manual/nav-BuildingNavMesh.html. Accessed 09 May 2023

27. Build new augmented reality experiences that seamlessly blend the digital and physical worlds. IARCorel. Google for Developers. https://developers.google.com/ar?hl=en. Accessed 12 Aug 2023

Analysis and Systematization of Vulnerabilities of Drone Subsystems

Maryna Kolisnyk[1,2(✉)] and Oleksandr Piskachov[1]

[1] National Aerospace University "Kharkiv Aviation Institute", Kharkiv, Ukraine
{m.kolisnyk,a.piskachev}@csn.khai.edu
[2] Vienna University of Technology (TU Wien), Vienna, Austria

Abstract. A drone is a software (SW) and hardware (HW) complex that has wireless data transfer technologies (Wi-Fi, LTE, 5G, Bluetooth, etc.) To transfer data, it uses various communication protocols: both specific and generally known.

Drones can perform complex tasks, but some cyber-attacks (such as Denial-of-Services - DoS) can lead to the failure of individual components of the drone and the entire system as a whole.

Guidelines for protecting against drone attacks are provided by many organizations that develop cybersecurity standards (NIST, CERT, CISA, etc.). Methods to prevent cyber-attacks can be used to drones, with adjustments to their parameters and architectural features. There are also recommendations for protecting drone components and also methods of communication protocols protection from cyber-attacks.

The authors of this study offer a comprehensive approach to the analysis of vulnerabilities of drone subsystems, which includes a system analysis of drone architecture, vulnerability analysis by different vulnerability databases, and their systematization.

Keywords: Vulnerabilities · Drone · System Analysis · Cyber-Attacks

1 Introduction

1.1 Motivation and Relevance of Research

Drones use communication protocols and embedded systems such as I2C, SPI, Serial, CAN, 1-wire, etc., operating systems (OS), hardware (HW) in the control device of the drone is similar to other digital systems. Therefore, drones can be affected by cyber-attacks that use the vulnerabilities of these HW and software (SW) subsystems of the drone in their mechanisms. Flight controllers used in UAVs are usually built using popular microcontrollers such as: Atmega, STM (ARM), Intel Movidius, sometimes low-voltage Intel/AMD x86/x64 microcontrollers (Fig. 1) [1, 2].

The scientific novelty of the obtained results lies in the fact that, based on the system analysis of the architecture of drones, the paper will perform an analysis of the vulnerabilities of their HW and SW subsystems, and their systematization according to the types of cyber-attacks.

© The Author(s), under exclusive license to Springer Nature Switzerland AG 2023
G. Antoniou et al. (Eds.): ICTERI 2023, CCIS 1980, pp. 65–81, 2023.
https://doi.org/10.1007/978-3-031-48325-7_6

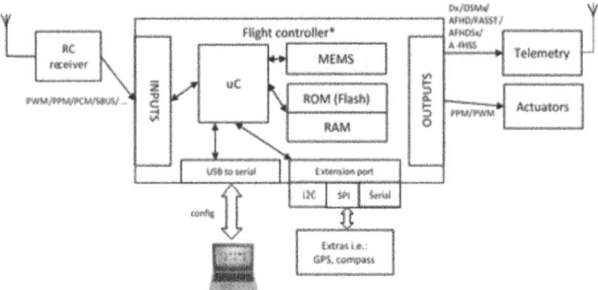

Fig. 1. The structure scheme of drone according to [1]

The practical significance of the obtained results is that the systematization of drone vulnerabilities according to the types of cyber-attacks will allow us to propose a method of prioritizing drone's vulnerabilities and a method of assessing and ensuring their dependability, which will increase the reliability and cybersecurity of drones.

1.2 Work Related Analysis

The issue of cybersecurity of drones is very relevant, especially during military operations. There are publications about communication methods and cybersecurity issues of drones. The report [1] describes decisions to organize IP access method to drones/ground stations based on the use of LTE and Wi-Fi technologies. Authors of the paper [2] described a high-level architecture multi-drone system consisting of quadrotors to develop a drone-based system with on-board sensors and built-in processing, detection, coordination, communication and networking functions. With different conditions and impact factors. An attacker can hack into a drone's video stream in order to gain access and steal confidential information, as described in the paper [3], and a threat analysis of the drone's security is performed. In paper [4], considering the main approaches to indoor collision avoidance, architecture, communication and routing protocols, Flying Ad-hoc Networks (FANET) is presented. The paper [5] presents some networking protocols for the UAV wireless networks. The paper [6] proposes the system of data transmission from a drone in the Internet of Things (DDC-IoT) for data collection centers and drones is considered in order to analyze the throughput and delay of data transmission. In [7] was analyzed a vulnerability in the micro-aircraft communication protocol (MAVLink) was described and a potential attack methodology leading to the failure of the tasks assigned to the drone. In [8], an encryption method is proposed that ensures the security of communication between the UAV and the GCS, based on the analysis of the vulnerabilities of the MAVlink protocol. Since a drone is a complex HW and SW device that connects to the Internet using appropriate data transmission technologies, it can be considered one of the network devices. And this means that it can be affected by all types of cyber-attacks that operate in a regular computer network [9, 10]. Similar measures to prevent cyber-attacks can be applied, with some amendments to the parameters and features of the drone structure. There are many organizations (NIST, CERT, etc.) offering guidance on how to defend against drone vulnerability attacks. There are also guidelines for protecting drone components and the communication protocols they use to transmit data separately from cyber-attacks. The authors of this study offer a comprehensive approach to analyzing the vulnerabilities of drone subsystems, based on their systematization.

The purpose of this study: analysis and systematization of the most serious vulnerabilities of drone subsystems and cyber-attacks that can affect them, and recommendations for their prevention, detection and mitigation.

2 Analysis of Drone's Subsystems Vulnerabilities

Vulnerabilities in the OS. The correct performance of drone tasks depends on the correct functioning of the OS. Microcontrollers and drone processors use the Linux OS. The [11] from 2021 indicates a large number of cyber-attacks specifically on the Linux OS [11].

EvilGnome spyware [12, 13] which is currently not included in all major antivirus security SW products, including features rare to most Linux malware. In the Linux kernel, a vulnerability signed in the National Vulnerability Database (NVD) as CVE-2020-14314, an out-of-memory read vulnerability was discovered in a way to access a wrongly indexed directory [14]. This vulnerability could allow a local user to fail the system and compromise system availability. Over the past 2 years, about 50 vulnerabilities of varying degrees of severity have been discovered in the Linux OS. According to the Linux OS Vulnerability Database, CVE-2020-16119 was discovered when the DCP protocol was injected into the OS kernel, a Denial of Service (DoS) attack (destruction of the OS) or the ability to execute malicious code, and vulnerable places lead to such consequences [15, 16].

Vulnerabilities CVE-2020-14314 [14] and CVE-2020-16120 [17], CVE-2020-14385 [18], CVE-2020-20285 [19], CVE-2020-25641 [20] help an attacker to perform successful DoS-attacks. Many malware targeting Linux OS mainly focus on attacks to create DoS botnets by hijacking vulnerable servers [14, 17–20]. The latest vulnerability of 2022 - CVE-2022-23222 [21] in the Linux kernel with severity level 7.2 gives possibility to a local users to change privileges [21]. There is a risk of complete disclosure of information, as a result of which all system files are disclosed, a complete violation of the integrity of the system occurs. An attacker can disable the entire cyberattack protection system and make the OS inoperable; there are no special access conditions or mitigating circumstances. Using the vulnerability does not require authentication and additional knowledge or skills. A large number of malware targeting Linux mainly focus on attacks to create DoS botnets by hijacking vulnerable servers. CVE-2022-2873 [22] and CVE-2023-2194 [23] - out-of-bounds vulnerabilities in the Linux kernel's SLIMpro I2C device driver [22, 23].

They could give a possibility to a local privileged user to crash the system (DoS) or remote code execution [23]. If the firewall in the OS fails, the risk of a successful attack on the drone increases, as a malfunctioning OS can cause the drone's systems to fail.

The drone works with a large number of communication protocols, such as: Hypertext Transfer Protocol (HTTP), Constrained Application Protocol (CoAP), Extensible Messaging and Presence Protocol (XMPP), Advanced Message Queuing Protocol (AMQP) and MQ Telemetry Protocol (MQTT). Discovery and Configuration Protocol (DCP), SSH File Transfer Protocol (SFTP), VPN, Simple Network Time Protocol (SNTP), Simple Network Management Protocol (SNMP), Hypertext Transfer Protocol (HTTP), Hypertext Transfer Protocol Secure (HTTPS), Secure Shell Protocol (SSH), Network Time Protocol (NTP), Real-time Transport Protocol (RTP), Real-time control Protocol (RTCP). Vulnerabilities in these protocols can lead to a successful attack on drone subsystems.

Vulnerabilities in the SSH Protocol. Over 39 SSH vulnerabilities led to successful cyber-attacks in 2022: Stored Cross-Site Scripting (XSS) Vulnerability Exploitable by an Attacker with General/Administrator Permissions (3 Vulnerabilities), DoS (11 Vulnerabilities), CSRF (4 Vulnerabilities), Remote Code Execution (7 vulnerabilities) (Fig. 2). CSRF vulnerability in Jenkins publisher plugin SCP 1.8 makes it possible to connect to an attacker-specified SSH server with fake credentials. CVE-2021-1378 [24, 25] and CVE-2021-1592 [26], vulnerabilities in the Cisco StarOS SSH service, were discovered in 2021 that could allow an unauthorized remote attacker to implement a DoS attack by exploiting a vulnerability related to a logic error when certain traffic conditions occur.

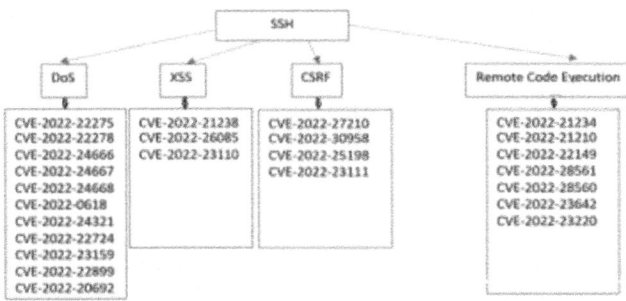

Fig. 2. Vulnerabilities in the SSH protocol

Vulnerabilities in the NTP Protocol. Network Time Protocol (NTP) is a network protocol for synchronizing the clocks of all servers and clients, used in drones. The NTP protocol has been found to be vulnerable to a number of attacks. The most frequently used were command injection (8 vulnerabilities in 2020–2022), XSS (2 vulnerabilities in 2020), DoS (3 vulnerabilities in 2020–2021). In 2022, vulnerabilities that facilitate brute force attacks (1 vulnerability) and remote command execution (3 vulnerabilities) appeared (Fig. 3).

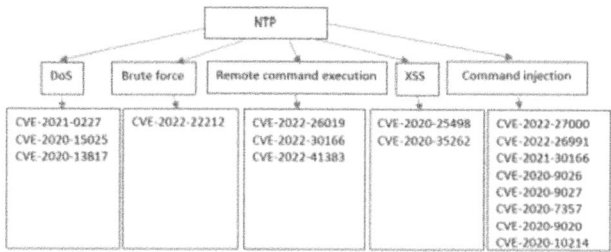

Fig. 3. Vulnerabilities of the NTP protocol

Several vulnerabilities led to successful DoS-attacks: CVE-2020-15025 allows remote attackers to cause DoS (memory consumption) [27]; CVE-2020-13817 [28] allows remote attackers to cause a DoS of drone (modification of system time or denial of service) by predicting transmission timestamps for use in forged packets; CVE-2020-11868 [29] - NTP is vulnerable to DoS; CVE-2022-27000 [30] gives a possibility to attackers to execute commands using fake request; CVE-2022-26019 [31] - an improper access control vulnerability allows a remote attacker to modify NTP GPS settings and overwrite files in the file system.

Real Time Protocol (RTP) Vulnerabilities. CVE-2018-0280 [32] - a bitstream processing vulnerability was detected in the Real-Time Transport Protocol (RTP) in Cisco Meeting Server. Exposure to an unauthenticated remote attacker leads to a DoS state (failure of audio and video services, failures in the media process) by sending a spoofed RTP bit stream when transmitting information from a drone.

Real-Time Control Protocol (RTCP) Vulnerabilities. All users who use PJMEDIA and receive incoming RTP/RTCP calls when operating the drone with PJSIP (using SIP, SDP, RTP, STUN, TURN, and ICE protocols) experience that incoming RTP/RTCP packets may exceed read access limits. In the NVD database, this is vulnerability CVE-2022-21722 [33].

VPN Vulnerabilities. The use of VPNs to improve drone cybersecurity has been proposed by several research groups (Fig. 4) [34]. In 2022, 21 VPN vulnerabilities have already been identified that can lead to various types of cyber-attacks, including DoS. VPNs are dangerous because they expose entire networks to threats such as malware, DDoS-attacks, and spoofing-attacks. A drone VPN network can be destroyed when an attacker penetrates a network through a compromised device. [34–36]. The most common vulnerabilities in 2022 that led to cyber-attacks were: DoS (7 vulnerabilities in 2020), SQL injection (1 vulnerability), Man-in-the-Middle (1 vulnerability), Buffer overflow (1 vulnerability), XSS (1 vulnerability), implementation of OS arguments and commands (3 vulnerabilities) (Fig. 5).

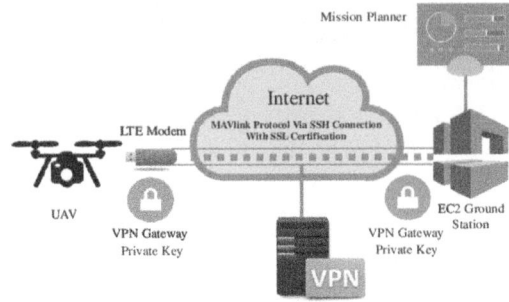

Fig. 4. Scheme of VPN implementation between drones according to [34]

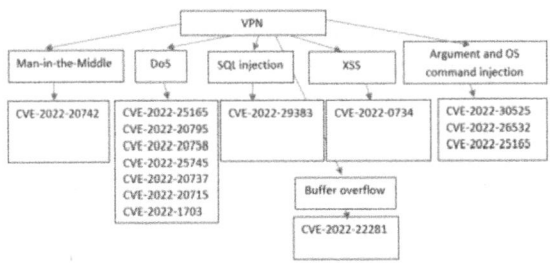

Fig. 5. Vulnerabilities in VPN in 2022

Random Access Memory (RAM) Vulnerabilities. It is known that a dynamic memory cell uses a capacitor and 3 transistors to store one bit of information. Capacitors lose charge over time, and a stored bit value of "1" (which may indicate a high charge) may change to a "0" (low charge). Every time a row of memory is activated for reading or writing (the bits are staggered in rows and columns), currents flowing inside the chip can cause the capacitors to discharge. The charge will flow faster in adjacent rows. By re-activating—or "injecting"—a row of memory (the "aggressor"), an attacker can cause bit errors in an adjacent row, also called a "victim" row. This bit error can be used to give attackers access to restricted areas of a computer system without relying on any software vulnerability. The vulnerability "Rowhammer" is a design flaw in the device's internal memory (DRAM) chips that creates a vulnerability that could allow an attacker to gain control of the drone.

Vulnerabilities in the SNMP Protocol. In 2022, a serious vulnerability appeared in the SNMP protocol [37], which leads to the failure of the drone when a DoS-attack is successfully implemented. In total, 5 protocol vulnerabilities were identified that can lead to successful DoS-attacks (Fig. 6).

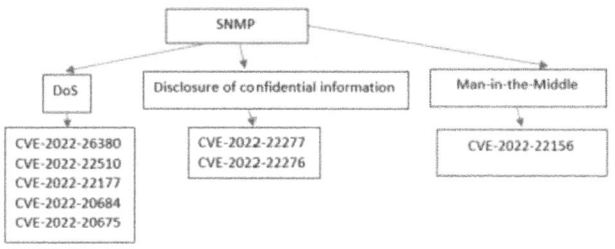

Fig. 6. Vulnerabilities in the protocol SNMP

Vulnerabilities in the MQTT Protocol. Vulnerabilities of the MQTT, CoAP protocols can be used to organize espionage, targeted attacks, intelligence by attackers [38]. An attacker can scan and gain access to vulnerable MQTT peripherals such as drones using IP web scanners and gain access to these devices. An unusual cache-hit vulnerability was discovered in this protocol: CVE-2022-0673 [39]—external schema file cache poisoning via directory traversal. The highest number of MQTT protocol vulnerabilities were exploited by attackers in 2022 and 2021 for successful DoS attacks (3 vulnerabilities), SSRF (1 vulnerability), Man-in-the-Middle (2 vulnerabilities), remote code execution (7 vulnerabilities) (Fig. 7).

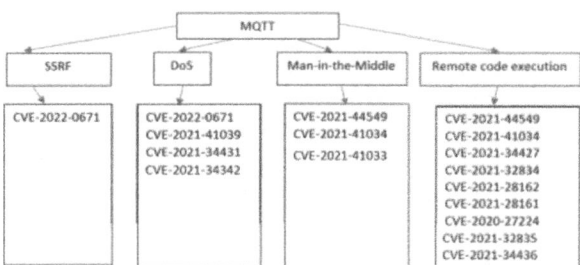

Fig. 7. Vulnerabilities in the protocol MQTT

Vulnerabilities in the CoAP Protocol. CVE-2020-3162 [40] - a Constrained Application Protocol (CoAP) vulnerability in Cisco IoT Field Network Director allows a remote attacker to trigger a DoS state on a drone. With insufficient inspection of incoming CoAP traffic, an attacker exploits the vulnerability by sending a fake CoAP packet to the drone, which causes the CoAP server to stop communicating with the drone.

Vulnerabilities in the XMPP Protocol. Drones use a client-side drone visual measurement tool [1]. The drone uses the camera recording module to transfer recorded data to the XMPP server and database.

The XMPP protocol is used for this purpose, and the data is updated and transmitted to the XMPP server in an optimal form. The XMPP protocol has vulnerabilities, shown in Fig. 8, which can lead to successful cyber-attacks.

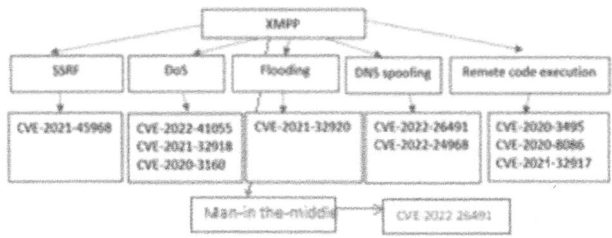

Fig. 8. Vulnerabilities in the protocol XMPP

Vulnerabilities in Processors. The processors used in drones can also contain vulnerabilities. Potential security vulnerability CVE-2022-25899 (vulnerability with severity 9.9) [41] in Intel®-supported Open AMT Cloud Toolkit allows an attacker to escalate privileges and influence the drone control process. There are also 2 of the most serious vulnerabilities that attackers can use to launch a successful cyber-attack. Meltdown (CVE-2017-5754), [42] allows a privileged attacker to read the entire memory of an attacked system via specially crafted executable code. An attacker must: gain physical access to the drone as an administrator, execute a specific program on the drone, read the protected data and send it back to the attacker. Specter [version 1: CVE-2017-5753, [43], version 2: CVE-2017-5715, [43] allows an attacker to read the memory of other processes using specially crafted executable code or dynamic code.

Vulnerabilities in the Protocol SFTP (Secure File Transfer Protocol). The drone can use the SFTP protocol to transfer data. SFTP protocol vulnerability CVE-2022-22899 [41] Core FTP/SFTP Server v2 Build 725 has been discovered to allow unauthenticated attackers to cause a DoS via a crafted packet via the SSH service.

Vulnerabilities in the Protocol UART (Universal Asynchronous Transmitter). CVE-2022-29402 [44]: vulnerability allows attackers to connect to the UART port via a serial connection and execute commands as the root user without authentication [44].

Vulnerabilities in the Protocol SPI. CVE-2021-26317 [45] – by exploiting this vulnerability, an attacker could control the protocol and modify the SPI (Serial Peripheral Interface) flash resulting in a potential arbitrary code execution [45].

Vulnerabilities in the Protocol Controller Area Network (CAN). CVE-2023-2166 – a vulnerability exists in the CAN protocol that an attacker can exploit to crash a drone or cause a DoS condition [46].

3 Systematization of Drone's Vulnerabilities by Severities and Cyber-Attacks

There are some organizations that collect vulnerability information, process it, and provide a severity score according to the Common Vulnerability Scoring System (CVSS). Tables 1, 2, 3, 4, 5, 6, 7, 8, 9, 10 and 11 and Figs. 9, 10, 11, 12, 13, 14 and 15 provide comparative tables listing vulnerabilities in various communication protocols and possible attacks against them, along with the severity of successful attacks on these vulnerabilities. The "other database" column means that information about vulnerabilities can be presented in such databases as SUSE, Red Hat, Greenbone, Talos, Open Source Vulnerability Database (OSVDB), Common Vulnerability Enumeration (CWE). The Open Web Application Security Project (OWASP) and others (Table 12).

Table 1. DoS-attacks

Protocol	Vulnerability	Severity in NVD	Severity in CVE DETAILS	Severity in other databases
MQTT	CVE-2022-0671	9.1	6.4	9.1
	CVE-2021-41039	7.5	5.0	7.5
	CVE-2021-34431	6.5	5.0	6.5
	CVE-2021-34432	7.5	5.0	7.5
XMPP	CVE-2022-41055	7.5	5.0	7.5
	CVE-2021-32918	7.5	5.0	7.5
	CVE-2020-31060	5.3	5.0	5.3
VPN	CVE-2022-25165	7.0	6.9	7.0
	CVE-2022-20795	7.5	5.1	5.0
	CVE-2022-20758	6.8	7.1	7.4
	CVE-2022-25745	9.8	5.5	7.6
	CVE-2022-20737	7.1	7.1	7.1
	CVE-2022-20715	7.5	7.8	8.6
	CVE-2022-1703	8.8	9.0	8.8
SNMP	CVE-2022-26380	7.5	7.0	7.5
	CVE-2022-22510	7.5	5.0	7.5
	CVE-2022-22177	7.5	5.0	7.5
	CVE-2022-20684	6.5	6.1	7.4
	CVE-2022-20675	5.3	5.0	5.3
NTP	CVE-2021-0227	7.5	4.0	7.5
	CVE-2020-15025	4.9	4.0	4.4
	CVE-2020-13817	7.4	5.8	5.9
SSH	CVE-2022-22275	7.5	5.0	7.5
	CVE-2022-22278	7.5	5.0	7.5
	CVE-2022-24666	7.5	5.0	7.5
	CVE-2022-24667	7.5	5.0	7.5
	CVE-2022-24668	7.5	5.0	7.5
	CVE-2022-0618	7.5	7.8	7.5
	CVE-2022-24321	7.5	5.0	7.5
	CVE-2022-22724	7.5	5.0	7.5
	CVE-2022-23159	6.5	4.0	4.8
	CVE-2022-22899	8.8	2.6	-
	CVE-2022-20692	6.5	6.8	7.7

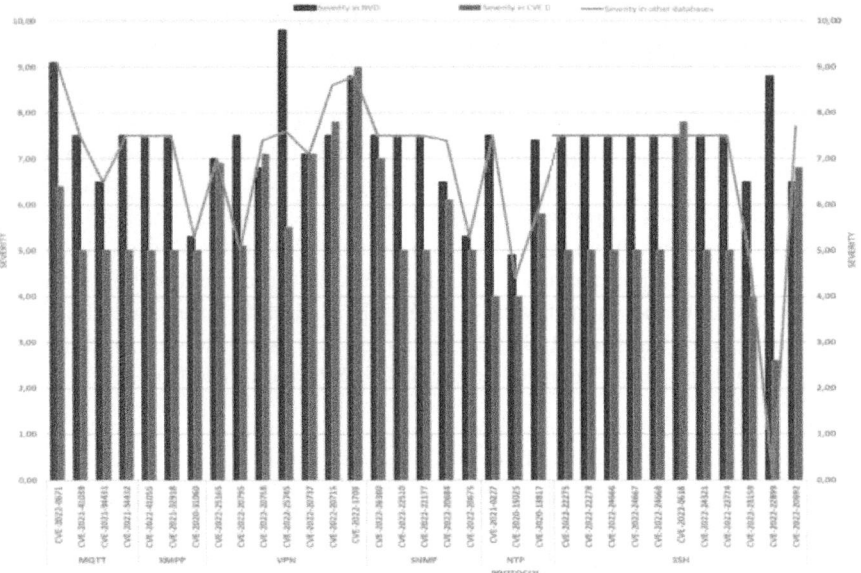

Fig. 9. Graphical dependence of the severity of the vulnerability on the types of vulnerability in the impact protocols DoS-attacks

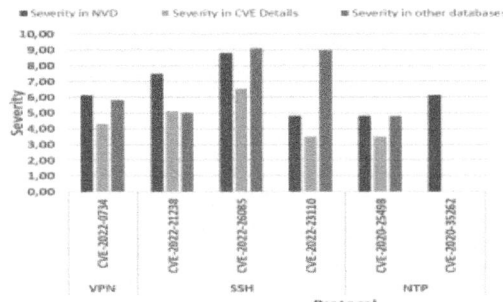

Fig. 10. Graphical dependences of the severity of vulnerabilities on the types of vulnerabilities in protocols when exposed XSS-attacks

Table 2. XSS-attacks

Protocol	Vulnerability	Severity in NVD	Severity in CVE DETAILS	Severity in other databases
VPN	CVE-2022-0734	6.1	4.3	5.8
SSH	CVE-2022-21238	7.5	5.1	5.0
	CVE-2022-26085	8.8	6.5	9.1
	CVE-2022-23110	4.8	3.5	-
NTP	CVE-2020-25498	4.8	3.5	4.8
	CVE-2020-35262	6.1	-	-

Table 3. SSRF-attacks

Protocol	Vulnerability	Severity in NVD	Severity in CVE DETAILS	Severity in other databases
XMPP	CVE-2021-45968	7.5	5.0	7.5
MQTT	CVE-2022-0671	9.1	6.4	6.4

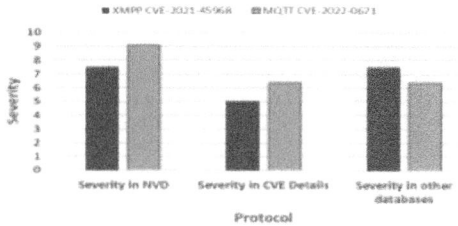

Fig. 11. Graphical dependences of the severity of vulnerabilities on the types of vulnerabilities in protocols and different databases under the influence of SSRF-attacks

Table 4. Flooding-attacks

Protocol	Vulnerability	Severity in NVD	Severity in CVE DETAILS	Severity in other databases
XMPP	CVE-2021-32920	7.5	7.8	7.5

Table 5. DNS-spoofing-attacks

Protocol	Vulnerability	Severity in NVD	Severity in CVE DETAILS	Severity in other databases
XMPP	CVE-2022-26491	5.9	6.4	5.9
	CVE-2022-24968	5.9	8.1	5.9

Table 6. Man-in-the-Middle-attacks

Protocol	Vulnerability	Severity in NVD	Severity in CVE DETAILS	Severity in other databases
VPN	CVE-2022-20742	7.4	5.8	7.4
SNMP	CVE-2022-22156	7.4	5.8	6.5
XMPP	CVE-2022-26491	5.9	6.4	5.9
MQTT	CVE-2022-44549	7.4	5.8	7.4
	CVE-2022-41034	8.1	6.8	8.1
	CVE-2022-41033	8.1	6.8	8.1

Table 7. Brute-force-attacks

Protocol	Vulnerability	Severity in NVD	Severity in CVE DETAILS	Severity in other databases
NTP	CVE-2022-22212	-	7.5	7.5

Table 8. Attacks to steal confidential information

Protocol	Vulnerability	Severity in NVD	Severity in CVE DETAILS	Severity in other databases
SNMP	CVE-2022-22277	5.3	5.0	5.3
	CVE-2022-22276	5.3	5.0	5.3

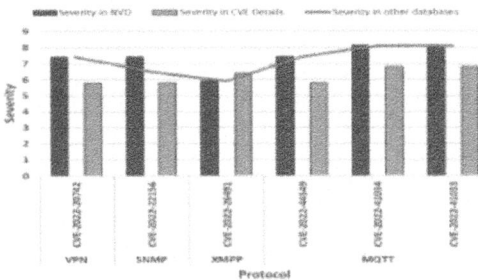

Fig. 12. Graphical dependences of the severity of vulnerabilities on the types of vulnerabilities in protocols in case of exposure to attacks Man-in-the-Middle

Table 9. SQL and others attacks of code injection

Protocol	Vulnerability	Severity in NVD	Severity in CVE DETAILS	Severity in other databases
VPN	CVE-2022-29383	9.8	7.5	9.8

Table 10. CSRF-attacks

Protocol	Vulnerability	Severity in NVD	Severity in CVE DETAILS	Severity in other databases
SSH	CVE-2022-27210	6.5	4.3	6.5
	CVE-2022-30958	8.8	6.8	8.8
	CVE-2022-25198	8.8	6.8	8.8
	CVE-2022-23111	4.3	4.3	4.3

Table 11. Argument injection-attacks and OS commands

Protocol	Vulnerability	Severity in NVD	Severity in CVE DETAILS	Severity in other databases
VPN	CVE-2022-30525	9.8	10.0	9.8
	CVE-2022-26532	7.8	7.3	7.8
	CVE-2022-25165	7.0	6.9	6.9
NTP	CVE-2022-27000	9.8	10	9.8
	CVE-2020-26991	9.8	7.5	6.6
	CVE-2020-9026	8.1	10.0	7.7
	CVE-2020-9027	5.4	10.0	8.1
	CVE-2022-7357 (1022)	5.4	3.5	8.1
	CVE-2022-9020	6.5	7.5	5.9
	CVE-2022-10214	5.9	9.0	6.4
	CVE-2022-30166	8.1	4.6	7.7

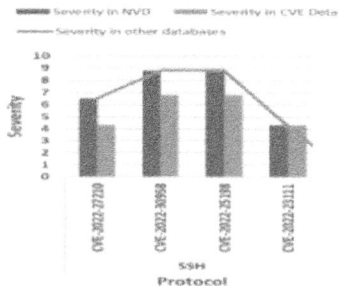

Fig. 13. Graphical dependences of criticality of vulnerabilities on types of vulnerabilities in protocols leading to CSRF-attacks

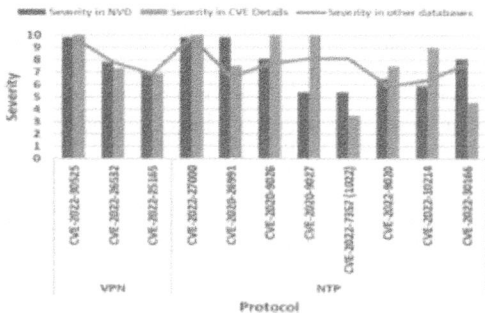

Fig. 14. Graphical dependences of the severity of vulnerabilities on the types of vulnerabilities in protocols, if the attacks of the injection of arguments and OS commands are affected

Table 12. Remote code-attacks and command execution

Protocol	Vulnerability	Severity in NVD	Severity in CVE DETAILS	Severity in other databases
MQTT	CVE-2021-44549	7.4	5.8	7.4
	CVE-2021-41034	8.1	6.8	8.1
	CVE-2021-34427	9.8	7.5	8.8
	CVE-2021-32834	9.9	6.5	9.9
	CVE-2021-28162	6.1	6.1	8.2
	CVE-2021-28161	7.5	5.0	7.5
	CVE-2020-27224	9.6	9.3	9.6
	CVE-2021-32835	9.9	9.9	9.9
	CVE-2021-34436	9.8	7.5	9.8
XMPP	CVE-2020-3495	8.8	8.8	9.9
	CVE-2020-8086	9.8	6.8	9.8
	CVE-2021-32917	5.3	4.3	5.3
NTP	CVE-2022-26019	8.8	8.5	8.8
	CVE-2022-30166	7.8	4.6	7.8
	CVE-2022-41383	5.5	-	-
SSH	CVE-2022-21234	8.8	6.5	9.1
	CVE-2022-21210	8.8	6.5	6.6
	CVE-2022-22149	8.8	6.5	9.1
	CVE-2022-28561	9.8	10.0	9.8
	CVE-2022-28560	9.8	10.0	9.8
	CVE-2022-23642	8.8	6.0	-
	CVE-2022-23220	7.8	7.2	-

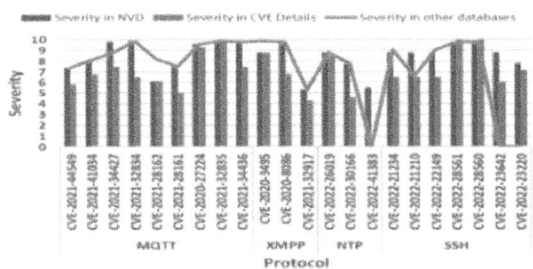

Fig. 15. Graphical dependences of the severity of vulnerabilities on the types of vulnerabilities in protocols in the case of an attack on remote code and command execution

Graphical dependencies show how the severity values of the same vulnerability differ, but in different vulnerability databases.

4   Conclusions

In this paper was realized the system analysis of drone's subsystems, and was made the analysis and the systematization of vulnerabilities in its subsystems. Were analyzed cyber-attacks, which can impact to these vulnerabilities.

The drone may be connected to the global Internet access network and may also be affected by attacks used in conventional computer networks. Cyber-attacks on a drone can be carried out when connected to its interfaces, using wireless Internet access technologies through a 4G/5G router or modem.

The each considered vulnerability is a hole in cybersecurity. From the moment a vulnerability is discovered to the moment a patch is installed for it, the minimum time should pass, since during this period of time attackers can successfully attack vulnerabilities known to them. Therefore, it is very important to update SW versions and fix vulnerabilities in a timely manner.

Drone uses VPN, but the number of VPN vulnerabilities has increased recently. Therefore, it is necessary to take measures for their additional cyber protection.

The conducted analysis showed that the vulnerabilities of drones are in the used general information transfer protocols, in HW and SW. The most of attacks on modern drones, which use vulnerabilities, are various types of DoS-attacks, Man-in-the-Middle-attacks and remote code execution attacks.

The analysis and the systematization of vulnerabilities and cyber-attacks on drones will allow drone manufacturers and users to propose new measures and update recommendations for ensuring cybersecurity for all drone components.

Further research will be aimed at developing a method for prioritizing drone vulnerabilities a method of evaluating and ensuring their dependability, which will increase the reliability and cybersecurity of drones.

References

1. Autonomian. UAV Data Transmission and Protocols, 92 p. https://robolabor.ee/img/cms/projektid/UAV%20Data%20Transmission%20and%20Communication%20Protocols.pdf. Accessed 04 Apr 2023
2. Yanmaz, E., Yahyanejad, S., Rinner, B., Hellwagner, H., Bettstetter, C.: Drone networks: communications, coordination, and sensing. Ad Hoc Netw. **68** (2017). https://doi.org/10.1016/j.adhoc.2017.09.001
3. Pleban, J., Band, R., Creutzburg, R.: Hacking and securing the AR.Drone 2.0 quadcopter - investigations for improving the security of a toy (2014). https://doi.org/10.1117/12.2044868
4. Menoret, S., Auburg, T., Nousi, V., Pitas Aristotle, I.: Drone communications. European Union's Horizon 2020 research and innovation programmed under grant agreement. No 731667 (MULTIDRONE)
5. Sawalmeh, A., Othman, N.: An overview of collision avoidance approaches and network architecture of unmanned aerial vehicles (UAVs). Int. J. Eng. Technol. **7** (2018). https://doi.org/10.14419/IJET.v7i4.35.27395
6. Aranzazu Suescun, C., Cardei, M.: Unmanned Aerial Vehicle Networking Protocols (2021). https://doi.org/10.18687/LACCEI2016.1.S.078
7. Aldeen, S., Yousra, Abdulhadi, H.: Data communication for drone-enabled internet of things. Indones. J. Electr. Eng. Comput. Sci. **22**, 1216. (2021). https://doi.org/10.11591/IJEECS.v22.i2.pp1216-1222
8. Kwon, Y.M., Yu, J., Cho, B.M., Eun, Y., Park, K.J.: Empirical analysis of MAVLink protocol vulnerability for attacking unmanned aerial vehicles. IEEE Access **6**, 43203–43212 (2018)
9. Khan, N.A., Jhanjhi, N.Z., Brohi, S.N., Almazroi, A.A., Almazroi, A.A.: A secure communication protocol for unmanned aerial vehicles. Comput. Mater. Continua **70**, 601–618 (2021). https://doi.org/10.32604/cmc.2022.019419
10. Aerosmart. UAV systems and solutions. Use-case. Drone Detection System (2022). https://www.aerosmart.ae/drone-detection-system/. Accessed 05 Apr 2023
11. Kaspersky. Endpoint Security for Linux. For workstations and servers. https://www.kaspersky.com/small-to-medium-business-security/endpoint-linux. Accessed 05 Apr 2023

12. Lee, M., Choi, G., Park, J., Cho, S.: Study of analyzing and mitigating vulnerabilities in uC/OS real-time operating system. In: 2018 Tenth International Conference on Ubiquitous and Future Networks (ICUFN), pp. 834–836 (2018). https://doi.org/10.1109/ICUFN.2018.8436965

13. Belding, G.: Malware Spotlight: EvilGnome (2020). https://resources.infosecinstitute.com/topic/malware-spotlight-evilgnome/. Accessed 10 Apr 2023

14. National Vulnerability Database. CVE-2020-14314. https://nvd.nist.gov/vuln/detail/CVE-2020-14314. Accessed 05 Apr 2023

15. National Vulnerability Database. CVE-2020-16119, https://nvd.nist.gov/vuln/detail/CVE-2020-16119. Accessed 10 Apr 2023

16. Linux RedHat. https://access.redhat.com/security/cve/cve-2020-16119. Accessed 10 Apr 2023

17. National Vulnerability Database. CVE-2020-16120, https://nvd.nist.gov/vuln/detail/CVE-2020-16120. Accessed 10 Apr 2023

18. National Vulnerability Database. CVE-2020-14385. https://nvd.nist.gov/vuln/detail/CVE-2020-14385

19. National Vulnerability Database. CVE-2020-20285. https://nvd.nist.gov/vuln/detail/CVE-2020-20285. Accessed 10 Apr 2023

20. National Vulnerability Database. CVE-2020-25641. https://nvd.nist.gov/vuln/detail/CVE-2020-25641. Accessed 10 Apr 2023

21. National Vulnerability Database. CVE-2022-23222. https://nvd.nist.gov/vuln/detail/CVE-2020-23222. Accessed 10 Apr 2023

22. RedHat. https://bugzilla.redhat.com/show_bug.cgi?id=2119048. Accessed 10 Apr 2023

23. RedHat. https://bugzilla.redhat.com/show_bug.cgi?id=2188396. Accessed 10 Apr 2023

24. MITRE. CVE-2021-1378. https://cve.mitre.org/cgi-bin/cvename.cgi?name=CVE-2021-1378. Accessed 10 Apr 2023

25. National Vulnerability Database. CVE-2021-1378. https://nvd.nist.gov/vuln/detail/CVE-2021-1378. Accessed 10 Apr 2023

26. National Vulnerability Database. CVE-2021-1592. https://nvd.nist.gov/vuln/detail/CVE-2021-1592. Accessed 10 Apr 2023

27. National Vulnerability Database. CVE-2020-15025. https://nvd.nist.gov/vuln/detail/CVE-2020-15025. Accessed 10 Apr 2023

28. National Vulnerability Database. CVE-2020-13817. https://nvd.nist.gov/vuln/detail/CVE-2020-13817. Accessed 10 Apr 2023

29. National Vulnerability Database. CVE-2020-11868. https://nvd.nist.gov/vuln/detail/CVE-2020-11868. Accessed 10 Apr 2023

30. National Vulnerability Database. CVE-2022-27000. https://nvd.nist.gov/vuln/detail/CVE-2022-27000. Accessed 10 Apr 2023

31. National Vulnerability Database. CVE-2022-26019. https://nvd.nist.gov/vuln/detail/CVE-2022-26019. Accessed 10 Apr 2023

32. National Vulnerability Database. CVE-2018-0280. https://nvd.nist.gov/vuln/detail/CVE-2018-0280. Accessed 10 Apr 2023

33. National Vulnerability Database. CVE-2022-21722. https://nvd.nist.gov/vuln/detail/CVE-2022-21722. Accessed 10 Apr 2023

34. Burleson-Davis, J.: 7 Common VPN Security Risks: The Not-So-Good, The Bad, and the Ugly, April 14, 2021. https://www.securelink.com/blog/vpnproblems/#:~:text=VPNs%20are%20insecure%20because%20they,network%20can%20be%20brought%20down. Accessed 10 Apr 2023

35. Aljehani, M., Inoue, M.: Communication and Autonomous Control of Multi-UAV System in Disaster Response Tasks (2017). https://doi.org/10.1007/978-3-319-59394-4_12

36. Rametta, C., Beritelli, F., Avanzato, R., Russo, M.: A smart VPN bonding technique for drone communication applications. In: 2019 15th International CONFERENCE on Distributed Computing in Sensor Systems (DCOSS), pp. 612–618 (2019). https://doi.org/10.1109/DCOSS.2019.00112
37. National Vulnerability Database. CVE-2022-22510. https://nvd.nist.gov/vuln/detail/CVE-2022-22510. Accessed 15 Apr 2023
38. Husnain, M., et al.: Preventing MQTT vulnerabilities using IoT-enabled intrusion detection system. Sensors **22**, 567 (2022). https://doi.org/10.3390/s22020567
39. National Vulnerability Database. CVE-2022-0673. https://nvd.nist.gov/vuln/detail/CVE-2022-0673. Accessed 21 Apr 2023
40. National Vulnerability Database. CVE-2020-3162. https://nvd.nist.gov/vuln/detail/CVE-2020-3162. Accessed 21 Apr 2023
41. National Vulnerability Database. CVE-2022-22899, https://nvd.nist.gov/vuln/detail/CVE-2022-22899. Accessed 21 Apr 2023
42. National Vulnerability Database. CVE-2017-5754. https://nvd.nist.gov/vuln/detail/CVE-2017-5754. Accessed 21 Apr 2023
43. National Vulnerability Database. CVE-2017-5753. https://nvd.nist.gov/vuln/detail/CVE-2017-5753. Accessed 21 Apr 2023
44. National Vulnerability Database. CVE-2022-29402. https://nvd.nist.gov/vuln/detail/cve-2022-29402. Accessed 05 May 2023
45. National Vulnerability Database. CVE-2021-26317. https://nvd.nist.gov/vuln/detail/CVE-2021-26317. Accessed 21 Apr 2023
46. National Vulnerability Database. CVE-2023-2166. https://nvd.nist.gov/vuln/detail/CVE-2022-2166. Accessed 21 Apr 2023

Helicopters Turboshaft Engines Parameters Identification Using Neural Network Technologies Based on the Kalman Filter

Serhii Vladov[1]([⊠]) [iD], Yurii Shmelov[1] [iD], Ruslan Yakovliev[1] [iD],
and Maryna Petchenko[2] [iD]

[1] Kremenchuk Flight College of Kharkiv National University of Internal Affairs, Peremohy Street, 17/6, Kremenchuk 39605, Ukraine
ser26101968@gmail.com
[2] Kharkiv National University of Internal Affairs, L. Landau Avenue, 27, Kharkiv 61080, Ukraine

Abstract. The work is a continuation of the research devoted to the development of a multidimensional Kalman filter connected at the output of the built-in helicopter's turboshaft engines mathematical dynamic model to improve the accuracy of helicopters turboshaft engines parameters identification and achieve high quality automatic control. The main difference is the use of radial basis functions neural networks, in which the multivariate Kalman filter is a training algorithm. The work illustrates well-known mathematical expressions underlying of optimal multidimensional filtering algorithms. The methods of mathematical modeling in the Matlab environment tested the proposed algorithms. The simulation results showed that the use of neural networks trained by the multidimensional Kalman matrix filter as part of the model of helicopters turboshaft engines built into the automatic control system allows achieving high indicators of the accuracy of identifying the parameters of helicopters turboshaft engines automatic control system – up to 0.9975, then in practical analogues.

Keywords: Helicopter Turboshaft Engine · Neural Network · Kalman Filter · Training · Accuracy

1 Introduction

Aircraft helicopters turboshaft engines (TE) are complex nonlinear systems, the characteristics of which have a significant scatter [1]. The quality of the built-in mathematical models of helicopters TE that are part of the automatic control system (ACS) [2] largely determines the quality of control and the possibility of using modern mathematical apparatus for the synthesis of ACS, as well as operating tools [3]. Satisfaction with the requirements for the reliability and quality of regulation of helicopters TE is possible only by expanding the functionality of the controls, in particular, endowing them with the ability to quickly adapt to changes in the characteristics of helicopters TE and external conditions. The change in the characteristics of helicopters TE is due to many reasons,

G. Antoniou et al. (Eds.): ICTERI 2023, CCIS 1980, pp. 82–97, 2023.
https://doi.org/10.1007/978-3-031-48325-7_7

the main of which are [4]: parameters technological spread due to tolerances for the manufacture and units' assembly; deviation in the similarity of modes under various external operating conditions; change in characteristics in the process of resource development (failures and units wear).

External interference is caused by a wide range of external destabilizing factors that affect the engine during operation, causing additional errors and reducing the resource: a wide range of operating temperatures ranging from −50 to 100 °C; mechanical shocks, linear acceleration and vibration corresponding to overloads of 10...15 tons or more; instability of ACS power supplies, and also due to the impact of pulses, which are almost twice the nominal value; electromagnetic interference; pressure pulsations with an amplitude of 20...90% of the upper limit; chemically aggressive impurities in the environment, etc.

Solving the task of ACS adaptation, as well as real-time condition monitoring and failure diagnostics, inevitably requires the use of identification methods using artificial intelligence tools (for example, neuron networks) [5]. In this case, the structure and accuracy of the applied mathematical model are determined by the nature of the problem for which they are applied.

2   Related Works

The process of ensuring the stability of engine parameters in all operating modes is one of the priority tasks at the helicopters TE ACS operation mode. At the same time, the ACS performs the following main functions: engine start automatic control, quick transition to other operating modes when controlling the engine or in case of a sharp change in external conditions, maintaining the specified engine operation mode or changing it in accordance with control programs, preventing the engine from entering dangerous operating modes. Of particular difficulty are the starting modes and engine transient modes operation under conditions of external and internal interference [6, 7].

To implement the above functions, a necessary condition is to obtain reliable data on the current parameters of helicopters TE at flight modes (in real time), such as fuel consumption, temperature, pressure, gas generator and free turbine rotor speeds, etc. [8, 9]. Modern helicopters TE ACS operates under interference conditions both in the built-in model channel (due to model inaccuracy and interference in the communication channel) and in the measurement channel (due to sensor error and communication channel interference). That's why an urgent task is to ensure the accuracy of parameter identification, taking into account calculated data obtained using the built-in model, and data from current on-board measurements. The identification accuracy is determined by the methods used [10, 11].

In connection with the foregoing, the aim of the work is to improve the accuracy of helicopters TE parameters identification by using neural network technologies, namely, training the multidimensional Kalman filter at the output of the built-in mathematical dynamic model of identification, develop on the basis of engine's dynamic and throttle characteristics, which makes it possible to identify the parameters and simulate the operation of the engine in stationary and dynamic modes [12, 13]. It should be noted that this study is a continuation of research in the field of development of helicopters

TE intelligent on-board automatic control systems, in which it is proposed to integrate a multidimensional Kalman filter [14].

3   Materials and Methods

Multidimensional filtering is carried out according to three helicopters TE thermogas-dynamic parameters, which are recorded on board the helicopter: n_{TC} – rotor r.p.m.; n_{FT} – free turbine rotor rotational speed; T_G – gas temperature in front of the compressor turbine, recorded on board of the helicopter, reduced to absolute parameters, according to the theory of gas-dynamic similarity [15] (Table 1).

Table 1. Fragment of the training sample during the operation of helicopters TE (on the example of TV3-117 TE) [15].

Number	n_{TC}	n_{FT}	T_G
1	0.932	0.929	0.943
2	0.964	0.933	0.982
3	0.917	0.952	0.962
4	0.908	0.988	0.987
5	0.899	0.991	0.972
6	0.905	0.987	0.979
7	0.923	0.972	0.983
8	0.948	0.966	0.989
9	0.962	0.952	0.997
...	...	...	...
256	0.953	0.973	0.981

According to [14], the recursive Kalman filter for the task posed is the most accurate and convenient in modeling, having the necessary properties of adaptation – self-correction in the process of filtering data. Adaptation is based on the application of a variable optimal Kalman coefficient obtained at the current moment when solving the problem of minimizing the mathematical expectation of the squared error of the identified parameter, taking into account the error at the previous moment (which determines the recurrence of the obtained relations) [14, 16]. Kalman filters are used for ergodic processes operating under noise conditions, characterized by a known time-independent dispersion and zero mathematical expectation [17].

To prove the ergodicity of the process, the feasibility of the Slutsky condition was checked, according to which the autocovariance function of the ergodic process should tend to zero as the lag value increases [14]. The research results showed the correctness of the hypothesis about the ergodicity of the observed processes. The analysis results for n_{TC}, n_{FT}, T_G are shown in Fig. 1, 2 and 3.

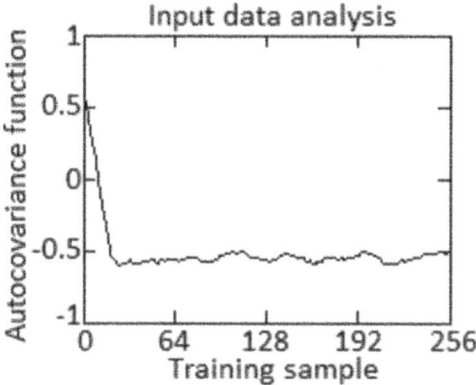

Fig. 1. Diagram of the autocovariance function for n_{TC} (author's research).

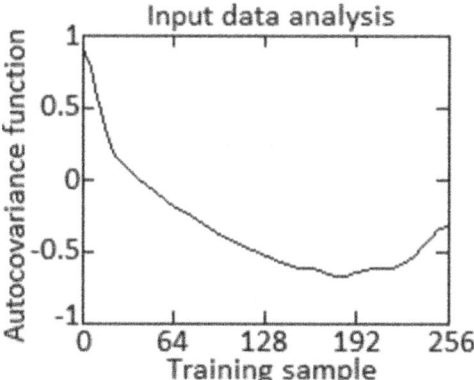

Fig. 2. Diagram of the autocovariance function for n_{FT} (author's research).

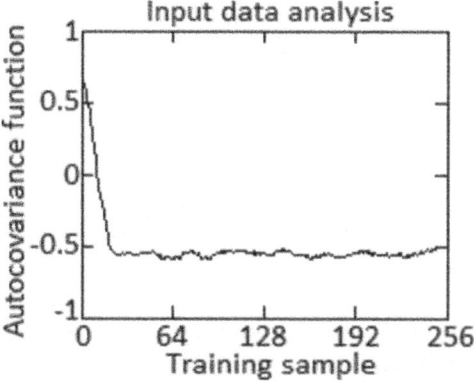

Fig. 3. Diagram of the autocovariance function for T_G (author's research).

According to [14], the analysis of real noises obtained from the data of flight tests of the TV3-117 engine showed that all of them are characterized by zero mathematical expectation, constant dispersion, and the same spectra in the study of one long-duration sample and several samples. Thus, the hypothesis of ergodicity of the processes under consideration was again confirmed, since the values of the mathematical expectation and variance are the same both in time and in the number of realizations. The application of the Pearson criterion showed the normal distribution of noise. All this allows us to draw a conclusion about the further possibility of using Kalman filters (including neural networks trained using the Kalman filter) for this class of processes [18]. In the identification problem using the optimal Kalman filtering, according to [14, 16], it is required at the current time to eliminate the error as much as possible both in the model channel (the predicted value of the identified parameter) and in the measurement channel (current sensor readings) for the four parameters under consideration. For this, a recursive matrix relation is used, which makes it possible to determine the matrix of the square of the covariance error over all coordinates [19]:

$$\mathbf{E}_k^2 = \mathbf{E}_{k-1}^2 + \sigma_\xi^2; \tag{1}$$

where $\mathbf{E}_k^2$ – column vector of the squared error of the covariance in the coordinates of the n_{TC}, n_{FT}, T_G at the k-th step, σ_ξ^2 – matrix of model variances in the coordinates of the n_{TC}, n_{FT}, T_G.

The solution of the *mathbf* $\mathbf{E}_k^2$ minimization problem allows us to determine the elements of the matrix of the Kalman coefficient:

$$\mathbf{K}_k = \mathbf{E}_k^2 \left[\mathbf{E}_k^2 + \sigma_\eta^2 \right]^{-1}; \tag{2}$$

where $\mathbf{K}_k$ – column vector of the Kalman coefficient for the coordinates of the n_{TC}, n_{FT}, T_G at the k-th step, is the variance matrix of the meter (sensor) in the matrix of model variances for the coordinates of the n_{TC}, n_{FT}, T_G.

Identification of parameters (obtaining an optimal estimate) is carried out through a column vector of Kalman coefficients, which determines in matrix form the ratio of the calculated (model) and measured components in the optimal values of the identified parameters:

$$\mathbf{X}_k^{opt} = \mathbf{X}_k (1 - \mathbf{K}_k) + \mathbf{K}_k \mathbf{Z}_k; \tag{3}$$

where $\mathbf{X}_k^{opt}$ – column vector of optimal estimates of coordinates n_{TC}, n_{FT}, T_G at the k-th step; $\mathbf{X}_k$ – column vector of model values of coordinates n_{TC}, n_{FT}, T_G, calculated at the k-th step; $\mathbf{Z}_k$ – column vector of coordinates n_{TC}, n_{FT}, T_G measured by sensors at the k-th step.

It should be noted that the problem solved using Kalman filtering is not a smoothing problem, but an identification problem [20]. The Kalman filter is not designed for smoothing data from the sensor, but is aimed at obtaining the closest to the real coordinates n_{TC}, n_{FT}, T_G, the values of their optimal estimates at the current time, obtained under conditions of external and internal interference in the channels of the built-in model and measurements and recorded in the column vector $\mathbf{X}_k^{opt}$.

In order to implement the considered method as part of the onboard neural network expert system for monitoring helicopters TE operational status [21] or in the modified closed onboard helicopters TE ACS [2, 15], as well as to increase the accuracy of helicopters TE parameters identification compared to [14], it is proposed to implement a multidimensional filter Kalman using neural networks.

Neural network training using the multidimensional Kalman filter [22] is the problem of estimating the true state of some unknown "ideal" neural network that provides zero mismatch, under states, in this case, the values of the weights of the neural network $w(k)$ are taken, and under the mismatch, the current error training $e(k)$. The dynamic training process of a neural network can be described by a pair of state-space equations, one of which is a process model representing the evolution of the weight vector under the influence of a random process $\xi(k)$, which is considered white noise with zero mathematical expectation and the known diagonal covariance matrix Q:

$$w(k + 1) = w(k) + \zeta(k). \tag{4}$$

The output equation is a linearized neural network model (5) at the k-th step, noisy with a random process $Q(k)$, which is considered to be white noise with zero mathematical expectation and a known diagonal covariance matrix R:

$$h(k) = \frac{\partial y(w(k), v(k), x(k))}{\partial w} + \zeta(k); \tag{5}$$

where $w(k)$ – neural network weights, $v(k)$ – neurons postsynaptic potentials, $x(k)$ – network input values. The instantaneous values of the derivatives $\frac{\partial y}{\partial w}$ are calculated using the backpropagation method. Mismatch $e(k)$ is calculated according to the expression:

$$e(k) = t(k) - \widetilde{y}(k); \tag{6}$$

where $t(k)$ – neural network target value, $\widetilde{y}(k)$ – neural network actual output, calculated according to the expression:

$$\widetilde{y}(k) = g\left(\sum_j w_j^{(2)} f\left(\sum_j w_{ji}^{(1)} x_i\right)\right); \tag{7}$$

where $w^{(1)}$ – hidden layer neurons weights, $f(\cdot)$ – hidden layer neurons activation functions, $w^{(2)}$ – output layer neurons weights, $g(\cdot)$ – output layer neurons activation functions.

Before training, the neural network goes through the initialization stage. The covariance matrices of measurement noise $R = \eta \cdot I$ and dynamic training noise $Q = \mu \cdot I$ are set, the size of the matrices is $L \times L$ and $N \times N$, respectively, where L – output neurons number, N – neural network weight coefficients number. The coefficient η has the meaning of the training rate, in this work according to [22] $\eta = 0.001$, the coefficient μ determines the measurement noise; in this work, according to [22], $\mu = 10^{-4}$. Also, a single covariance matrix P of size $N \times N$ and a zero-measurement matrix H of size $L \times N$ are set at the initialization stage. The training stage is performed online, the neural network correction weights is sequentially performed for each example of the training sample. At the k-th step, the following actions are performed:

1) A neural network output new value $\widetilde{y}(k)$ is calculated according to (4), a "forward pass" of the neural network is performed.

2) A "reverse pass" of the neural network is performed: the derivatives are calculated using the backpropagation method $\frac{\partial \widetilde{y}}{\partial w_i}$, $i = \overline{1, N}$. This is done using the same technique as in the error backpropagation method, but the local gradients for the output neurons are set not to the current error $e(k)$, but to the constant 1, which, with all the same calculations, provides the neural network outputs Jacobians values $\frac{\partial \widetilde{y}}{\partial w}$ instead of gradients $\frac{\partial (e(k))^2}{\partial w}$ because $\frac{\partial (e(k))^2}{\partial w} = 2e(k)\frac{\partial y}{\partial w}$. Observation matrix $H(k)$ is formed:

$$H(k) = \left[\frac{\partial \widetilde{y}}{\partial w_1} \ \frac{\partial \widetilde{y}}{\partial w_2} \ \cdots \ \frac{\partial \widetilde{y}}{\partial w_N} \right]^T. \tag{8}$$

3) The current error of the network operation $e(k)$ is determined according to (6), a deviation matrix $E(k)$ of size $1 \times L$ is formed:

$$E(k) = [e(k)]. \tag{9}$$

4) New values of the neural network weights $w(k + 1)$ and correlation matrix $P(k + 1)$ are calculated according to the expressions:

$$K(k) = P(k)H(k)^T \left[H(k)P(k)H(k)^T + R \right]^{-1}; \tag{10}$$

$$P(k + 1) = P(k) - K(k)H(k)P(k) + Q; \tag{11}$$

$$w(k + 1) = w(k) + K(k)e(k); \tag{12}$$

where $K(k)$ – Kalman gain matrix, its dimension is $N \times L$.

Actions 1 – 4 are performed for all elements of the training sample. The correlation matrix P updated at each clock cycle contains second-order information about the error surface, which provides the extended Kalman filter method with an advantage over first-order training methods such as gradient descent and its modifications.

4   Experiment

Based on [14], a neural network of radial basis functions (RBF) with 5 inputs is taken as a neural network, three of which are responsible for the input parameters n_{TC}, n_{FT}, T_G, and two specify the model error and the measurement error (Fig. 4).

It is known that RBF-networks model an arbitrary non-linear function using only one intermediate layer, thereby eliminating the need to decide on the number of layers [23, 24]. Secondly, the parameters of the linear combination in the output layer can be optimized using well-known linear optimization methods that are fast and do not suffer from local minimum that interfere with training using the backpropagation algorithm.

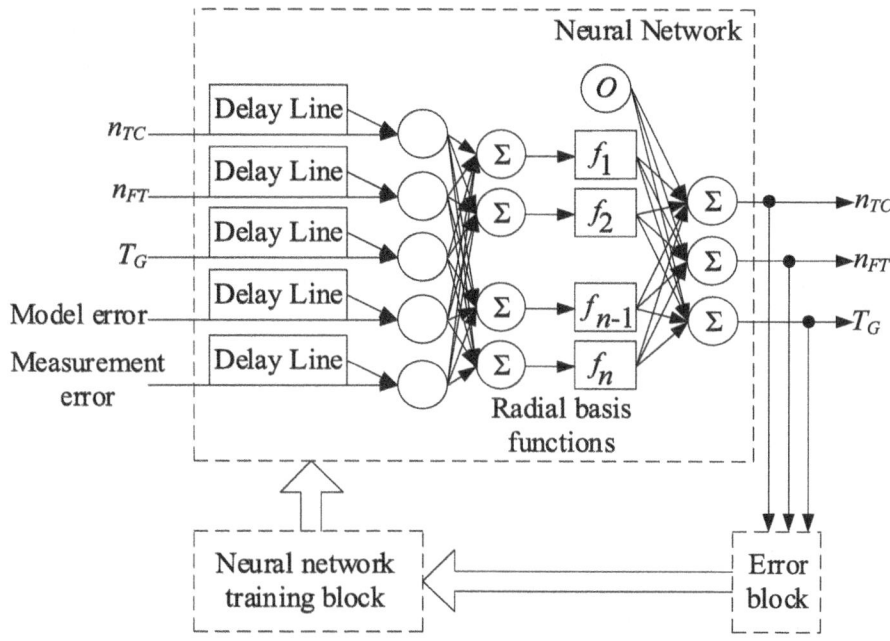

Fig. 4. Neural network diagram (author's research).

Therefore, the RBF-network trains very quickly – an order of magnitude faster than using the backpropagation algorithm.

In order to determine the optimal number of neurons in the hidden layer, an experimental dependence $E = f(N)$ was constructed, shown in Fig. 5, where E – neural network training error; N – number of neurons in the hidden layer (it is assumed that the number of neurons in the input layer is 5, in the output layer is 3) [25].

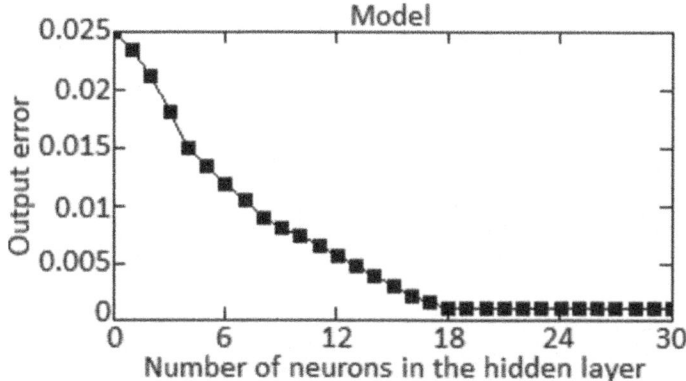

Fig. 5. Neural network training error diagram the number of neurons in the hidden layer (author's research).

As can be seen from Fig. 5, with 18 neurons in the hidden layer, the smallest training error of the neural network is achieved, that is, the optimal structure of the neural network is 5–18–3.

The neural network training was carried out on a personal computer with an AMD Ryzen 5 5600 6-Core Processor 3.50 GHz CPU of the Zen 3 architecture, 32 GB of DDR-4 RAM and an Nvidia GeForce RTX 3060 GPU times compared to training on the CPU. To control of neural network training, we used the accuracy metric and the loss function, which was chosen as categorical entropy [26]. The experiment was carried out in laboratory conditions. At the same time, it was proved that under the conditions of on-board implementation, this method can be easily implemented using the 64-bit Intel Neural Compute Stick 2 neuroprocessor [27]. Diagrams of the dependence of the change in these values on the number of the epoch of network training are shown in Fig. 6.

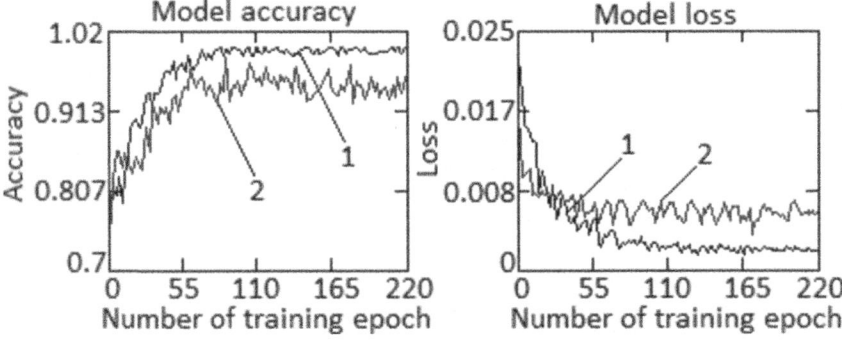

Fig. 6. Diagrams of changes in accuracy and loss when a neural network training: 1 – train, 2 – test (author's research).

As can be seen from Fig. 6, the accuracy indicator approaches one, and loss indicator – tends to zero, which indicates the high accuracy of the model and its minimal error [28, 29].

Table 2 contains the values of training and testing errors, as well as the training time, for various training algorithms, from which it can be seen that for the task at hand, the RBF-network training algorithm based on the multidimensional Kalman filter significantly outperforms other RBF-network training algorithms in terms of convergence rate (number of epochs) and identification accuracy.

Table 2. RBF-network training results (author's research).

Training algorithm	Identification error	Number of training epochs	Training time, s
Backpropagation algorithm	0.034	283	14
Quasi Newton algorithm [30]	0.031	275	12
Q-learning [31]	0.029	272	12
Genetic algorithm combination method [32]	0.027	268	11
Expectation-maximization algorithm [33]	0.026	264	11
Gradient algorithm [34]	0.018	247	8
Multidimensional Kalman filter [14, 22]	0.009	220	3.25

5   Results

Similarly, to [14], the task of developing multidimensional Kalman filtering algorithms using neural network technologies was solved on the basis of a model experiment in the Simulink interactive environment, which allows building dynamic models of the researched control objects based on block diagrams in the form of directed diagrams [34, 35].

Its main advantages are the variety of built-in libraries, including those included in the Matlab environment, visibility and ease of modeling, the ability to monitor the system operational status in real time, and the convenience of an interface that makes it easy to influence the designed algorithm and model experiment [36].

The task of developing an algorithm is reduced to modeling mathematical expressions (1)–(3) to calculate the specified values at each step. Calculations in the multidimensional Kalman filtering algorithms were based on the dispersions of the model and sensors along the coordinates n_{TC}, n_{FT}, T_G, obtained by statistical processing at flight test data of the TV3-117 TE.

The generalized block diagram of the model of the multidimensional Kalman filter, which performs real-time simultaneous identification by three coordinates of n_{TC}, n_{FT}, T_G, is shown in Fig. 7.

A detailed functional diagram of filtering along one coordinate is taken from [14] and is shown in Fig. 8.

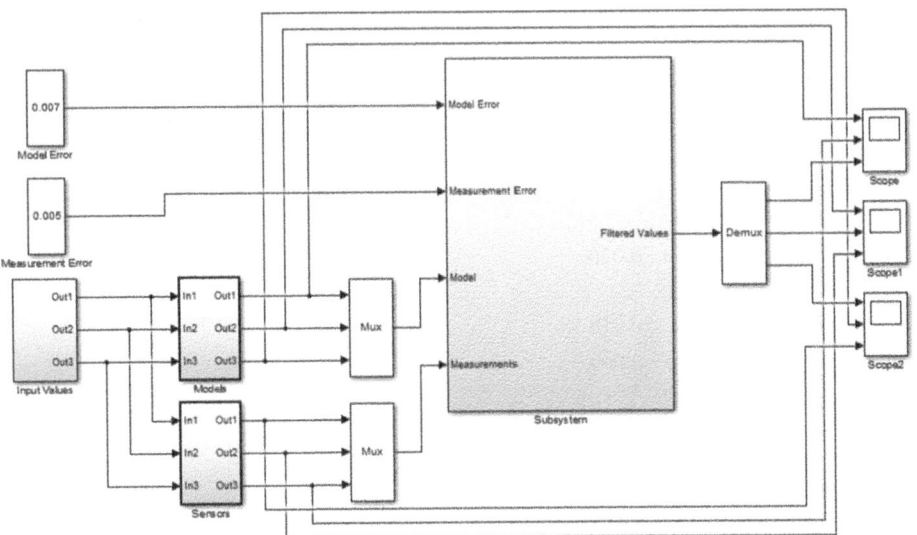

Fig. 7. Generalized block diagram of the multidimensional Kalman filter in Simulink (author's development based on [14]).

Similarly, to [14], in the model experiment, the values of the coordinates of the column vector of the readings of the sensors Z_k under noise conditions were modeled by superimposing several types of noise on the calculated values: gaussian noise; real noise extracted from experimental data obtained during flight tests of the TV3-117 engine for the considered modes and types of input signals; several high frequency sinusoids; combined noise obtained by superimposing several high-frequency sinusoids on real noise. The results of filtering under the action of combined noises when applying input signals that provide a change in the identified values in the entire operating range for various input signals (acceleration – operating mode – reset) along one coordinate are shown in Fig. 9, 10, where: 1 – change in time of the coordinates of the column vector $\mathbf{X}_k$ – calculated (model) values of n_{TC}, n_{FT}, T_G; 2 – change in time of the coordinates n_{TC}, n_{FT}, T_G of the column vector $\mathbf{Z}_k$ – measured by the sensors of the values of the coordinates n_{TC}, n_{FT}, T_G; 3 – change in time of the coordinates of the column vector $\mathbf{X}_k^{opt}$ – optimal estimates of the coordinates n_{TC}, n_{FT}, T_G.

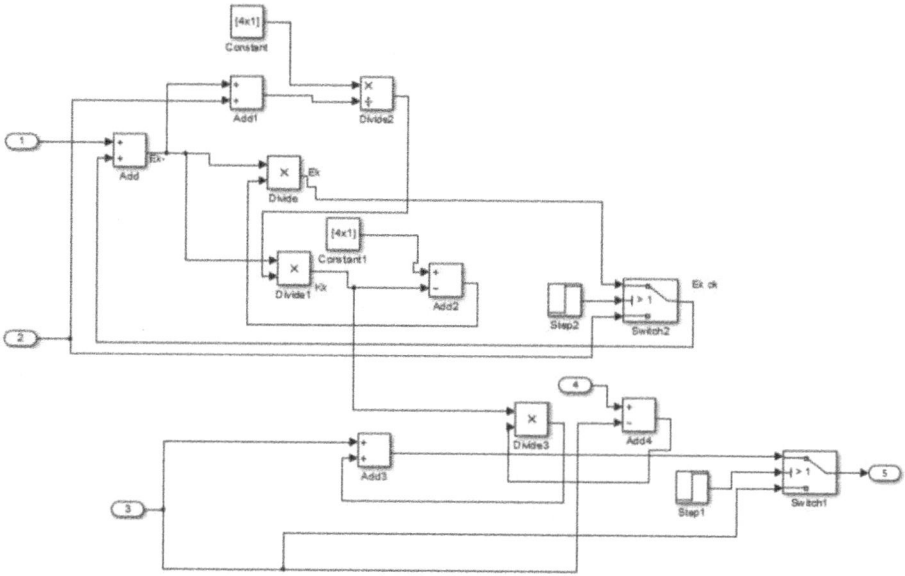

Fig. 8. Single coordinate filtering block diagram in Simulink [14]: 1 – model error, 2 – measurement error, 3 – model, 4 – measurement, 5 – filtered value.

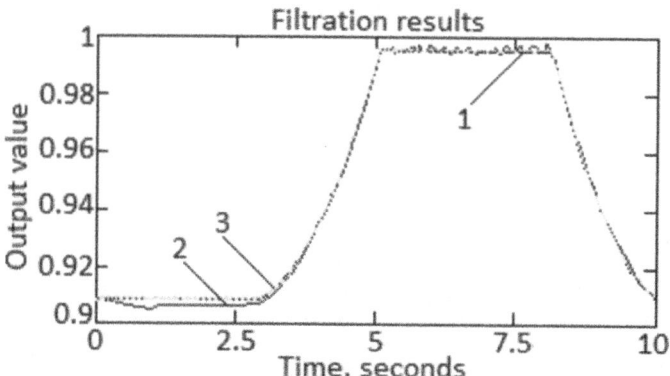

Fig. 9. The results of applying the multidimensional Kalman filter in the injectivity modes (reset under conditions of combined noise) (author's research).

6 Discussions

Researches have shown that the relative error of the signals at the output of the multidimensional Kalman filter neural network model for all researched coordinates n_{TC}, n_{FT}, T_G does not exceed 0.25%, which corresponds to the specified technical requirements for the accuracy of identification algorithms. The results obtained indicate a two-fold decrease in the error of signals in the output of the multidimensional Kalman filter

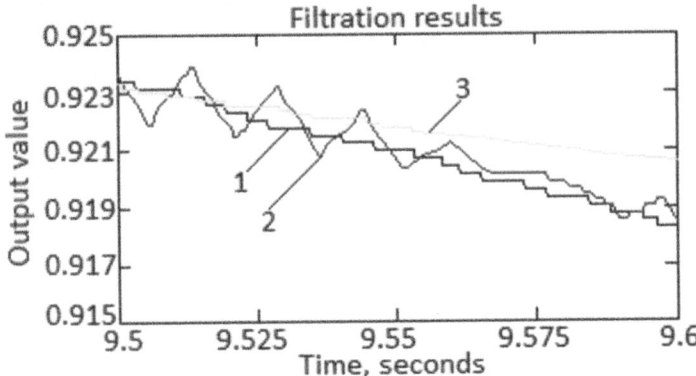

Fig. 10. The results of applying the multidimensional Kalman filter in the operating mode under conditions of combined noise (author's research).

neural network model compared to the results obtained in [14]. The developed the multidimensional Kalman filter neural network model operates both in static and dynamic modes under the influence of "hard" external and internal disturbances in a wide range of helicopter TE operating modes.

Table 3 shows the results of a comparative analysis of the identification of the parameters of TV3-117 TE at three coordinates n_{TC}, n_{FT}, T_G by determining errors of the 1st and 2nd kind. It follows from Table 3 that the use of the multidimensional Kalman filter neural network reduces errors by an average of 20% compared to the results obtained in [14].

Table 3. Identification 1st and 2nd kind errors calculation results (author's research).

Identification method based on multidimensional Kalman filter	Probability of error in parameter identification					
	n_{TC}		n_{FT}		T_G	
	Type 1st error	Type 2nd error	Type 1st error	Type 2nd error	Type 1st error	Type 2nd error
Without the use of neural networks [14]	0.84	0.63	0.85	0.64	0.79	0.58
Using neural networks	0.62	0.38	0.63	0.41	0.56	0.33

7 Conclusions

1. The method for helicopters turboshaft engines parameters identification using the multidimensional Kalman filter has been further developed, which differs from the existing one in that due to the use of radial basis functions neural network, it has

increased the accuracy of helicopters turboshaft engines parameters identification with an accuracy of 0.9975.

2. The method of radial basis functions neural network training has been improved, which, due to the use of a training algorithm based on the multivariate Kalman filter, has reduced the identification error from 0.034 to 0.009 (3.4 to 0.9%), the number of training epochs from 283 to 220, the training time from 14 to 3.25 s, which is critical in terms of on-board implementation.

3. It is proved that the errors of the 1st and 2nd implementations of the method for helicopters turboshaft engines parameters identification using the multidimensional Kalman filter neural network model did not exceed 0.63% and 0.41%, respectively, while for the classical method (multidimensional Kalman filter direct application [14]) they amounted to 0.85% and 0.64%, respectively. The obtained results prove that the use of the multidimensional Kalman filter neural network filter will allow 25% more accurate helicopters turboshaft engines parameters identification at the helicopter flight mode.

4. The prospect for further research is testing the developed multidimensional Kalman filter neural network model in the closed control loop of the modified closed onboard helicopters turboshaft engines automatic control system [2, 15], the onboard neural network expert system for monitoring helicopters turboshaft engines operational status [21] and a comparative analysis of the results obtained with the results of applying the element-by-element model and bench and flight test data in similar modes.

References

1. Zeng, J., Cheng, Y.: An ensemble learning-based remaining useful life prediction method for aircraft turbine engine. IFAC-PapersOnLine **53**(3), 48–53 (2020). https://doi.org/10.1016/j.ifacol.2020.11.00

2. Vladov, S., Shmelov, Y., Yakovliev, R.: Helicopters aircraft engines self-organizing neural network automatic control system. In: The Fifth International Workshop on Computer Modeling and Intelligent Systems (CMIS-2022), 12 May 2022, Zaporizhzhia, Ukraine. CEUR Workshop Proceedings, vol. 3137, pp. 28–47 (2022). https://doi.org/10.32782/cmis/3137-3

3. Yildirim, M.T., Kurt, B.: Aircraft gas turbine engine health monitoring system by real flight data. Int. J. Aerosp. Eng. 1–12 (2018). https://doi.org/10.1155/2018/9570873. https://www.hindawi.com/journals/ijae/2018/9570873/

4. Pang, S., Li, Q., Ni, B.: Improved nonlinear MPC for aircraft gas turbine engine based on semi-alternative optimization strategy. Aerosp. Sci. Technol. **118**, 106983 (2021). https://doi.org/10.1016/j.ast.2021.106983

5. Shen, Y., Khorasani, K.: Hybrid multi-mode machine learning-based fault diagnosis strategies with application to aircraft gas turbine engines. Neural Netw. **130**, 126–142 (2020). https://doi.org/10.1016/j.neunet.2020.07.001

6. Filippone, A., Zhang, M., Bojdo, N.: Validation of an integrated simulation model for aircraft noise and engine emissions. Aerosp. Sci. Technol. **89**, 370–381 (2019). https://doi.org/10.1016/j.ast.2019.04.008

7. Cai, C., Wang, Y., Fang, J., Chen, H., Zheng, Q., Zhang, H.: Multiple aspects to flight mission performances improvement of commercial turbofan engine via variable geometry adjustment. Energy **263**(A), 125693 (2023). https://doi.org/10.1016/j.energy.2022.125693

8. de Voogt, A., St. Amour, E.: Safety of twin-engine helicopters: risks and operational specificity. Saf. Sci. **136**, 105169 (2021). https://doi.org/10.1016/j.ssci.2021.105169

9. Li, Y., Xuan, Y.: Thermal characteristics of helicopters based on integrated fuselage structure/engine model. Int. J. Heat Mass Transf. **115**(A), 102–114 (2017). https://doi.org/10.1016/j.ijheatmasstransfer.2017.07.038

10. Sheng, H., et al.: Research on dynamic modeling and performance analysis of helicopter turboshaft engine's start-up process. Aerosp. Sci. Technol. **106**, 106097 (2020). https://doi.org/10.1016/j.ast.2020.106097

11. Infante, V., Freitas, M.: Failure analysis of compressor blades of a helicopter engine. Eng. Fail. Anal. **104**, 67–74 (2019). https://doi.org/10.1016/j.engfailanal.2019.05.024

12. Chen, Y.-Z., Tsoutsanis, E., Wang, C., Gou, L.-F.: A time-series turbofan engine successive fault diagnosis under both steady-state and dynamic conditions. Energy **263**(D), 125848 (2023). https://doi.org/10.1016/j.energy.2022.125848

13. Andoga, R., Fozo, L., Schrotter, M., Ceskovic, M.: Intelligent thermal imaging-based diagnostics of turbojet engines. Appl. Sci. **9**(11), 2253 (2019). https://doi.org/10.3390/app9112253

14. Kuznetsova, T., Likhacheva, Y., Gubarev, E., Yakushev, A.: The identification of gas turbine engine parameters by the multidimensional Kalman filter. Electr. Eng. Inf. Technol. Control Syst. **10**, 114–126 (2014)

15. Vladov, S., Shmelov, Y., Yakovliev, R.: Modified helicopters turboshaft engines neural network on-board automatic control system using the adaptive control method. In: ITTAP'2022: 2nd International Workshop on Information Technologies: Theoretical and Applied Problems, 22–24 November 2022, Ternopil, Ukraine. CEUR Workshop Proceedings, vol. 3309, pp. 205–229 (2022). (ISSN 1613-0073)

16. Karvonen, T., Bonnabel, S., Moulines, E., Sarkka, S.: On stability of a class of filters for nonlinear stochastic systems. SIAM J. Control Optimiz. **58**(4), 2023–2049 (2020). https://doi.org/10.1137/19M1285974

17. Jia, C., Hu, J.: Variance-constrained filtering for nonlinear systems with randomly occurring quantized measurements: recursive scheme and boundedness analysis. Adv. Differ. Equ. **2019**, 53 (2019). https://doi.org/10.1186/s13662-019-2000-0. https://d-nb.info/1180488768/34

18. Long, Z., Bai, M., Ren, M., Liu, J., Yu, D.: Fault detection and isolation of aeroengine combustion chamber based on unscented Kalman filter method fusing artificial neural network. Energy **1**, 127068 (2023). https://doi.org/10.1016/j.energy.2023.127068

19. Kellermann, C., Ostermann, J.: Estimation of unknown system states based on an adaptive neural network and Kalman filter. Proc. CIRP **99**, 656–661 (2021). https://doi.org/10.1016/j.procir.2021.03.089

20. Togni, S., Nikolaidis, T., Sampath, S.: A combined technique of Kalman filter, artificial neural network and fuzzy logic for gas turbines and signal fault isolation. Chin. J. Aeronaut. **34**(2), 124–135 (2021). https://doi.org/10.1016/j.cja.2020.04.015

21. Shmelov, Y., Vladov, S., Klimova, Y., Kirukhina, M.: Expert system for identification of the technical state of the aircraft engine TV3-117 in flight modes. In: System Analysis & Intelligent Computing: IEEE First International Conference on System Analysis & Intelligent Computing (SAIC), 08–12 October, pp. 77–82 (2018). https://doi.org/10.1109/SAIC.2018.8516864

22. Chernodub, A.: Training of neuroemulators with use of pseudoregularization for model reference adaptive neurocontrol. Artif. Intell. **4**, 602–614 (2012)

23. Gyamfi, K.S., Brusey, J., Gaura, E.: Differential radial basis function network for sequence modelling. Expert Syst. Appl. **189**, 115982 (2022). https://doi.org/10.1016/j.eswa.2021.115982

24. Sideratos, G., Hatziargyriou, N.D.: A distributed memory RBF-based model for variable generation forecasting. Int. J. Electr. Power Energy Syst. **120**, 106041 (2020). https://doi.org/10.1016/j.ijepes.2020.106041

25. Vladov, S., Shmelov, Y., Yakovliev, R., Petchenko, M., Drozdova, S.: Neural network method for helicopters turboshaft engines working process parameters identification at flight modes. In: 2022 IEEE 4th International Conference on Modern Electrical and Energy System (MEES), Kremenchuk, Ukraine, 20–22 October 2022, pp. 604–609 (2022). https://doi.org/10.1109/MEES58014.2022.10005670

26. Puchkov, A., Dli, M., Lobaneva, E., Vasilkova, M.: Predicting an object state based on applying the Kalman filter and deep neural networks. Softw. Syst. **32**(3), 368–376 (2019). https://doi.org/10.15827/0236-235X.127.368-376

27. Vladov, S., Shmelov, Y., Yakovliev, R.: Modified method of identification potential defects in helicopters turboshaft engines units based on prediction its operational status. In: 2022 IEEE 4th International Conference on Modern Electrical and Energy System (MEES), Kremenchuk, Ukraine, 20–22 October 2022, pp. 556–561 (2022). https://doi.org/10.1109/MEES58014.2022.10005605

28. Vladov, S., Shmelov, Y., Yakovliev, R. Methodology for control of helicopters aircraft engines technical state in flight modes using neural networks. In: The Fifth International Workshop on Computer Modeling and Intelligent Systems (CMIS-2022), 12 May 2022, Zaporizhzhia, Ukraine, CEUR Workshop Proceedings, vol. 3137, pp. 108–125 (2022). https://doi.org/10.32782/cmis/3137-10). (ISSN 1613-0073)

29. Yoo, H.-M., Park, M.-K., Park, B.-G., Lee, J.-H.: Effect of layer-specific synaptic retention characteristics on the accuracy of deep neural networks. Solid-State Electron. **200**, 108570 (2023). https://doi.org/10.1016/j.sse.2022.108570

30. Saneifard, R., Jafarian, A., Ghalami, N., Measoomy Nia, S.: Extended artificial neural networks approach for solving two-dimensional fractional-order Volterra-type integro-differential equations. Inf. Sci. **612**, 887–897 (2022). https://doi.org/10.1016/j.ins.2022.09.017

31. Shamlitskiy, Y., Popov, A., Saidov, N., Moiseeva, K.: Transport stream optimization based on neural network learning algorithms. Transp. Res. Proc. **68**, 417–425 (2023). https://doi.org/10.1016/j.trpro.2023.02.056

32. Wei, W., Chuan, J.: A combination forecasting method of grey neural network based on genetic algorithm. Proc. CIRP **109**, 191–196 (2022). https://doi.org/10.1016/j.procir.2022.05.235

33. Hu, X., Wei, X., Gao, Y., Liu, H., Zhu, L.: Variational expectation maximization attention broad learning systems. Inf. Sci. **608**, 597–612 (2022). https://doi.org/10.1016/j.ins.2022.06.074

34. Pattnaik, M., Badoni, M., Tatte, Y.: Application of Gaussian-sequential-probabilistic-inference concept based Kalman filter for accurate estimation of state of charge of battery. J. Energy Storage **54**, 105244 (2022). https://doi.org/10.1016/j.est.2022.105244

35. Lu, F., Gao, T., Huang, J., Qiu, X.: Nonlinear Kalman filters for aircraft engine gas path health estimation with measurement uncertainty. Aerosp. Sci. Technol. **76**, 126–140 (2018). https://doi.org/10.1016/j.ast.2018.01.024

36. Cordeiro, R.A., Azinheira, J.R., Moutinho, A.: Actuation failure detection in fixed-wing aircraft combining a pair of two-stage Kalman filters. IFAC-PapersOnLine **53**(2), 7144–7749 (2020). https://doi.org/10.1016/j.ifacol.2020.12.825

Modification of Fuzzy TOPSIS Based on Various Proximity Coefficients Metrics and Shapes of Fuzzy Sets

Dmytro Hapishko[1] , Ievgen Sidenko[1]([⊠]) , Galyna Kondratenko[1] ,
Yuriy Zhukov[3] , and Yuriy Kondratenko[1,2]

[1] Petro Mohyla Black Sea National University, 68th Desantnykiv Str., 10, Mykolaiv 54003,
Ukraine
gapishko99@gmail.com, {ievgen.sidenko,halyna.kondratenko,
yuriy.kondratenko}@chmnu.edu.ua
[2] Institute of Artificial Intelligence Problems, Mala Zhytomyrs'ka Str., 11/5, Kyiv 01001,
Ukraine
[3] C-Job Nikolayev, Artyleriyska Str., 17/6, Mykolaiv 54006, Ukraine
yuriy.zhukov@nuos.edu.ua

Abstract. The relevance of this work lies in the need for qualitative evaluation
of innovative projects, which are characterized by various technical and economic
indicators, and, as a rule, significant capital investments are required for their
implementation. The use of multi-criteria methods, in particular the modified
Fuzzy TOPSIS, will allow to increase the accuracy of decision-making when
choosing the optimal innovative project. The purpose of the work is to select
the optimal innovative project through the use of the developed software application based on the modified Fuzzy TOPSIS method, as well as to investigate
the influence of the parameters of the specified method on the decision-making
result. To modify the Fuzzy TOPSIS method, the Euclidean distance, the Hamming distance, and the Chebyshev distance were used to determine the proximity
coefficients. Another element of the modification of the Fuzzy TOPSIS method
was the ability to calculate results using various fuzzy numbers, such as triangular, trapezoidal, and Gaussian. A comparative analysis of the obtained results
with other multi-criteria decision-making methods was carried out. A data entry
template for experts has also been created and the results presented in a clear and
easy-to-read format.

Keywords: Fuzzy TOPSIS · DSS · MCDM · Fuzzy Sets · Proximity
Coefficients · Innovative Projects

1 Introduction

Today, the optimal choice and application of multi-criteria decision-making (MCDM)
methods in various tasks, especially in fuzzy conditions, is quite relevant and difficult.
This is due to the fact that not all methods are modified to solve problems with fuzzy
data. Even those in which this possibility is implemented do not have enough shapes of
fuzzy sets to describe the input data and metrics to evaluate the best decision [1–3].

© The Author(s), under exclusive license to Springer Nature Switzerland AG 2023
G. Antoniou et al. (Eds.): ICTERI 2023, CCIS 1980, pp. 98–113, 2023.
https://doi.org/10.1007/978-3-031-48325-7_8

Choosing optimal innovative projects, as one of the MCDM tasks, has become quite popular and widespread in any field of activity. Today, there is a large number of relevant projects with different (quantitative and qualitative) initial data, therefore it was decided to propose a modification of the Fuzzy TOPSIS (Fuzzy Technique for Order of Preference by Similarity to Ideal Solution) method, which has proven itself as an effective compromise method for solving MCDM problems, to integrate it into the developed system for choosing optimal innovative projects, as well as to investigate the influence of its parameters on the final result. The relevance of this work lies in the need for qualitative evaluation of innovative projects, which are characterized by various technical and economic indicators, and, as a rule, significant capital investments are required for their implementation. The use of MCDM methods, in particular the modified Fuzzy TOPSIS, will allow to increase the accuracy of decision-making when choosing the optimal innovative project [2, 4–9].

The goal of the work is to modify the Fuzzy TOPSIS method for solving the MCDM problems in the fuzzy conditions and variability of the shapes of fuzzy sets, in particular, when choosing the optimal innovative project. To solve the set goal, it is necessary to perform the following tasks: (a) overview of the current state of the MCDM problems for the evaluation and selection of innovative projects; (b) analysis of recent research and publications; (c) overview of multi-criteria decision-making methods; (d) modification of the Fuzzy TOPSIS method, modeling and analysis of expert evaluations of innovative projects according to defined criteria; (e) software implementation of the system for choosing an optimal innovative project. To modify the Fuzzy TOPSIS method, the Euclidean distance, the Hamming distance, and the Chebyshev distance were used to determine the proximity coefficients. Another element of the modification of the Fuzzy TOPSIS method was the ability to calculate results using various shapes of fuzzy sets, such as triangular, trapezoidal, and Gaussian.

2 Related Works and Problem Statement

Innovative projects are a set of actions aimed at creating and promoting new high-tech products on the market that have limited resources. Usually, as indicated in the name, such projects are characterized by innovative approaches in the field where they are applied in order to solve the problems of enterprises or areas to which these projects were directed [10–12].

Due to the positive results that such innovative projects often give, their popularity and necessity continue to grow. For example, there are several Ukrainian sites (https://imzo.gov.ua/osvitni-proekti/, https://mon.gov.ua/ua/nauka/inn ovacijna-diyalnist-ta-transfer-tehnologij/innovacijni-proekti/derzhavnij-reyestr-inn ovacijnih-proektiv) that work with innovative projects. But in order for investors and participants not to lose their funds, it is necessary to evaluate the effectiveness of the innovative project when financing it.

The main stages of evaluation of innovative projects are (a) expert assessment of the significance of projects for global and national economic projects; (b) determination of performance indicators for the purpose of finding investors; (d) determination of effectiveness after consideration of the funding scheme [10, 11].

Another important factor when working on innovative projects is risk. When working with projects, it is important to take into account many details, such as inaccurate or incomplete information about the conditions and measures of project implementation, which create the risk of deviation of expected results from real ones [11, 12].

Therefore, in order to take into account all criteria for evaluating innovative projects, and possible variants of quantitative and qualitative data, it will be appropriate and convenient to use such a method as Fuzzy TOPSIS. It belongs to MCDM methods [2, 4, 8], the goal of which is to find the optimal alternative based on similarity to the ideal solution. The idea of the TOPSIS method is to find for the selected alternative not only the smallest distance to the positive ideal solution (PIS) but also the largest distance to the negative ideal solution (NIS). However, this method is used only when the expert evaluates the criteria and alternatives in the form of clear estimates (clear numbers). In real modern conditions, input data is not only quantitative but also qualitative, so there is a need to evaluate criteria and alternatives in a fuzzy form, which is more convenient for an expert. In this case, the Fuzzy TOPSIS method is ideal for this and allows the use of linguistic terms with the triangular shape of fuzzy sets.

The authors of the publication [13] consider the problem related to the results of the analysis of trends in the development of the world economy, which shows that trade is heading toward high-tech products. And with this in mind, the countries that produce and export this product will have significant growth in the field of industry. Therefore, investing in these technologies is quite a profitable choice. But since investing in such a field requires paying attention to a number of important factors, such as risks (production, environmental, market, technological, financial, and management). Therefore, in this study, the authors, using the Fuzzy TOPSIS method, try to estimate investment risks. Taking into account the assessments of experts in the field of risks, and applying to them a decision support system (DSS), it was concluded that the most influential is financial risk, followed by market risk, then technological risk, management risk, then production risk, and the last one is environmental risk.

The authors [14] utilize a set of criteria and sub-criteria to evaluate the cars and provide a comprehensive analysis of the decision-making process. They also apply the TOPSIS algorithm to rank the cars according to their suitability for families. The study offers insights into how the TOPSIS algorithm can be applied in real-world scenarios and demonstrates the effectiveness of this approach in decision-making. Overall, this publication contributes to the field of decision-making by showcasing how the TOPSIS algorithm can be applied to select the best family car.

The publication [15] proposes a new approach for prioritizing strategies using a combination of the Fuzzy Analytic Hierarchy Process (AHP) and Fuzzy TOPSIS methods. The authors explain how their proposed model can be used to deal with the uncertainties and ambiguities that are typically associated with strategic decision-making. They also provide a case study to demonstrate the effectiveness of their approach in real-world scenarios. Overall, this publication offers a new method for prioritizing strategies that can account for the complexity and ambiguity of decision-making scenarios in this area.

The publication [16] presents a methodology to select the best cloud service provider using the Fuzzy TOPSIS. The authors argue that cloud computing has become a popular solution for organizations due to its various advantages such as scalability, cost efficiency, and flexibility. However, selecting the most appropriate cloud service provider from a range of options can be challenging. The proposed methodology is based on the identification of criteria and sub-criteria, as well as the assignment of weights and ratings to each one of them, which are then used to calculate the overall performance score of each provider. The fuzzy TOPSIS algorithm is used to rank the providers based on their performance scores. The proposed methodology is applied to a real-world case study to evaluate the performance of four popular cloud service providers. The results show that the proposed methodology can help decision-makers to select the most appropriate cloud service provider by taking into consideration multiple criteria and sub-criteria in a fuzzy environment.

The publication [17] proposes a framework for selecting sustainable suppliers based on a multi-criteria approach that integrates compensatory fuzzy AHP and TOPSIS methods. The paper highlights the increasing importance of sustainability in supply chain management and the need for a comprehensive and objective approach to selecting sustainable suppliers. The authors first present a literature review on the subject, followed by a detailed description of the proposed framework, which consists of three stages: (1) criteria identification, (2) criteria weighting using the fuzzy AHP method, and (3) supplier ranking using the fuzzy TOPSIS method. The proposed framework aims to provide a systematic and structured approach for selecting sustainable suppliers, while also considering the trade-offs between different criteria.

The authors pf publication [18] propose a new approach to solving the supplier selection problem. The authors introduce the concept of interval-valued intuitionistic fuzzy sets (IVIFS), which allows for a more accurate representation of the uncertainty and ambiguity inherent in the decision-making process. The proposed method combines the TOPSIS approach with IVIFS to provide a more comprehensive evaluation of the available alternatives. The paper presents a step-by-step procedure for applying the proposed method to the supplier selection problem, including the determination of the weights of the criteria and the ranking of the suppliers. The authors also provide a numerical example to demonstrate the effectiveness and applicability of the proposed approach.

In the humanitarian sector, it is essential to ensure the availability and timely delivery of critical supplies. This study [19] proposes a decision-making framework for selecting supply partners in continuous aid humanitarian supply chains using fuzzy AHP and fuzzy TOPSIS methods. The proposed framework addresses the challenges of supply partner selection by considering multiple criteria, including quality, reliability, price, delivery, and social responsibility. The study uses linguistic variables to capture the decision-makers preferences and applies interval type-2 fuzzy sets to manage the uncertainty and vagueness associated with the decision-making process. The proposed methodology is demonstrated using a case study of a non-governmental organization operating in the Syrian refugee crisis.

The publication [20] proposes a new approach for selecting alternatives based on linear programming and extended fuzzy TOPSIS under the framework of dual hesitant fuzzy sets. The proposed method is used for decision-making problems in which the decision criteria are uncertain and the decision-makers express their preferences in a hesitant manner. The proposed approach first transforms the dual hesitant fuzzy decision matrix into an interval-valued hesitant fuzzy decision matrix and then solves the linear programming problem to obtain the relative weights of the criteria. Finally, the extended fuzzy TOPSIS method is used to rank the alternatives based on the weighted criteria.

The publication [21] focuses on the application of two well-known MCDM techniques, namely Fuzzy AHP and Fuzzy TOPSIS for optimizing the electro-discharge machining (EDM) process. The article describes the integration of Fuzzy AHP and Fuzzy TOPSIS methods to identify the optimal machining parameters that maximize the material removal rate and minimize surface roughness. The study uses the dual representation of hesitant fuzzy sets to better reflect the uncertainty and vagueness in decision-making. The article presents a case study where the proposed method is applied to optimize the EDM process parameters for the machining of AISI D2 tool steel.

Having analyzed all the above-mentioned publications, it can be seen that the Fuzzy TOPSIS method is quite widely used in many different areas. This shows its real usefulness in various tasks, in particular when choosing innovative projects. At the same time, it should be noted that in existing research and development there is no possibility of choosing the shape of fuzzy sets when evaluating alternatives and criteria, and there is also no possibility of changing the metric for choosing the optimal solution.

3 Modification of Fuzzy TOPSIS to Solve the Problem

Having a problem that includes a set of alternatives and a set of criteria, multi-criteria decision-making methods are a good choice for solving this type of problem. Since each decision is evaluated according to many criteria, it becomes necessary to combine the obtained evaluations into one global criterion [2, 4, 22, 23].

Fuzzy ARAS, Fuzzy TOPSIS, Fuzzy COPRAS, Fuzzy VIKOR, Fuzzy AHP, Fuzzy PROMETHEE, Fuzzy ELECTRE, Fuzzy SAW are examples of multi-criteria decision-making methods. Next, several of the methods defined above are described [7, 22–25].

The Fuzzy ARAS method is aimed at choosing the optimal alternative based on several criteria. The ranking of alternatives is determined by calculating the degree of usefulness of each alternative. The degree of optimality of the alternative is considered to be the ratio of the sum of the normalized and weighted points of the criteria of the considered alternative to the sum of the normalized and weighted points of the criteria of the optimal alternative. The Fuzzy ARAS method compares the value of the utility function of each alternative with the value of the optimal utility function [26].

The Fuzzy AHP method of finding fuzzy weights is based on the direct fuzzification of Saati's method. This method applies the comparison of criteria in pairs with a defined measurement scale to find priority criteria. The main input data in the Fuzzy AHP method are expert ratings. This fact creates an element of subjectivity when making decisions. Additionally, a big plus of this method is taking into account the reliability of data with limits of inconsistency. However, significant uncertainty in the assessment can significantly affect the accuracy of the results. Taking into account all the above-mentioned elements, a method of fuzzy analytical process was created [2, 4].

Fuzzy VIKOR consists in determining the compromise solution and the best solution among all alternatives. A compromise solution is a comparison of the coefficient of proximity to the ideal alternative. Also, each alternative is evaluated by all criteria [27].

After a review and comparison of multi-criteria decision-making methods, the Fuzzy TOPSIS method was chosen for further work, as it is better suited to the chosen task of evaluation and selection of the optimal innovative project. After that, the possibility of modifying this method was considered, based on the specifics of the problem and input data. It was decided to make a modification using other metrics for calculating the coefficient of proximity (closeness) to the ideal solution. Therefore, several distances were chosen, namely: the Euclidean distance, the Hamming distance; the Chebyshev distance [5, 7, 8].

To demonstrate the operation of the modified method in solving the task of evaluation and selection of an innovative project, the following alternatives (projects) were determined, which were presented at the start-up project competition: MPBoard, Pixium, GUPY Services, Snager, SIFmeter, Holo Media System, Interview.top, InstaAdver, Pillars of Light, E-Cup, Elxy, Electro teacher, Cyberstick, Econd. The following criteria are defined for the evaluation of relevant projects: relevance and social significance, project idea; project issues; specificity, significance of the achievement of project results; economic efficiency and business model; investment attractiveness of the project; current situation; decision options; advantages over competitors; potential clients; market reach. The survey of experts was conducted after the demonstration for them of relevant innovative projects and their thorough review and discussion with other specialists. The criteria were selected taking into account the previous successful practices of evaluating innovative projects and various scientific publications.

When implementing the selected Fuzzy TOPSIS method, data for calculations were collected at each step. Thus, the first step was to define linguistic terms for criteria and alternatives (Table 1) [5, 13–15].

Table 1. Linguistic terms for criteria and alternatives.

Linguistic terms	Fuzzy triangular numbers	Fuzzy trapezoidal numbers	Fuzzy Gaussian numbers
Very low (VL)	(0.0, 0.0, 0.1)	(0.0, 0.0, 0.017, 0.15)	([0, 0.3], 0, 0.07, 0.01)
Low (L)	(0.0, 0.1, 0.3)	(0.02, 0.15, 0.18, 0.32)	([0, 0.47], 0.17, 0.07, 0.01)
Medium low (ML)	(0.1, 0.3, 0.5)	(0.18, 0.32, 0.35, 0.48)	([0.03, 0.6], 0.33, 0.07, 0.01)
Medium (M)	(0.3, 0.5, 0.7)	(0.35, 0.48, 0.52, 0.65)	([0.2, 0.8], 0.5, 0.07, 0.01)
Medium high (MH)	(0.5, 0.7, 0.9)	(0.52, 0.65, 0.68, 0.82)	([0.37, 0.97], 0.7, 0.07, 0.01)
High (H)	(0.7, 0.9, 1.0)	(0.68, 0.82, 0.85, 0.98)	([0.53, 1], 0.83, 0.07, 0.01)
Very high (VH)	(0.9, 1.0, 1.0)	(0.85, 0.98, 1, 1)	([0.7, 1], 1, 0.07, 0.01)

After that, the criteria (see Fig. 1) and alternatives (see Fig. 2) were assessed by experts in linguistic form. The experts were specialists in the field of innovative projects.

Criteria evaluation						
	Expert 1	Expert 2	Expert 3	Expert 4	Expert 5	Expert 6
Relevance and social significance, project idea	VH	H	VH	VH	VH	H
Project issues	H	MH	M	MH	M	H
Specificity, significance of the achievement of project results	M	MH	MH	MH	H	H
Economic efficiency and business model	M	M	ML	MH	H	MH
Investment attractiveness of the project	MH	ML	M	MH	M	H
Current situation	MH	L	M	MH	M	H
Decision options	MH	MH	MH	M	MH	H
Advantages over competitors	M	M	MH	ML	M	MH
Potential clients	MH	L	MH	H	MH	H
Market reach	VH	M	M	H	MH	H

Fig. 1. Evaluation of criteria by experts in linguistic form.

Alternatives evaluation						
	MPBoard					
	Expert 1	Expert 2	Expert 3	Expert 4	Expert 5	Expert 6
Relevance and social significance, project idea	G	F	G	G	G	G
Project issues	G	F	MG	G	MG	MG
Specificity, significance of the achievement of project results	MG	MP	MP	F	MG	G
Economic efficiency and business model	MG	F	MP	F	F	MG
Investment attractiveness of the project	MG	MG	G	F	F	MG
Current situation	MG	MG	G	F	F	F
Decision options	MG	F	MG	F	MG	MG
Advantages over competitors	MG	P	MG	MP	F	F
Potential clients	F	P	F	F	F	F
Market reach	F	P	MP	MP	F	F

Fig. 2. Evaluation of alternatives by experts in linguistic form.

Then, the estimates were converted into fuzzy numbers. This modification of the method makes it possible to use three shapes of fuzzy sets: triangular, trapezoidal, and Gaussian. Next, some calculations using the appropriate shapes will be illustrated (see Fig. 3 and Fig. 4) [5, 14, 28].

Transform the linguistic terms into fuzzy numbers						
	Criterion					
Relevance and social significance, project idea	(0,9; 1; 1)	(0,7; 0,9; 1)	(0,9; 1; 1)	(0,9; 1; 1)	(0,9; 1; 1)	(0,7; 0,9; 1)
Project issues	(0,7; 0,9; 1)	(0,5; 0,7; 0,9)	(0,3; 0,5; 0,7)	(0,5; 0,7; 0,9)	(0,3; 0,5; 0,7)	(0,7; 0,9; 1)
Specificity, significance of the achievement of project results	(0,3; 0,5; 0,7)	(0,5; 0,7; 0,9)	(0,5; 0,7; 0,9)	(0,5; 0,7; 0,9)	(0,7; 0,9; 1)	(0,7; 0,9; 1)
Economic efficiency and business model	(0,3; 0,5; 0,7)	(0,3; 0,5; 0,7)	(0,1; 0,3; 0,5)	(0,5; 0,7; 0,9)	(0,7; 0,9; 1)	(0,5; 0,7; 0,9)
Investment attractiveness of the project	(0,5; 0,7; 0,9)	(0,1; 0,3; 0,5)	(0,3; 0,5; 0,7)	(0,5; 0,7; 0,9)	(0,3; 0,5; 0,7)	(0,7; 0,9; 1)
Current situation	(0,5; 0,7; 0,9)	(0; 0,1; 0,3)	(0,3; 0,5; 0,7)	(0,5; 0,7; 0,9)	(0,3; 0,5; 0,7)	(0,7; 0,9; 1)
Decision options	(0,5; 0,7; 0,9)	(0,5; 0,7; 0,9)	(0,5; 0,7; 0,9)	(0,3; 0,5; 0,7)	(0,5; 0,7; 0,9)	(0,7; 0,9; 1)
Advantages over competitors	(0,3; 0,5; 0,7)	(0,3; 0,5; 0,7)	(0,5; 0,7; 0,9)	(0,1; 0,3; 0,5)	(0,3; 0,5; 0,7)	(0,5; 0,7; 0,9)
Potential clients	(0,5; 0,7; 0,9)	(0; 0,1; 0,3)	(0,5; 0,7; 0,9)	(0,7; 0,9; 1)	(0,5; 0,7; 0,9)	(0,7; 0,9; 1)
Market reach	(0,9; 1; 1)	(0,3; 0,5; 0,7)	(0,3; 0,5; 0,7)	(0,7; 0,9; 1)	(0,5; 0,7; 0,9)	(0,7; 0,9; 1)
Transform the linguistic terms into fuzzy numbers						
	MPBoard					
Relevance and social significance, project idea	(0,7; 0,9; 1)	(0,3; 0,5; 0,7)	(0,7; 0,9; 1)	(0,7; 0,9; 1)	(0,7; 0,9; 1)	(0,7; 0,9; 1)
Project issues	(0,7; 0,9; 1)	(0,3; 0,5; 0,7)	(0,5; 0,7; 0,9)	(0,7; 0,9; 1)	(0,5; 0,7; 0,9)	(0,5; 0,7; 0,9)
Specificity, significance of the achievement of project results	(0,5; 0,7; 0,9)	(0,1; 0,3; 0,5)	(0,1; 0,3; 0,5)	(0,3; 0,5; 0,7)	(0,5; 0,7; 0,9)	(0,7; 0,9; 1)
Economic efficiency and business model	(0,5; 0,7; 0,9)	(0,3; 0,5; 0,7)	(0,1; 0,3; 0,5)	(0,3; 0,5; 0,7)	(0,3; 0,5; 0,7)	(0,5; 0,7; 0,9)
Investment attractiveness of the project	(0,5; 0,7; 0,9)	(0,5; 0,7; 0,9)	(0,7; 0,9; 1)	(0,3; 0,5; 0,7)	(0,3; 0,5; 0,7)	(0,5; 0,7; 0,9)
Current situation	(0,5; 0,7; 0,9)	(0,5; 0,7; 0,9)	(0,7; 0,9; 1)	(0,3; 0,5; 0,7)	(0,3; 0,5; 0,7)	(0,3; 0,5; 0,7)
Decision options	(0,5; 0,7; 0,9)	(0,3; 0,5; 0,7)	(0,5; 0,7; 0,9)	(0,3; 0,5; 0,7)	(0,5; 0,7; 0,9)	(0,5; 0,7; 0,9)
Advantages over competitors	(0,5; 0,7; 0,9)	(0; 0,1; 0,3)	(0,5; 0,7; 0,9)	(0,1; 0,3; 0,5)	(0,3; 0,5; 0,7)	(0,3; 0,5; 0,7)
Potential clients	(0,3; 0,5; 0,7)	(0; 0,1; 0,3)	(0,3; 0,5; 0,7)	(0,3; 0,5; 0,7)	(0,3; 0,5; 0,7)	(0,3; 0,5; 0,7)
Market reach	(0,3; 0,5; 0,7)	(0; 0,1; 0,3)	(0,1; 0,3; 0,5)	(0,1; 0,3; 0,5)	(0,3; 0,5; 0,7)	(0,3; 0,5; 0,7)

Fig. 3. Evaluation of criteria and alternatives by experts in the form of fuzzy triangular numbers.

Since the database (DB) of alternatives and criteria is quite large, the authors have demonstrated only a part of the DB and calculations. The user can always view all data and calculations in a file created automatically during system operation.

Transform the linguistic terms into fuzzy numbers					
	Criterion				
Relevance and social significance, project idea	(0,85; 0,98; 1; 1)	(0,68; 0,82; 0,85; 0,9)	(0,85; 0,98; 1; 1)	(0,85; 0,98; 1; 1)	(0,85; 0,98; 1; 1)	(0,68; 0,82; 0,85; 0,98)
Project issues	(0,68; 0,82; 0,85; 0,98)	(0,52; 0,65; 0,68; 0,82)	(0,35; 0,48; 0,52; 0,65)	(0,52; 0,65; 0,68; 0,82)	(0,35; 0,48; 0,52; 0,65)	(0,68; 0,82; 0,85; 0,98)
Specificity, significance of the achievement of project results	(0,35; 0,48; 0,52; 0,65)	(0,52; 0,65; 0,68; 0,82)	(0,52; 0,65; 0,68; 0,82)	(0,52; 0,65; 0,68; 0,82)	(0,68; 0,82; 0,85; 0,98)	(0,68; 0,82; 0,85; 0,98)
Economic efficiency and business model	(0,35; 0,48; 0,52; 0,65)	(0,35; 0,48; 0,52; 0,65)	(0,18; 0,32; 0,35; 0,48)	(0,52; 0,65; 0,68; 0,82)	(0,68; 0,82; 0,85; 0,98)	(0,52; 0,65; 0,68; 0,82)
Investment attractiveness of the project	(0,52; 0,65; 0,68; 0,82)	(0,18; 0,32; 0,35; 0,48)	(0,35; 0,48; 0,52; 0,65)	(0,52; 0,65; 0,68; 0,82)	(0,35; 0,48; 0,52; 0,65)	(0,68; 0,82; 0,85; 0,98)
Current situation	(0,52; 0,65; 0,68; 0,82)	(0,02; 0,15; 0,18; 0,32)	(0,35; 0,48; 0,52; 0,65)	(0,52; 0,65; 0,68; 0,82)	(0,35; 0,48; 0,52; 0,65)	(0,68; 0,82; 0,85; 0,98)
Decision options	(0,52; 0,65; 0,68; 0,82)	(0,52; 0,65; 0,68; 0,82)	(0,52; 0,65; 0,68; 0,82)	(0,35; 0,48; 0,52; 0,65)	(0,52; 0,65; 0,68; 0,82)	(0,68; 0,82; 0,85; 0,98)
Advantages over competitors	(0,85; 0,48; 0,52; 0,65)	(0,35; 0,48; 0,52; 0,65)	(0,52; 0,65; 0,68; 0,82)	(0,18; 0,32; 0,35; 0,48)	(0,35; 0,48; 0,52; 0,65)	(0,52; 0,65; 0,68; 0,82)
Potential clients	(0,52; 0,65; 0,68; 0,82)	(0,02; 0,15; 0,18; 0,32)	(0,52; 0,65; 0,68; 0,82)	(0,68; 0,82; 0,85; 0,98)	(0,52; 0,65; 0,68; 0,82)	(0,68; 0,82; 0,85; 0,98)
Market reach	(0,85; 0,98; 1; 1)	(0,35; 0,48; 0,52; 0,65)	(0,35; 0,48; 0,52; 0,65)	(0,68; 0,82; 0,85; 0,98)	(0,52; 0,65; 0,68; 0,82)	(0,68; 0,82; 0,85; 0,98)
Transform the linguistic terms into fuzzy numbers					
	MPBoard				
Relevance and social significance, project idea	(0,68; 0,82; 0,85; 0,98)	(0,35; 0,48; 0,52; 0,65)	(0,68; 0,82; 0,85; 0,98)	(0,68; 0,82; 0,85; 0,98)	(0,68; 0,82; 0,85; 0,98)	(0,68; 0,82; 0,85; 0,98)
Project issues	(0,68; 0,82; 0,85; 0,98)	(0,35; 0,48; 0,52; 0,65)	(0,52; 0,65; 0,68; 0,82)	(0,68; 0,82; 0,85; 0,98)	(0,52; 0,65; 0,68; 0,82)	(0,52; 0,65; 0,68; 0,82)
Specificity, significance of the achievement of project results	(0,53; 0,65; 0,68; 0,82)	(0,18; 0,32; 0,35; 0,48)	(0,18; 0,32; 0,35; 0,48)	(0,35; 0,48; 0,52; 0,65)	(0,35; 0,48; 0,52; 0,65)	(0,52; 0,65; 0,68; 0,82)
Economic efficiency and business model	(0,52; 0,65; 0,68; 0,82)	(0,35; 0,48; 0,52; 0,65)	(0,18; 0,32; 0,35; 0,48)	(0,35; 0,48; 0,52; 0,65)	(0,35; 0,48; 0,52; 0,65)	(0,35; 0,48; 0,52; 0,65)
Investment attractiveness of the project	(0,52; 0,65; 0,68; 0,82)	(0,52; 0,65; 0,68; 0,82)	(0,68; 0,82; 0,85; 0,98)	(0,35; 0,48; 0,52; 0,65)	(0,52; 0,65; 0,68; 0,82)	(0,52; 0,65; 0,68; 0,82)
Current situation	(0,52; 0,65; 0,68; 0,82)	(0,52; 0,65; 0,68; 0,82)	(0,68; 0,82; 0,85; 0,98)	(0,35; 0,48; 0,52; 0,65)	(0,35; 0,48; 0,52; 0,65)	(0,52; 0,65; 0,68; 0,82)
Decision options	(0,52; 0,65; 0,68; 0,82)	(0,35; 0,48; 0,52; 0,65)	(0,52; 0,65; 0,68; 0,82)	(0,35; 0,48; 0,52; 0,65)	(0,52; 0,65; 0,68; 0,82)	(0,52; 0,65; 0,68; 0,82)
Advantages over competitors	(0,52; 0,65; 0,68; 0,82)	(0,02; 0,15; 0,18; 0,32)	(0,52; 0,65; 0,68; 0,82)	(0,18; 0,32; 0,35; 0,48)	(0,35; 0,48; 0,52; 0,65)	(0,35; 0,48; 0,52; 0,65)
Potential clients	(0,35; 0,48; 0,52; 0,65)	(0,02; 0,15; 0,18; 0,32)	(0,35; 0,48; 0,52; 0,65)	(0,35; 0,48; 0,52; 0,65)	(0,35; 0,48; 0,52; 0,65)	(0,35; 0,48; 0,52; 0,65)
Market reach	(0,35; 0,48; 0,52; 0,65)	(0,02; 0,15; 0,18; 0,32)	(0,18; 0,32; 0,35; 0,48)	(0,18; 0,32; 0,35; 0,48)	(0,35; 0,48; 0,52; 0,65)	(0,35; 0,48; 0,52; 0,65)

Fig. 4. Evaluation of criteria and alternatives by experts in the form of fuzzy trapezoidal numbers.

The next step is to average the evaluations of the criteria and alternatives (see Fig. 5).

Averaged evaluations of criteria and alternatives												
	Criterion			MPBoard			Pixium			GUPY Services		
Relevance and social significance, project idea	0,83	0,97	1,00	0,63	0,83	0,95	0,57	0,77	0,92	0,70	0,90	1,00
Project issues	0,50	0,70	0,87	0,53	0,73	0,90	0,43	0,63	0,83	0,63	0,83	0,97
Specificity, significance of the achievement of project results	0,53	0,73	0,90	0,37	0,57	0,75	0,43	0,63	0,83	0,40	0,60	0,78
Economic efficiency and business model	0,40	0,60	0,78	0,33	0,53	0,73	0,37	0,57	0,77	0,23	0,43	0,63
Investment attractiveness of the project	0,40	0,60	0,78	0,47	0,67	0,85	0,43	0,63	0,82	0,40	0,60	0,78
Current situation	0,38	0,57	0,75	0,43	0,63	0,82	0,33	0,53	0,73	0,40	0,60	0,80
Decision options	0,50	0,70	0,88	0,43	0,63	0,83	0,37	0,57	0,77	0,53	0,73	0,90
Advantages over competitors	0,33	0,53	0,73	0,28	0,47	0,67	0,27	0,47	0,67	0,43	0,63	0,83
Potential clients	0,48	0,67	0,83	0,25	0,43	0,63	0,23	0,43	0,63	0,47	0,67	0,83
Market reach	0,57	0,75	0,88	0,18	0,37	0,57	0,30	0,50	0,70	0,57	0,77	0,90

Fig. 5. Averaging the evaluations of criteria and alternatives.

Next, the averaged evaluations of the alternatives were normalized and a weighted normalized matrix of the averaged evaluations of the alternatives was constructed. Next, FPIS and FNIS are found. After that, the distance of each alternative to the FPIS and to the FNIS was calculated (see Fig. 6).

After that, the proximity coefficient was calculated and the alternatives were ranked. The top 3 projects have the following proximity coefficient indicators: Cyberstick is 0.5901 (first place), Holo Media System is 0.5211 (second place), GUPY Services is 0.5204 (third place).

The distance of each alternative to the FPIS and to the FNIS, as well as the proximity coefficient using fuzzy trapezoidal numbers, were also calculated. The top 3 projects have the following proximity coefficient indicators: Cyberstick is 0.7177 (first place), GUPY Services is 0.6634 (second place), Holo Media System is 0.6630 (third place).

The distance of each alternative to the FPIS and to the FNIS, as well as the proximity coefficient using fuzzy Gaussian numbers, were also calculated. The top 3 projects have the following proximity coefficient indicators: Cyberstick is 0.4105 (first place), GUPY Services is 0.3784 (second place), Holo Media System is 0.3775 (third place).

According to the obtained results, the innovative Cyberstick project is the most prioritized when choosing for investors in various shapes of fuzzy sets.

The distance of each alternative to the FPIS														
	FPIS (A+)													
Relevance and social significance, project idea	0,30	0,34	0,25	0,37	0,32	0,32	0,42	0,34	0,44	0,53	0,34	0,27	0,27	0,34
Project issues	0,39	0,44	0,35	0,48	0,35	0,38	0,39	0,39	0,47	0,54	0,43	0,39	0,35	0,39
Specificity, significance of the achievement of project results	0,47	0,43	0,45	0,51	0,49	0,45	0,43	0,41	0,41	0,49	0,47	0,51	0,37	0,41
Economic efficiency and business model	0,45	0,43	0,50	0,47	0,45	0,36	0,43	0,43	0,47	0,54	0,48	0,54	0,40	0,48
Investment attractiveness of the project	0,40	0,42	0,43	0,50	0,46	0,40	0,46	0,48	0,45	0,57	0,43	0,51	0,35	0,48
Current situation	0,40	0,45	0,42	0,45	0,45	0,42	0,40	0,39	0,40	0,52	0,40	0,50	0,33	0,42
Decision options	0,45	0,49	0,40	0,45	0,47	0,42	0,45	0,44	0,45	0,53	0,49	0,49	0,37	0,45
Advantages over competitors	0,48	0,48	0,41	0,44	0,41	0,44	0,48	0,41	0,44	0,53	0,46	0,42	0,34	0,42
Potential clients	0,53	0,53	0,40	0,55	0,45	0,43	0,43	0,41	0,45	0,45	0,38	0,41	0,35	0,45
Market reach	0,60	0,52	0,36	0,40	0,34	0,38	0,48	0,38	0,45	0,48	0,38	0,36	0,33	0,38
Total	4,47	4,53	3,97	4,61	4,19	3,98	4,38	4,08	4,43	5,17	4,27	4,41	3,44	4,24
The distance of each alternative to the FNIS														
	FNIS (A-)													
Relevance and social significance, project idea	0,51	0,47	0,57	0,45	0,50	0,50	0,40	0,48	0,37	0,30	0,48	0,54	0,54	0,48
Project issues	0,45	0,40	0,51	0,35	0,51	0,47	0,46	0,46	0,37	0,29	0,42	0,45	0,51	0,46
Specificity, significance of the achievement of project results	0,38	0,44	0,40	0,34	0,35	0,41	0,44	0,45	0,45	0,36	0,39	0,34	0,49	0,45
Economic efficiency and business model	0,37	0,39	0,30	0,35	0,37	0,48	0,39	0,39	0,35	0,26	0,32	0,26	0,42	0,32
Investment attractiveness of the project	0,43	0,41	0,39	0,31	0,35	0,43	0,35	0,33	0,37	0,23	0,39	0,29	0,50	0,33
Current situation	0,37	0,31	0,35	0,31	0,31	0,35	0,37	0,39	0,37	0,24	0,37	0,26	0,46	0,35
Decision options	0,40	0,35	0,46	0,40	0,38	0,45	0,40	0,42	0,40	0,31	0,35	0,35	0,50	0,39
Advantages over competitors	0,28	0,28	0,38	0,34	0,37	0,34	0,28	0,38	0,34	0,23	0,30	0,36	0,46	0,35
Potential clients	0,29	0,29	0,44	0,27	0,38	0,40	0,40	0,43	0,38	0,37	0,46	0,42	0,50	0,37
Market reach	0,26	0,35	0,52	0,47	0,54	0,51	0,40	0,51	0,42	0,39	0,51	0,52	0,57	0,50
Total	3,74	3,70	4,31	3,58	4,06	4,34	3,90	4,23	3,83	2,98	3,99	3,80	4,96	3,99

Fig. 6. The distance of each alternative to the FPIS and to the FNIS using the Euclidean distance.

Further, calculations were carried out with a change in the distance metric to the ideal solution and proximity coefficients with various shapes of fuzzy sets.

For convenient use of the modified Fuzzy TOPSIS method when solving various problems, a corresponding DSS was created (see Fig. 7).

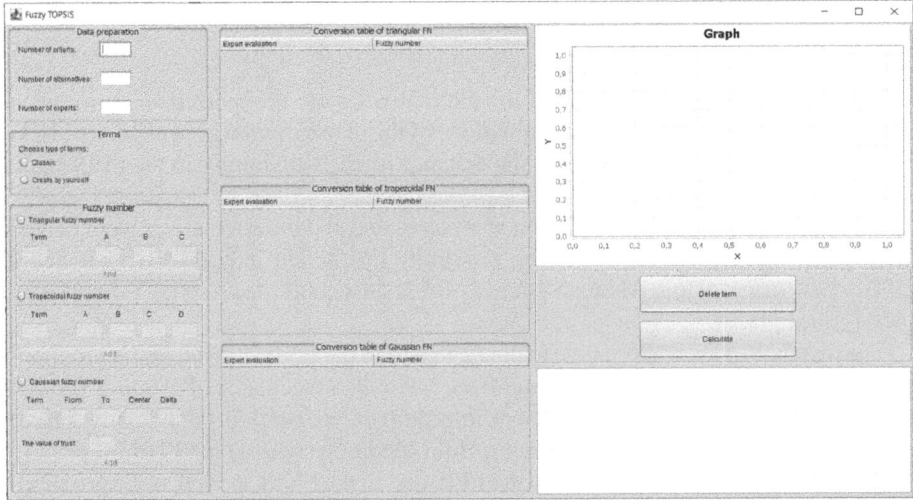

Fig. 7. The interface of the developed DSS using the proposed modified Fuzzy TOPSIS method.

This DSS provides an opportunity to consider existing linguistic terms (see Fig. 8) or add your own linguistic terms (see Fig. 9).

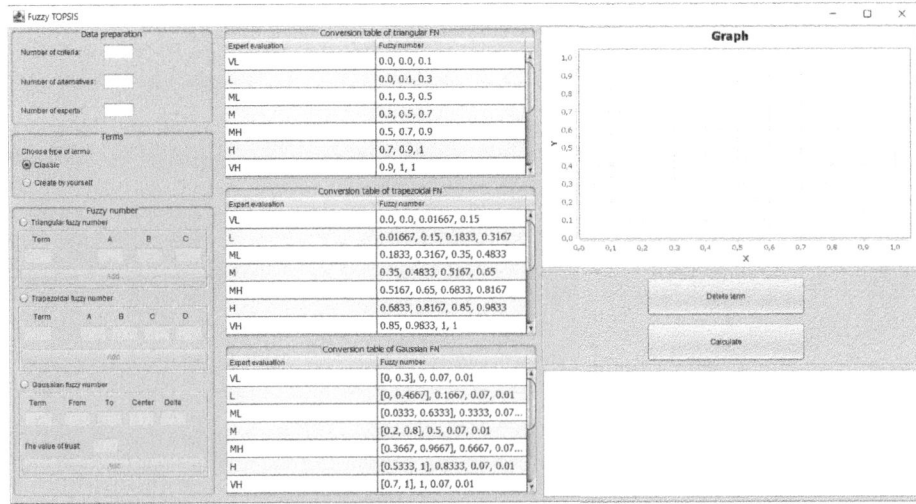

Fig. 8. Consider existing linguistic terms.

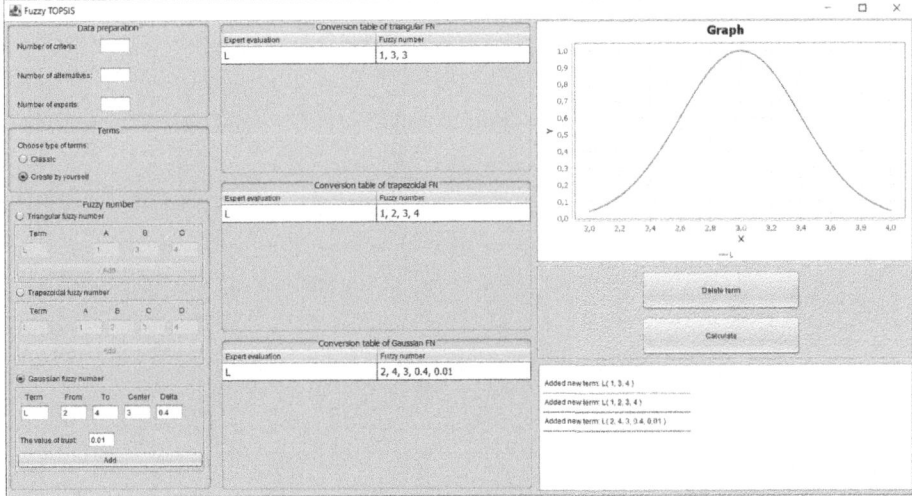

Fig. 9. Add your own linguistic terms.

A small console window has been added to the lower right, which provides information about the performed actions (see Fig. 9). Validation of entered data is also created. This DSS provides an opportunity to enter only correct data. Thus, only numeric data can be entered in the fields describing the number of criteria, alternatives, and experts. The same check is made for fuzzy number input fields.

When creating linguistic terms, it is possible to consider the corresponding fuzzy numbers on the graph (see Fig. 10).

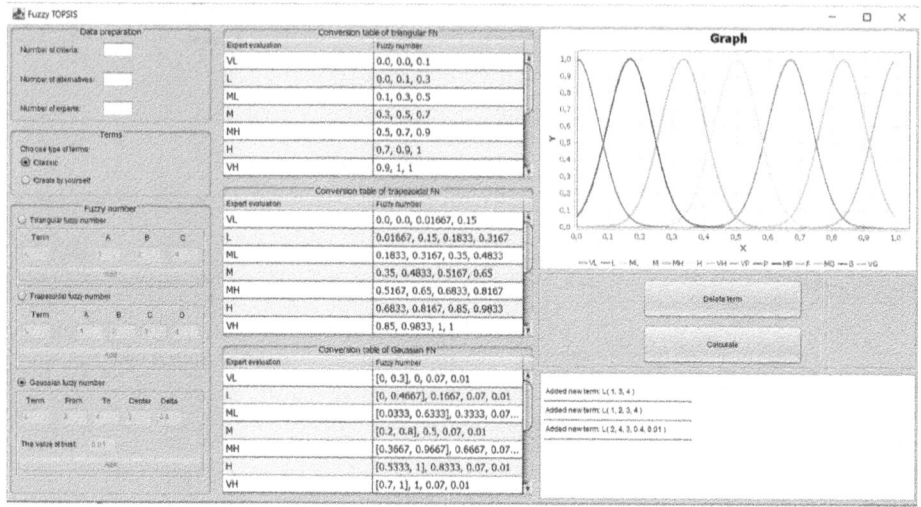

Fig. 10. Visualization of fuzzy Gaussian numbers.

Also, for convenient viewing of created terms, the option to highlight the selected term has been added (see Fig. 11).

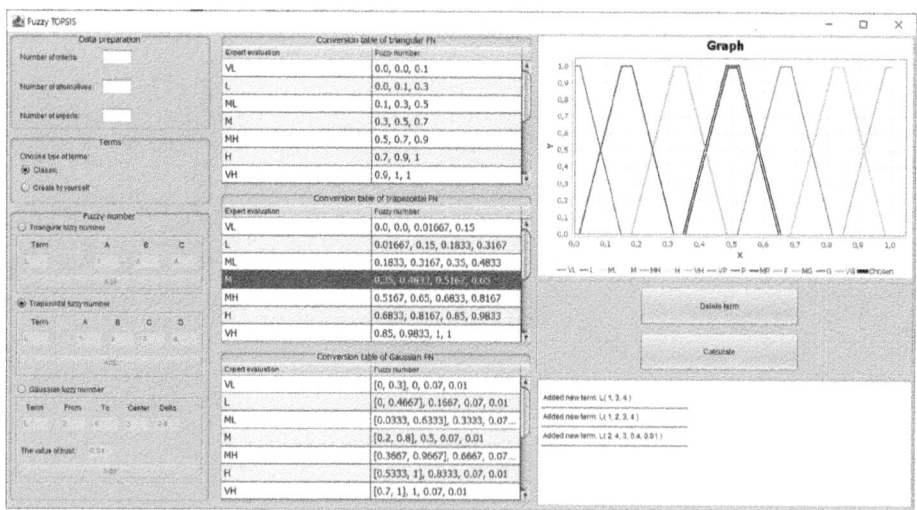

Fig. 11. Selection of the chosen term.

In addition, the possibility of data correction is provided (see Fig. 12). A template was also created that reflects the correctness of data entry into the database (see Fig. 13).

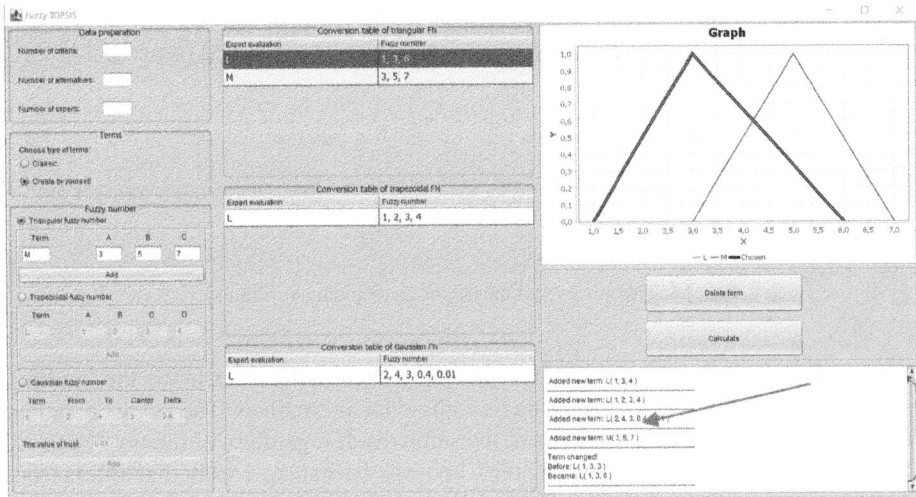

Fig. 12. Correction of the chosen term.

Criteria					
	Criterion 1	Criterion 2	Criterion 3	Criterion 4	Criterion 5
Expert 1	VH	H	M	M	MH
Expert 2	H	MH	MH	M	ML
Expert 3	VH	M	MH	ML	M

Expert 1					
	Criterion 1	Criterion 2	Criterion 3	Criterion 4	Criterion 5
Alternative 1	G	G	MG	MG	MG
Alternative 2	G	MG	MG	F	F
Alternative 3	G	G	MG	MP	G
Alternative 4	MG	MG	MG	F	F

Expert 2					
	Criterion 1	Criterion 2	Criterion 3	Criterion 4	Criterion 5
Alternative 1	F	F	MP	F	MG
Alternative 2	F	MG	MP	F	MG
Alternative 3	G	G	F	F	F
Alternative 4	F	F	MG	F	MG

Fig. 13. Data entry template.

Warnings have also been added when you try to make calculations without entering all the necessary data. So, for example, when you try to make calculations without defining the number of criteria and alternatives, an error message is issued. Another warning appears when trying to add an existing linguistic term. Also, if the file with the input data is missing or damaged, a corresponding warning is issued (see Fig. 14).

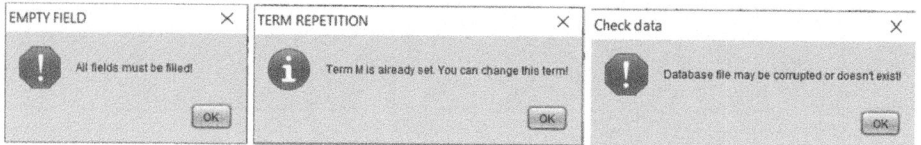

Fig. 14. Warnings and messages during data validation.

According to the results obtained by calculating the Euclidean distance, the resulting ranking has very close values, differing by individual alternatives. The results obtained when using the Hamming distance have the same result regardless of the type of fuzzy number. According to the results using the Chebyshev distance, a result that differs from the previous results was obtained. The results are significantly worse than when using other distances.

In the work, additional calculations were made for the selection of an innovative project using the Fuzzy ARAS method. The results obtained using the Fuzzy ARAS method agree with the results obtained using the modified Fuzzy TOPSIS method in selecting the best solution. A small difference in the obtained results may be precisely due to taking into account the largest distance to the fuzzy negative ideal solution by the Fuzzy TOPSIS method [5, 27, 29, 30].

Analyzing the obtained data, it can be seen that in general, the application of the classical method of calculating the distance to the ideal solution and the proximity coefficient, the Euclidean distance and the Hamming distance have the same ranking results. This shows that using different approaches can achieve the same result. However, there are still minimal differences, for example, the Euclidean and Hamming distances have a difference in the ranking of the second and third alternatives, which can be important if it is planned to choose several innovative projects for investment. It can also be observed that the results of the classical calculation of the proximity coefficient and the results using the Euclidean distance have the same results. The results using the Hamming distance differ slightly from the other metrics, but are still practically similar, which indicates that the Fuzzy TOPSIS method using the classical distance, Euclidean distance or Hamming distance gives very close results, so it does not make a fundamental difference which one to use. It is worth noting that the results obtained when applying the Chebyshev method for distance calculation are significantly different from the results obtained when applying other distances. Especially bad results were obtained when using fuzzy Gaussian numbers when calculating the distance using the Chebyshev metric. That is, the use of the Chebyshev distance is inappropriate for the Fuzzy TOPSIS method [5, 16, 27].

Summarizing all the above analysis, it can be concluded that the Fuzzy TOPSIS method is well optimized for the use of many forms of fuzzy sets and metrics for calculating the distance to the ideal solution and proximity coefficients. Therefore, the choice remains entirely up to the expert. As a result of the calculations, in most cases the best alternative was the Cyberstick. Therefore, this innovative project will be the most profitable choice for investors.

Taking into account the above-mentioned comparison of the application results of the Fuzzy TOPSIS method modifications, in particular the change of the distance in the metrics of the proximity coefficients and the forms of the fuzzy sets, it can be concluded that they have different effects on the final result of choosing the best solution. This influence is significantly felt when the distance to the best solution is changed in the metrics of proximity coefficients. Therefore, it is advisable to use Euclidean and Hamming distances with different forms of fuzzy sets. The results of the evaluation of the relevant projects by experts also showed a similar ranking of solutions, in which the innovative Cyberstick project is the most prioritized.

4 Conclusions

Within the framework of this work, the following tasks were successfully completed: (a) a review of the current state of MCDM problems for evaluation and selection of innovative projects was carried out; (b) analyzed the latest research and publications; (c) a review of multi-criteria decision-making methods was conducted; (d) the Fuzzy TOPSIS method was modified; (e) the software implementation of the optimal innovation project selection system has been developed.

To modify the Fuzzy TOPSIS method, the Euclidean distance, the Hamming distance, and the Chebyshev distance were used to determine the proximity coefficients. Another element of the modification of the Fuzzy TOPSIS method was the ability to calculate results using various fuzzy numbers, such as triangular, trapezoidal, and Gaussian. A comparative analysis of the obtained results with other multi-criteria decision-making methods was carried out. A data entry template for experts has also been created and the results presented in a clear and easy-to-read format.

References

1. Velimirović, L.Z., Janjić, A., Velimirović, J.D.: Multi-criteria Decision Making for Smart Grid Design and Operation. Springer, Singapore (2023)
2. Munier, N., Hontoria, E., Jiménez-Sáez, F.: Strategic Approach in Multi-criteria Decision Making. A Practical Guide for Complex Scenarios. Springer, Cham (2019)
3. Sahoo, L., Senapati, T., Yager, R.: Real Life Applications of Multiple Criteria Decision Making Techniques in Fuzzy Domain. Springer, Singapore (2023)
4. Singh, N., Singh, K.R.: Application of TOPSIS – a multi criteria decision making approach in surface water quality assessment. In: Haq, I., Kalamdhad, A.S., Dash, S. (eds.) Environmental Degradation: Monitoring, Assessment and Treatment Technologies. Springer, Cham (2022). https://doi.org/10.1007/978-3-030-94148-2_20
5. Martynova, L., Kondratenko, G., Sidenko, I., Kondratenko, Y.: Application of fuzzy TOP-SIS method in group decision-making for ranking political parties. In: IEEE International Conference on Advanced Trends in Information Theory (ATIT), pp. 384–388 (2019)
6. Skarga-Bandurova, I., Nesterov, M., Kovalenko, Y.: A group decision support technique for critical IT infrastructures. In: Zamojski, W., Mazurkiewicz, J., Sugier, J., Walkowiak, T., Kacprzyk, J. (eds.) Theory and Engineering of Complex Systems and Dependability. AISC, vol. 365, pp. 445–454. Springer, Cham (2015). https://doi.org/10.1007/978-3-319-19216-1_42

7. Borysenko, V., Kondratenko, G., Sidenko, I., Kondratenko, Y.: Intelligent forecasting in multi-criteria decision-making. In: CEUR Workshop Proceedings. 3rd International Workshop on Computer Modeling and Intelligent Systems, CMIS 2020, vol. 2608, pp. 966–979. National University "Zaporizhzhia Polytechnic", Zaporizhzhia, Ukraine (2020)
8. Nazim, M., Wali Mohammad, C., Sadiq, M.: Analysis of fuzzy AHP and fuzzy TOPSIS methods for the prioritization of the software requirements. In: Kulkarni, A.J. (ed.) Multiple Criteria Decision Making. SSDC, vol. 407, pp. 79–90. Springer, Singapore (2022). https://doi.org/10.1007/978-981-16-7414-3_4
9. Chumachenko, D., Meniailov, I., Bazilevych, K., Chumachenko, T.: On intelligent decision making in multiagent systems in conditions of uncertainty. In: XIth International Scientific and Practical Conference on Electronics and Information Technologies (ELIT), Lviv, Ukraine, pp. 150–153 (2019). https://doi.org/10.1109/ELIT.2019.8892307
10. Rebiy, E.Y., Boris, O.A., Lepyahova, E.N.: Selection of innovative projects for technical and technological development within network management structure environment at industrial enterprises. In: Popkova, E.G., Ostrovskaya, V.N., Bogoviz, A.V. (eds.) Socio-economic Systems: Paradigms for the Future. SSDC, vol. 314, pp. 709–719. Springer, Cham (2021). https://doi.org/10.1007/978-3-030-56433-9_75
11. Topka, V.V.: The cross-impact analysis of innovative projects in a portfolio. J. Comput. Syst. Sci. Int. **58**, 736–746 (2019). https://doi.org/10.1134/S1064230719050149
12. D, G.Y., Cui, J., Bekun, F.V.: Ecological risks and innovative-investment projects. Environ. Sci. Pollut. Res. **30**, 33124–33132 (2023). https://doi.org/10.1007/s11356-022-24405-7
13. Sadeghi, M.E., Nozari, H., Dezfoli, H.K., Dezfoli, M.K.: Ranking of different of investment risk in high-tech projects using TOPSIS method in fuzzy environment based on linguistic variables. J. Fuzzy Extension Appl. **2**(3), 246–258 (2021)
14. Sarkar, A., Ghosh, A., Karmakar, B., Shaikh, A., Mondal, S.P.: Application of fuzzy TOPSIS algorithm for selecting best family car. In: International Conference on Decision Aid Sciences and Application (DASA), Sakheer, Bahrain, pp. 59–63 (2020). https://doi.org/10.1109/DASA51403.2020.9317175
15. Abadchi, A.R., Cyrus, K.M.: Designing a model for prioritization of strategies using combined fuzzy AHP and fuzzy TOPSIS. In: International Conference on Electrical, Computer and Energy Technologies (ICECET), Prague, Czech Republic, pp. 1–6 (2022). https://doi.org/10.1109/ICECET55527.2022.9872894
16. Kalaiarasan, C., Venkatesh, K.A.: Cloud service provider selection using fuzzy TOPSIS. In: IEEE International Conference for Innovation in Technology (INOCON), Bangluru, India, pp. 1–5 (2020). https://doi.org/10.1109/INOCON50539.2020.9298207
17. Gupta, A.K.: Framework for the selection of sustainable suppliers using integrated compensatory fuzzy AHP-TOPSIS multi-criteria approach. In: IEEE International Conference on Industrial Engineering and Engineering Management (IEEM), Kuala Lumpur, Malaysia, pp. 0772–0775 (2022). https://doi.org/10.1109/IEEM55944.2022.9989663
18. Tiwari, A., Lohani, Q.M.D., Muhuri, P.K.: Interval-valued intuitionistic fuzzy TOPSIS method for supplier selection problem. In: IEEE International Conference on Fuzzy Systems (FUZZ-IEEE), Glasgow, UK, pp. 1–8 (2020) https://doi.org/10.1109/FUZZ48607.2020.9177852
19. Venkatesh, V.G., Zhang, A., Deakins, E., et al.: A fuzzy AHP-TOPSIS approach to supply partner selection in continuous aid humanitarian supply chains. Ann. Oper. Res. **283**, 1517–1550 (2019). https://doi.org/10.1007/s10479-018-2981-1
20. Sindhu, M.S., Rashid, T.: Selection of alternative based on linear programming and the extended fuzzy TOPSIS under the framework of dual hesitant fuzzy sets. Soft. Comput. **27**, 1985–1996 (2023). https://doi.org/10.1007/s00500-022-07173-x
21. Roy, T., Dutta, R.K.: Integrated fuzzy AHP and fuzzy TOPSIS methods for multi-objective optimization of electro discharge machining process. Soft. Comput. **23**, 5053–5063 (2019). https://doi.org/10.1007/s00500-018-3173-2

22. Kondratenko, Y., Kondratenko, G., Sidenko, I.: Multi-criteria decision making and soft computing for the selection of specialized IoT platform. In: Chertov, O., Mylovanov, T., Kondratenko, Y., Kacprzyk, J., Kreinovich, V., Stefanuk, V. (eds.) ICDSIAI 2018. AISC, vol. 836, pp. 71–80. Springer, Cham (2019). https://doi.org/10.1007/978-3-319-97885-7_8

23. Kondratenko, Y., Kondratenko, G., Sidenko, I.: Multi-criteria decision making for selecting a rational IoT platform. In: IEEE 9th International Conference on Dependable Systems, Services and Technologies (DESSERT), Kiev, Ukraine, pp. 147–152 (2018). https://doi.org/10.1109/DESSERT.2018.8409117

24. Lavrynenko, S., Kondratenko, G., Sidenko, I., Kondratenko, Y.: Fuzzy logic approach for evaluating the effectiveness of investment projects. In: IEEE 15th International Scientific and Technical Conference on Computer Sciences and Information Technologies (CSIT), pp. 297–300 (2020)

25. Kondratenko, Y., Kondratenko, G., Sidenko, I., Taranov, M.: Intelligent information system for investment in uncertainty. In: 10th IEEE International Conference on Intelligent Data Acquisition and Advanced Computing Systems: Technology and Applications (IDAACS), pp. 216–221 (2019)

26. Heidary Dahooie, J., Estiri, M., Zavadskas, E.K., et al.: A novel hybrid fuzzy DEA-Fuzzy ARAS method for prioritizing high-performance innovation-oriented human resource practices in high tech SME's. Int. J. Fuzzy Syst. **24**, 883–908 (2022). https://doi.org/10.1007/s40815-021-01162-2

27. Cheng, S., Jianfu, S., Alrasheedi, M., et al.: Correction to: a new extended VIKOR approach using q-Rung Orthopair fuzzy sets for sustainable enterprise risk management assessment in manufacturing small and medium-sized enterprises. Int. J. Fuzzy Syst. **23**, 2504 (2021). https://doi.org/10.1007/s40815-021-01178-8

28. Kryvyi, S.L., Opanasenko, V.M., Zavyalov, S.B.: Algebraic operations on fuzzy sets and relations in automata interpretation implemented by logical hardware. Cybern. Syst. Anal. **58**, 649–659 (2022). https://doi.org/10.1007/s10559-022-00497-4

29. Tirmikcioglu, N.: ARAS method in picture fuzzy environment for the selection of catering firm. In: Kahraman, C., Tolga, A.C., Cevik Onar, S., Cebi, S., Oztaysi, B., Sari, I.U. (eds.) Intelligent and Fuzzy Systems (INFUS). LNNS, vol. 504, pp. 481–488. Springer, Cham (2022). https://doi.org/10.1007/978-3-031-09173-5_57

30. George, J., Badoniya, P., Xavier, J.F.: Comparative analysis of supplier selection based on ARAS, COPRAS, and MOORA methods integrated with fuzzy AHP in supply chain management. In: Sachdeva, A., Kumar, P., Yadav, O.P., Tyagi, M. (eds.) Recent Advances in Operations Management Applications. LNME, pp. 141–156. Springer, Singapore (2022). https://doi.org/10.1007/978-981-16-7059-6_13

Development of the Professional Competence of Bachelors in Preschool Education Through Online Interaction

Tatyana Ponomarenko⬤, Olena Kovalenko⬤, Tetiana Shynkar(✉)⬤,
Larysa Harashchenko⬤, and Tetiana Holovatenko⬤

Boris Grinchenko Kyiv University, Kyiv, Ukraine
{t.ponomarenko,o.kovalenko,t.shynkar,l.harashchenko,
t.holovatenko}@kubg.edu.ua

Abstract. The article discusses the issue of pre-service professional training of preschool teachers on the example of the course Workshop on Children's Play Activities for students of the first (Bachelor's) level of higher education of the educational and professional program 012.00.01 Preschool Education. This paper focuses on the ways of developing the professional competence of Bachelors of Preschool Education through online interaction. The place of the course Workshop on Children's Play Activities in the e-learning system of Borys Grinchenko Kyiv University is presented. The article describes the structure of the electronic online training course, the logic of thematic unit mapping of the course. Authors suggest effective services and resources for online interaction of participants of the educational process, namely: during lectures (Google Meet, Google Chat, Google Hangouts, Google Classroom, Webex, Zoom, Lucidspark, Agile, Canva, Mentimeter, Kahoot, Piktochart); seminar sessions (Hangouts, Zoom, Easel.ly Padlet, Piktochart, AnswerGarden, Slides, Prezi, Learningapps, Socratic); and practical classes (Microsoft Teams, Migo, Whiteboard, Agile Workflows, Class Dojo, Mural, GoAnimate, GIFAnimate, Slasstools, Learningapps, Triventy, Yumpu, Emaze, Genially, Flippity, Wordwall, Kahoot). Authors claim that usage of information and communication technologies is effective for facilitating online interaction, increasing students' engagement with theoretical material, and making the preparation to seminars and practical classes easier for students. Using the suggested services and resources will allow the organization of dynamic and engaging classes, as well as promoting active interaction with students even in online setting.

Keywords: Online Interaction · Professional Competence · Information and Communication Technologies · Play Activities · Forms

1 Introduction

The experience of organizing the educational process in a distance form during the COVID-19 pandemic assisted higher education institutions in adapting to new realities – the introduction of Martial Law in Ukraine. The period of dynamic digital transformation, and the availability of digital technologies and tools allow professionals to

G. Antoniou et al. (Eds.): ICTERI 2023, CCIS 1980, pp. 114–127, 2023.
https://doi.org/10.1007/978-3-031-48325-7_9

discover new ways of developing distance learning. However, completing tasks in the online format and the summative assessment in the form of an online test is expected to lead to difficulties for students of specialty 012 Preschool Education during direct interaction with children. According to the educational and professional program 012.00.01 Preschool Education of the first (Bachelor's) level of higher education for students at Borys Grinchenko Kyiv University (Ukraine), a graduate should develop a number of professional competencies, including the ability to support the play activities of young children during early childhood. This competence is primarily formed on the content of the course Workshop on Children's Play Activities and is also integrated into the study of other courses on teaching methods. Therefore, preparing students of the first (Bachelor's) level of specialty 012 Preschool Education for organizing the leading activity (play) of preschool children in an online setting is a considerable challenge.

2   Analysis of Publications

The analysis of psychological and pedagogical literature on play activities of preschool children gives grounds for concluding that this issue is relevant and important. The study of didactic principles and methodological approaches, which were developed in the course of historical development on the issue of play activities of preschool children, emphasizes their theoretical significance and is the basis for further justification of psychological and pedagogical support of the child's play.

The theoretical and methodological foundations of supporting the play activities of preschool children are presented in the works of contemporary scholars O. Bezsonova, H. Bielienka, N. Havrysh, N. Kudykina, T. Ponimanska, T. Pirozhenko, K. Karasova, O. Kornieieva, K. Krutii, O. Staienna and others.

Further study of the research problem requires identification of the main mechanisms of formation of professional competence of Bachelors of Preschool Education in online setting. The results of the literature analysis indicate that raising the quality of training of future educators is a priority for the system of higher education. Much research on the issue of using information and communication technologies in the educational process of higher education institutions has been done.

Psychological and pedagogical aspects of the use of information technologies in the educational process are discussed in the research of V. Bykov, O. Buinytska, S. Vasylenko, L. Varchenko-Trotsenko, A. Hurzhii, V. Lapinskyi, N. Morze, O. Spirin and others. Researchers of the issue of effective use of information and communication technologies in the educational process of higher education institutions S. Vasylenko and N. Morze studied common innovative technologies, ICT, and best pedagogical practices, and did comparative research of digitalization across EU universities [1].

The issue of forming digital competence was studied by O. Buinytska and S. Vasylenko, who have developed a corporate standard for the digital competence of a university instructor and mapped detailed skills of instructors in accordance with the levels of digital competence formation and areas of its application [2]. Recently, a group of researchers from Borys Grinchenko Kyiv University conducted research on the use of digital technologies for formative assessment [3]. The issue of developing methodological approaches to research of various types of tools used in online setting raises scholarly discussions. O. Torubara and E. Kleino made a classification of the issues of using ICT in higher education [4]. M. Miastkovska, I. Kobylianska and N. Vasazhenko analyzed the shortcomings of the use of information and communication technologies in higher education institutions [5]. The issues of using ICT in higher education are a certain focal point combining the personal and social aspects of choosing a teaching career.

The results of the analysis of theory and practice show that scientific and practical conferences are systematically held at various levels on the implementation of e-learning and distance learning in higher education institutions. In particular, CEUR. Workshop Proceedings and ICTERI: International Conference on ICT in Research, Education and Industrial Applications provides an opportunity for scholars and practitioners to get acquainted with high-quality developments, innovations in the field of ICT and use them in their professional activities [6–8].

International research on the integration of digital technologies in preschools shows that ICT is used in various ways: to enrich and transform existing curriculum and practices; to enhance children's cultural literacy and narrow the gap for young immigrant children; to keep children busy; to communicate and document preschool practices [9]. As Hernwall argues, it is important for preschool teachers to have a high level of digital proficiency and know how to support the child's development by means of ICT [10]. At the same time, it is also important for instructors to know how preschoolers use technology in their families to bridge the gap between various patterns used by children in different settings. According to Zevenbergen and Logan, the majority of preschool children use technology for some educational activities under the supervision of adults (79.54%), non-educational activities (59.90%), and drawing (48.92%) [11]. As a result, it is important for preschool teachers to develop a diverse repertoire of gamified activities using ICT for their learners. Authors believe that ICT has the potential not only for overall child development but to transform children's play activities in particular.

The purpose of the study is to prove the feasibility and positive impact of using ICT in online interaction to form students' competence in supporting the play activities of children of early and preschool age on the example of the Workshop on Play Activities course.

3 Discussion

We designed a multi-stage experiment on the research issue. The first step was administering a pre-test. We were guided by the following research questions: 1) identifying student readiness to develop the play competence of preschool children and 2) identifying the attitude of students to incorporating various tools of online interaction, as well as finding out which services students usually use during online interaction.

3.1 Students' Readiness to Develop the Play Competence of Preschool Children

The authors have identified the following criteria and indicators of students' preparedness for the formation of play competence in preschool children:

- motivational and emotional criteria (indicators: motivation to use play as the main type of activity of children; tolerance to the child's opinion and replicating lived experiences in games; emotionally correct reaction to changes in the rules of the game and behaviour of the child; motivation to demonstrate values of group solidarity while playing games (humanity, responsibility, justice, self-control, friendliness, positive communication, tolerance); emotional support of game activities, using expressive means during the game (emotional and expressive movements, facial expressions, pantomimic, the timbre of voice, role-playing, etc.); responsible attitude to children's choice and performance of the role; readiness to reflect on the game);
- cognitive criteria (indicators: knowledge about various types, names of games and their content; understanding actions in creative games and rule-based games; ability to involve children in a common game; the formation of role-based ways of behaviour in games; following the rules and etiquette of communication while playing games; understanding the possibilities of the game environment; identifying reasons and consequences of successful and unsuccessful games);
- action-based criteria (indicators: the ability to apply methodological tools of supporting game activities of preschool children; demonstrating creativity, critical and logical thinking, ability to make non-standard decisions in various situations in games; following the rules of a role-play (admitting, decision-making, equity, support, etc.); creative usage of the game environment and designing the game field).

The authors used observations, individual interviews, analysis of student artefacts, and surveying. While conducting the pre-test to study students' preparedness to form the play competence of preschool children, observation was used as the main research method. We observed students during their practical training in kindergartens. In particular, we focused on how students supported children's play. The data obtained from the observation indicate that 85% of students know the types of children's plays, characteristics of children's plays in different ages (78%), methods and techniques of supporting children's play activities (64%), recommendations of educational programs for children's play in a different age (74%), etc. However, not all students can apply their theoretical knowledge into practice (36%).

Students demonstrated good skills of organizing and conducting outdoor games with various levels of activity for children, didactic games, and activities from different sections of the curriculum during organized forms of interaction with children. A majority of students (63%) suggested children playing games that they themselves played in childhood, which raised the interest of children, causing sympathy and establishing a certain level of affinity.

The most difficult thing for students was to support story-based role-playing and creative games of children. Young children suggested some students (43%) to play the roles of «dad», «grandmother», «dog», «parrot» etc. The unwillingness and reluctance of a number of students to play along with the children and to take on the offered roles have actually led to stopping the game at this stage. Students explain their reluctance in playing along with children at their request by noticing that «… you have to be an actor».

Another issue raising our concern during the observations was the inability of students to apply techniques to support children's play into practice. These techniques include conversations with children about a new literary work (video, excursion, walk, something seen on weekends, etc.); making an attribute for games and introducing it to the group for further use; showing children and directly using substitute items, etc.

The results of our observations give grounds to the conclusion that most children, when left without the guidance of an adult, are not able to organize independent activities and make meaning of it: they wander, push, sort and throw toys, etc. Thus, supporting children's play, directing it, developing it by introducing a new attribute, substituting items, and introducing new characters is important for young children, yet difficult for students (43%). In individual interviews and reflections, students admit that they are not ready to «rise to the level of children» and «completely immerse themselves in the play with children», because «we are adults, and they are little children».

The pre-test aiming at identifying the students' preparedness to form the play competence of preschool children has also comprised administering an anonymous survey of students. It included the following questions: «What approaches to classifying children's games do you know?», «What is common and different in the structure and support of didactic games and didactic activities?», «Name and characterize the methods and techniques of supporting children's play», «Choose the games that you know how to support», and «Personally for you, what is the most difficult thing when supporting children's plays?».

The analysis of the results of an anonymous survey of students in numerical data has coincided with the results of our observations. When answering the question «What approaches to classifying children's games do you know?», 85% of students gave correct answers, 12% of respondents gave incorrect answers, and 3% of respondents refused to answer. When answering the question «What is common and different in the structure and support of didactic games and didactic activities?», 64% of students gave correct answers, 23% of respondents gave incorrect answers and 13% of respondents did not give an answer. When answering the question «Name and characterize the methods and techniques of supporting children's play», 79% of students gave correct answers to a theoretical question, and 15% of respondents made mistakes.

Students' answers to the question «Choose the games that you know how to support» actually mirrored our observations during pedagogical practice: 43% of students chose role-plays and indicated specific difficulties they have in supporting them; more than 30% of students indicated that they do not know how to support them but want to learn; 12% of respondents did not want to answer this question.

In the answer to the survey question «Personally for you, what is the most difficult thing when supporting children's plays?», the majority of students (64%) have chosen «Applying theoretical knowledge into the practice of working with children»; 18% of respondents reported it being an approach/method to support children's play; 10% of students noted that the most difficult thing for them is to adapt their language use to the level of child's language; 8% of respondents noted their unwillingness «to take the role of an actor» in children's games.

Overall, the results of the ability of students to form play competence of preschool children during the pre-test are the following: 43% of participants demonstrated insufficient level, 46 respondents demonstrated sufficient level, and 11 pre-service teachers demonstrated high level.

Due to Russian full-scale invasion in Ukraine, we have been living and studying under a state of «delayed danger» for two consecutive years, being continuously exposed to the possibility of bombing and shelling. As a result, the educational process at Grinchenko University is being conducted online. The scope of the pre-test was also guided by our attempt to identify the attitude of students toward incorporating various tools of online interaction, as well as finding out which services students usually use during online interaction. To answer this research question, we used such research methods as interview, survey, and analysis of student artefacts.

The following questions were used to conduct an anonymous survey: «What are the digital tools that you use most often when learning our course and why?»; «What are the digital tools that you think are most helpful for learning our course and why?»; «What are the digital tools that you think are unnecessary in learning our course, and why?»; «Give your suggestions for adding new digital tools to our e-learning course».

In the response to the first question of the survey «What are the digital tools that you use most often when learning our course and why?», 92% of students indicated Google Meet, Zoom, Padlet; 5% of students chose AnswerGarden, Slides, Google Chat, and 3% of students did not answer.

In the response to the first question of the survey «What are the digital tools that you think are most helpful for learning our course and why?», 86% of respondents mentioned Learningapps, Genially, Flippity, Padlet, Kahoot; 15% of students replied with GoAnimate, Genially, Flippity. When justifying their answers, students prioritized the user-friendliness of the interface and free-of-charge usage of tools, the availability, and scope of useful material, as well as bright photos and videos provided to incorporate in children's games.

Students' answers to the third question «What are the digital tools that you think are most helpful for learning our course and why?» were distributed as follows: 88% of respondents indicated there were no such tools in the e-learning course, 15% of students named Genially, Canva, Kahoot; Flippity, 7% of respondents refused to give an answer. Explaining their answers, students mentioned the difficulties with logging in, limited time to use (trial period), and numerous explanations in English, etc.

When answering the fourth question «Give your suggestions for adding new digital tools to our e-learning course», 89% of students indicated that no changes to the e-learning course are needed. In their opinion, the most important thing is a fast victory over enemies and a return to the offline educational process in brick-and-mortar classrooms; 11% of students suggested incorporating the following online tools in the e-learning course: Class Dojo, modifications of mBot, VR-zone, Lucidspark, and Agile; 2% of students refused to suggest anything.

The analysis of the results of the pre-test of the experiment became the basis for conducting the intervention stage of the experiment.

3.2 Syllabus Outline to Teach Using ICT within the Course Workshop on Children's Play Activities

We The educational and professional program 012.00.01 Preschool Education for the first (Bachelor's) level of higher education at Borys Grinchenko Kyiv University (Ukraine) [12] provides for the study of the Workshop on Children's Play Activities course. This is a 180-h (6 credits) course. According to the structural and logical scheme of the educational and professional program, students study the educational component during the 7th and 8th semesters. The summative control is in the form of a credit.

The purpose of the Workshop on Children's Play Activities course is to develop the professional competence of future specialists in the field of preschool education through the study of theoretical and methodological foundations of play-based activities of preschool children at an educational institution and in the family.

Main objectives of the course:

– formation of students' skills in developing basic personality traits in children of early and preschool age through play activities;
– formation of knowledge about the peculiarities of organizing play activities for children of early and preschool age;
– formation of skills to provide pedagogical support for the play activities of preschool children using pedagogical methods and approaches.

The objectives of the course provide for the formation of the integral competence. Integral competence is the ability to solve complex specialized tasks and practical problems in the field of Preschool Education aiming at the development, teaching and fostering children of early and preschool age. It provides for the use of general psychological and pedagogical theories and professional methods of Preschool Education, and is characterized by complexity and uncertainty of conditions;

General competence (GC)-5. Ability to ensure the quality of work performed: plan, make predictions and anticipate the consequences of personal professional and innovative activities;

Special competence (SC)-2. Ability to develop basic personality traits in children of early and preschool age (independence, creativity, initiative, freedom of behaviour, self-awareness, self-esteem, and self-respect);

SC-6. Ability to teach children of early and preschool age the skills of conscious compliance with socially recognized moral, ethical norms and rules of behaviour;

SC-13. Ability to organize and manage play activities of children of early and preschool age.

This course is taught using online resources provided by the University. Each student has their personal student account at Borys Grinchenko Kyiv University. The e-learning system on the Moodle platform hosts a developed e-learning course Workshop on Children's Play Activities. The developed mobile application to the system allows you to do and submit the tasks offline. Online classes are organized using the services Google Meet (for Education), Google Chat, Google Hangouts, Google Classroom, Webex (Enterprise), and Zoom. To enhance the user experience and educational experience of students with the course, the information block of the e-learning course contains a syllabus, assessment criteria for each type of task, a course map, a glossary, and recommended sources, including internet resources. Taking into account the results of the pre-test of the experiment, we improved the content of the educational component Workshop on Children's Play Activities by enhancing it with a greater variety of digital ICT tools.

When developing the scope of material for lectures, seminars and practical classes, we took into account a number of characteristic features of plays of preschool children, namely: the shallowness and simplicity of plots; copying plots of popular movies and cartoons; unwillingness of children to stick to the rules of the game; the inability and unwillingness of children to use role-playing speech and act out role-plays; enacting roles of fictional television characters (Spider-Man, ghosts, etc.). The enormous gap between the children's play and the lives of their close circle of adults may indicate that social life and gradual transition into the adult world are no longer the content of children's plays, as it was envisioned by the classical psychological concept of children's play. As a result, the development of the plot of games in preschool age affects the overall mental and personal development of children.

The research results prove that modern children are not able to organize meaningful independent activities without an adult's support. Most children lack imagination, creativity, and independent thinking. And since preschool age is a sensitive period for the formation of these important qualities, it is highly unlikely they will be formed at a more mature age. Lack of imagination and creativity also affects the communicative development of children. Preschoolers, who do not know how to play, are not able to participate in meaningful communication, collaboration, and lack conflict resolution skills. As a result, children become more aggressive, alienated, and hostile towards their peers.

Due to the above-mentioned, in this course we focus the attention of our students on the importance of «pedagogical support» of preschool children's games, as opposed to the outdated concept of «leadership» in children's games. Pedagogical support is aligned with the person-oriented approach in organizing educational interaction with children.

The e-learning course provides theoretical material for each lecture, a visual aid in the form of presentation, and a list of additional resources. If absent, the student can work with these materials independently. If necessary, they can get an individual consultation using the Forum resource in the e-learning course or by e-mail. It is effective to use such digital tools and resources as Lucidspark (a virtual whiteboard where you can create tasks, projects, collaborate and discuss the project), Agile (a resource for planning the study of the material); Canva, Mentimeter, Kahoot, Piktochart (creating presentations, demonstrating lectures, surveys, built-in templates, and animation to diversify the theoretical material). Implementing these resources and tools does not cause any trouble for students, aids better understanding of the content, and creates positive emotional attitudes in students towards the learning process.

The content of seminars and practical classes, recommended resources with active hyperlinks, recommendations for doing tasks and assessment criteria are available in the e-learning course. Seminars are held using such digital tools as: Hangouts, Zoom, Easel.ly, Padlet, Piktochart, AnswerGarden, Slides, Prezi, Learningapps, Socratic, etc. These services enhance students' experience with the e-learning course, raise their interest in studying theoretical material, and make the process of preparing for classes easier for students.

During the seminar classes, students learn the methodological foundations of the Play Theories; the structure of preschool children's play activities and stages of its development; classification of preschool children's games; program requirements for pedagogical support of preschool children's play activities; features of diagnostics, planning, and support of children's play activities. We engage all students in the group participation during the classes with the use of interactive teaching methods (such as Microphone,

World Cafe, Socrates Dialogue, Debate, Aquarium, Brainstorming, Six Pairs of Shoes, Vernissage, Take a position, etc.).

According to the Memorandum of Understanding between the Ministry of Education and Science of Ukraine and The LEGO Foundation On Cooperation in the Field of Education and Science [13], the special course Comprehensive Development of the Child Through Play is integrated into this course in two modules: Play in Contemporary Setting; Play as a Tool and Mechanism of Educational Activities in the Preschool Education System (60 h). The purpose of the special course is to prepare pre-service teachers to introducing the play as a universal approach to interacting with preschool children, ensure the comprehensive development of children and continuity between preschool and primary stages of education in the context of the implementation of the State Standard of Preschool Education and the Concept of the New Ukrainian School. These modules include only practical classes delivered in the form of trainings organized using the Google Meet (for Education) and Webex (Enterprise). The usage of digital tools and resources, such as Microsoft Teams, Miro (used for teamwork, chatting and file sharing); Whiteboard (for visualizing ideas); Agile Workflows (easy project planning, clear definition of deadlines in projects); Class Dojo (a convenient tool for evaluating students in real time); Mural (allows you to visualize collaboration in the form of diagrams as a result improving the ability to make joint decisions and implement common ideas) proved its effectiveness in this course. These digital tools and resources do not require advanced skills of using the PC and can be easily integrated with meetings in Google Meet, Webex, or Zoom.

According to the State Standard of Preschool Education [14], plays of preschool children fall into two categories: games organized by an adult and free amateur game of a child. In the final two content modules of the course on Workshop on Children's Play Activities, students learn to create algorithms for various types of plays, engage with children during the playtime, and employ indirect methods and techniques to support and encourage children's plays. In this course, students develop team projects (Playbook, Playday, Day without Toys, Playdate, etc.), and learn how to create content for online games using digital tools and resources (GoAnimate, GIFAnimate, Slasstools, Learningapps, Triventy, Yumpu, Emaze, Genially, Flippity, Wordwall, Kahoot, etc.). Implementing these resources and tools creates favourable environment for developing logical and creative thinking; looking for feasible ways to support children's games even when this task is not explicitly given; developing an information culture.

When developing the content of seminars and practical classes for students, we incorporated materials developed as a result of the cooperation of the Ministry of Education and Science of Ukraine with The LEGO Foundation (LEGO construction program (2010) [15]; Curriculum for educating children of early and preschool age EDUCATION & CARE (2021) [16]; Curriculum for child development from 2 to 6 years, and methodological recommendations Infinite World of Play with LEGO (2016) [17]; Curriculum «Intellectual Mosaic» (2022) [18]; Supplemental curriculum Creators of the Future (2022) [19]). Students are asked to familiarize themselves with these programs, compare them, highlight life hacks, etc.

To develop information and digital, subject and methodological competencies of the participants of the educational process, the Center for Innovative Educational Technologies (ICR class) [20] was created based on the Faculty of Pedagogical Education. Students work there as a part of the study of the educational component Workshop on Children's Play Activities. The ICR class was created as part of the implementation of the international scientific project Modernization of Higher Education Using Innovative Teaching Tools (MoPED) funded by the Erasmus + program. Three working zones of the ICR class (STEM-Lab, IT-Lab, and VR-zone) create opportunities for the use of innovative teaching methods (IBL, PBL, and PrBL); integrated learning and competence approach in implementing elements of STEAM education in play activities for preschool children; digital tools for accompanying play activities; 5E research learning model; technologies of flipped classroom and blended learning; virtual and augmented reality software, etc. For example, when studying the topic Play of Young Children and its Pedagogical Support, we use VR-zone to teach pre-service teachers using virtual and augmented reality. While working with virtual and augmented reality students use textbooks and manuals with built-in mini lessons, explore the development of play activities of young children, and game attributes for young children. IT-Lab is a zone for working with information technologies, allowing you to create a team project Play Space for Young Children, develop a project plan and presentation materials about the project implementation. While studying the module Peculiarities of Organizing Various Types of Games for Preschool Children in the STEM-Lab zone, students create content of online games for preschool children in the digital space. Using the existing Makeblock laboratory (robotics) STEM Classroom Kit mBot, students create engaging and exciting games for direct interaction with preschoolers.

A majority of end-of-module assessments are in the form of practice-oriented questions (pedagogical scenarios) that require thorough knowledge of the methodology for organizing support for play activities of children of early and preschool age. For example: «A teacher saw in a neighbouring preschool institution how children were playing fishing games in an interesting way. To introduce this game in her group, she made her own fishing accessories and offered children the theme of the game. The game did not work out well and the teacher always had to tell the children what to do next. Why do you think the play didn't work out. How can you make sure that all children actively participate in the game and can perform both main and secondary roles? Suggest an algorithm of preparing the suggested game»; «Plan the preparatory stage of teacher's interaction (forms, methods, techniques) with children aiming to enrich the game experience by organizing a role-play «Theatre» for children of the senior group of kindergarten. Make sure to provide for educational interaction between the teacher and children on getting acquainted with theatre jobs and plan artistic activities to create sets, costumes, and posters» etc.

Ways of Implementing ICT in the Workshop on Children's Play Activities course is given in image. (Fig. 1).

Individual module assessments are in the form of tests. The test has 25 questions of different levels of difficulty. The tests are developed on the Moodle platform. We also use tools for implementing gamification elements, interactive activities (crosswords, puzzles, test tasks etc.) using H5P. Module tests include various types of tasks: multiple

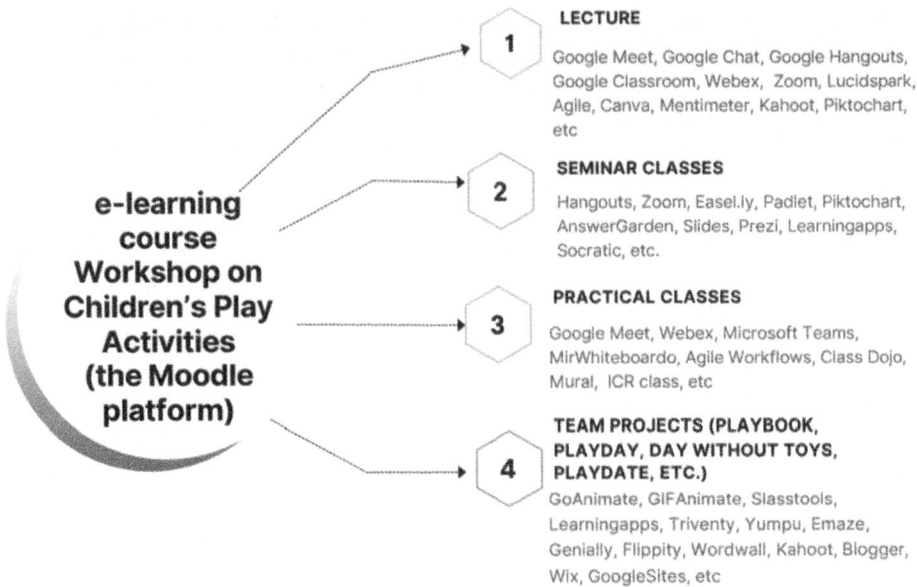

Fig. 1. The level of formation of artistic competence

choice questions, true/false questions, matching activity, providing a short answer, drag and drop, and labelling the image. Self-study work involves creating a Play-portfolio hosted on the Blogger, Wix, or GoogleSites platform.

3.3 Research Results

During the intervention stage of the experiment, the same set of methods of research as during the pre-test were used. They were tailored to research the peculiarities of using various digital tools and resources in the educational process. The results of the analysis of students' responses to the first question of the survey showed that when studying the course, all students (100% in contrast to 92% of students in the pre-test) who took part in the study, use such digital tools as Google Meet, Google Chat, Google Hangouts, Google Classroom, Webex (Enterprise), Zoom, and Padlet. Also, 100% of students indicated that they started using the following digital tools and resources: Lucidspark, Agile; Canva, Mentimeter, Kahoot, Piktochart, Easel.ly Piktochart, Prezi, Socratic, Microsoft Teams, Migos, Whiteboard, Agile Workflows, Class Dojo, Mural, GoAnimate, GIFAnimate, Slasstools, Learningapps, Triventy, Yumpu, Emaze, Genially, Flippity, Wordwall, Kahoot, etc. The number of students using AnswerGarden and Slides has increased to 9% (it was 5% in the pre-test). There were no students who did not wish to answer the question. Observations of students' activities during the study of this course confirmed the results of the survey.

In the answers to the second question of the survey, 92% of students (against 86% in the pre-test) indicated the following digital tools as the most effective services for mastering this course: Learningapps, Genially, Flippity, Padlet, Kahoot; and 21% of respondents (versus 15% in the pre-test) named GoAnimate, Genially, and Flippity. At the same time, 96% of students ranked the following digital tools and resources as effective: Lucidspark, Agile; Canva, Mentimeter, Kahoot, Piktochart, Easel.ly Piktochart, Prezi, Socratic, Microsoft Teams, Migos, Whiteboard, Agile Workflows, Class Dojo, Mural, GoAnimate, GIFAnimate, Slasstools, Learningapps, Triventy, Yumpu, Emaze, Genially, Flippity, Wordwall, Kahoot, etc.

In the answers to the third question of the survey, 100% of students (against 88% in the pre-test) indicated that no irrelevant services were found in the e-learning course, and 4% of students (against 15% in the pre-test) indicated the complexity of using such services as Genially, Canva, Kahoot, and Flippity, still mentioning difficulties with logging in, limited time to use (trial period), and numerous explanations in English, etc. On the fourth question, 96% of students indicated that no changes to the e-learning course are needed, and only 4% of students suggested adding such digital tools as: Class Dojo, modifications of mBot, and VR-zone.

Analysis of the results of the survey and observation of students during the practical training in preschool institutions allowed us to state that 96% of students (against 85% in the pre-test) know the types of children's games; 89% of students (against 78% in the pre-test) are aware of the features of games of early and preschool age; 87% of students (against 64% in the pre-test) know the methods and techniques of psychological and pedagogical support of children's games; 93% of students take into account the recommendations of educational programs when organizing games of children of different ages (against 74% in the pre-test), etc. The number of students who suggested children's games that they themselves played in childhood increased to 75% (against 63% in the pre-test). The unwillingness and reluctance of a number of students to perform certain roles in the game decreased from 43% (in the pre-test) to 29% (in the post-test). Thus,

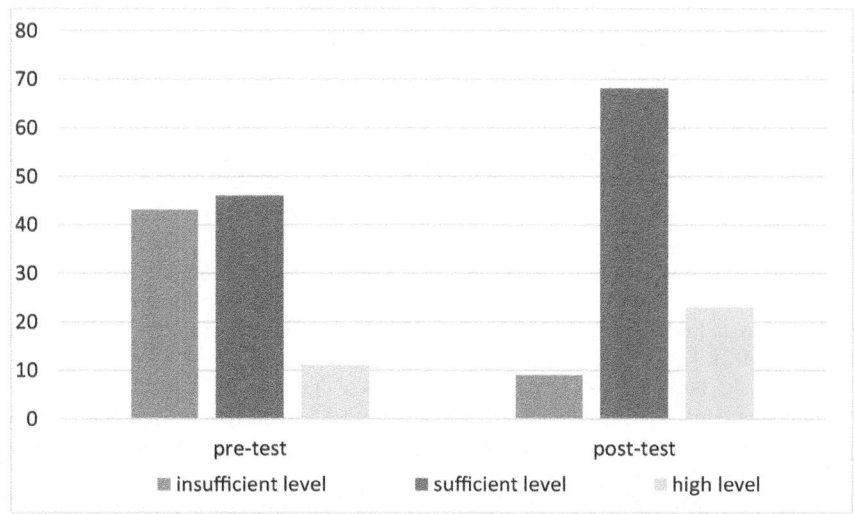

Fig. 2. The Results of Students' Preparedness for the Formation of Play Competence of Preschool and early-age children

the results of the post-test confirmed the effectiveness of the intervention stage of the experiment. Overall, the results on the ability of students to form play competence of preschool and early childhood children at the post-test were the following: insufficient level – 9% of students demonstrated insufficient level, 68% of respondents demonstrated sufficient level and 23% of pre-service teachers have demonstrated a high level. The data with the results of the pre-test and post-test is presented on Fig. 2.

4   Conclusion

The Teaching in the blended and distance form is necessary due to a number of reasons. Notwithstanding the situation in the country, the graduate should develop a complex of professional competencies. Therefore, universities should address the challenges of preparing students of the first (Bachelor's) level of specialty 012 Preschool Education in online setting. An effective tool in distance education is the use of information and communication technologies that help to raise the interest of students in studying theoretical material, make the process of preparing students for seminars and practical classes easier. Using the offered services and resources allows organizing interactive, interesting classes and promote active interaction with students even in online setting. The prospects for further research are researching the possibilities of using ICTs in the implementation of the educational process for future teachers of preschool educational institutions in the conditions of blended and distance learning.

References

1. Morze, N., Vasylenko, S.: Report 1. Innovative learning and best practices: Ukrainian universities. Open educational e-environment of a modern university: special issue «Pedagogical higher education of Ukraine: analysis and research», pp. 1–68 (2020)
2. Buinytska, O., Vasylenko, S.: Corporate standard for university lecturer's digital competence. Electron. Sci. Prof. J. «Open Educational E-Environment of Modern University» 12, 1–20 (2022). https://doi.org/10.28925/2414-0325.2022.121. Accessed 21 Feb 2023
3. Morze, N., Vember, V., Boiko, M.: Using of digital technologies for formative assessment. Electron. Sci. Prof. J. «Open Educational E-Environment of Modern University» 202–214 (2019). https://doi.org/10.28925/2414-0325.2019s19. Accessed 21 Feb 2023
4. Torubara, O., Kleino, Ye.: The specific use of personal computers in the study process in out-of-school educational institutions. Sci. Notes Ser. Pedagogy 2, 56–62 (2016)
5. Miastkovska, M., Kobylianska, I., Vasazhenko, N.: Analysis of disadvantages of using information and communication technology in higher education institutions. Health Safety Pedagogy 4(2), 173–179 (2020)
6. Bielienka, A., Polovina, O., Kondratets, I., Shynkar, T., Brovko, K.: The use of ICT for training future teachers: an example of the course on «Art Education of Preschool Children». In: ICTERI-2021: 16th International Conference on ICT in Research, Education and Industrial Applications, vol. XX, pp. 361–370 (2021)
7. Bielienka, A., Shynkar, T., Kovalenko, O.: ICT as a tool for improving the quality of methodical work in the first link of the system of education. In: ICTERI-2020: 15th International Conference on ICT in Research, Education and Industrial Applications, vol. I, pp. 334–341 (2020)

8. Iatsyshyn, A., Kovach, V., Lyubchak, V., Zuban, Y., Piven, A., Sokolyuk, O., et al.: Application of augmented reality technologies for education projects preparation. CEUR Workshop Proc. **2643**, 134–160 (2020)
9. Masoumi, D.: Preschool teachers' use of ICTs: towards a typology of practice. Contemp. Issues Early Child. **16**(1), 5–17 (2015)
10. Hernwall, P.: «We have to be professional» – Swedish preschool teachers' conceptualisation of digital media. Nordic J. Digit. Lit. **11**(1), 5–23 (2016)
11. Zevenbergen, Logan, H.: Computer use by preschool children: rethinking practice as digital natives come to preschool. Australas. J. Early Childh. **33**(1), 37–44 (2008). https://doi.org/10.1177/183693910803300107
12. Educational and professional programme 012.00.01 Preschool education of the first (bachelor's) level of higher education for students of Borys Grinchenko Kyiv University. https://kubg.edu.ua/images/stories/Departaments/vstupnikam/pi/OPP_bak_DO_012.00.01.pdf. Accessed 18 Feb 2023
13. Memorandum between the Ministry of Education and Science and the Lego Foundation on Cooperation in Education and Science. https://mon.gov.ua/ua/news/memorandum-mizh-mon-i-lego-foundation-shodo-spivpraci-u-sferi-osviti-ta-nauki. Accessed 18 Feb 2023
14. State Standard of Preschool Education (new edition) (2021). https://ezavdnz.mcfr.ua/book?bid=37876. Accessed 27 Feb 2023
15. Pekker, T., Holota N., Tereshchenko, O., Reznichenko, I.: Curriculum for the development of preschool children's constructive abilities «LEGO Construction», 121 p., Kyiv (2010)
16. Voronov, V., Kovalchuk, K., Pikanova, N., et al.: Early childhood education programme «Education and Care», 130 p., Kyiv: FOP V.B. Ferenets (2021)
17. Roma, O., Blyzniuk, V., Boruk, O.P.: Curriculum for development of children aged 2 to 6 and methodological recommendations. The Infinite World of LEGO Games, 140 p., The LEGO Foundation (2016)
18. Zdanevych, L., Pisotska, L., Myskova, N., et al.: Intellectual mosaic. Supplementary curriculum of intellectual development of children of early and preschool age, 175 p., Kyiv (2022)
19. Roma, O., Malevych, H., Piskova, I., et al.: Supplementary curriculum creators of the future and methodological recommendations, 109 p., The LEGO Foundation (2022)
20. Centre for Innovative Educational Technologies Innovative Classroom (ICR). https://fpo.kubg.edu.ua/struktura/inshi-pidrozdily/tsentr-innovatsiinykh-osvitnikh-tekhnolohii-icr-klas.html. Accessed 26 Feb 2023

Design and Implementation of Didactic Process Based on Simulation

Paweł Plaskura[✉][iD]

Academy in Piotrków Trybunalski, Słowackiego 114/118, 97-300 Piotrków
Trybunalski, Poland
pawel.plaskura@gmail.com

Abstract. The problem of designing didactic processes has not been
solved to this day, although some specific issues are considered. The article presents a generalized approach model based on the reference ADDIE
model including the organisational level to the design and implementation of the didactic process. The model uses Bloom's taxonomy and
Gardner's Theory of Multiple Intelligences. The model has been modified towards a dynamic design oriented on quality, efficiency and adaptive
learning. A competency-based decomposition is used. The lowest level of
decomposition is the activities level where the information flows as well
as the learning and forgetting are taken into account. The competences-
objectives-activities-data linking is discussed. The level of activities is
represented in the form of an electrical-like network. The network represents differential equations describing dynamic learning and forgetting as
well as the structure and information flows of the didactic process. The
element models can be easily extended. The networks can be simulated
and optimized. The results can be used during the design process. The
simulation result as well as the structure of network equations enables
inference. The article presents the results of simulations of some aspects
of didactic processes including scheduling and sequencing, gap effect,
workload, predicting achievements and effectiveness.

Keywords: Didactical Process Modelling · Didactical Process Design
Model · Didactical Process Simulation · Learner-Centered Design ·
Personalized Learning

1 Introduction

Designing of the highly efficient didactic processes is still an unresolved problem. The design of didactic processes began with the development of systematic
training programs, later called *Instructional Systems Design* (ISD). This was
the result of post-war research by the U.S. Army to achieve a more effective
and manageable way to create training programs [22]. A series of ISD models
were developed and implemented in the late 1960s [17,27]. Almost all developed
models were based on the general ADDIE process [1, p.433] which consists of five
phases: *analysis of training requirements and needs; definition of education and*

G. Antoniou et al. (Eds.): ICTERI 2023, CCIS 1980, pp. 128–143, 2023.
https://doi.org/10.1007/978-3-031-48325-7_10

training requirements; *development of objectives and tests*; *planning, creating and validating training*; *conducting and evaluation of training*. All processes and products of individual phases are constantly assessed in the context of quality (meeting the needs). The assessment ensures continuous improvement.

Today, a more comprehensive approach is required, considering the need to develop different types of intelligence. The need to solve these problems depends on the level of education. Especially in early school and preschool education, developing multiple intelligences is required. It is different, for example, in vocational education, where the acquisition of specific skills is needed.

> The article aims to present the author's model for designing and conducting the didactic process based on competency decomposition which incorporates the Theory of Multiple Intelligences [18–20], Bloom's taxonomy [5] as well as techniques of modelling and simulation.

The use of simulation enables the design of processes used at the stage of more ergonomic didactic processes and the conduct of adaptive didactic processes. In this work, we will focus on a general approach that takes into account the development of different types of intelligence. The issues of profiling goals and achievements will be discussed using Bloom's taxonomy in the cognitive sphere. The project approach will be based on a modified ADDIE model embedded in an organizational context [45].

1.1 Practical Problems

The application of the presented techniques can be difficult in real situations due to several problems that occur [46], among others: openness to innovations; low level of digital literacy among teachers; lack of adequate support for teachers. Design of high efficient didactic processes is difficult due to the:

- incorrect content design causing learning problems: cognitive overload (e.g. incorrect representation of information in drawings), large amount of detail, lack of knowledge structure, difficulties with interdisciplinary connections;
- incorrect schedule - too large time intervals between classes can cause forgetfulness, because it is beyond the limit of forgetting;
- difficult detection of repetition of similar issues within different courses, where synchronisation of didactic units and subjects based on teacher cooperation and the modular approach can be used;
- difficulties or lack of use of efficiency improvement techniques - e.g. short breaks during classes or several short practice sessions [9,25].

Designing didactic processes with the use of competency-based engineering techniques may be a way to solve some of these problems. The most important is modelling the process of information flow, collection (learning) and loss (forgetting) during the learning process. They give us insight into many phenomena and enable their analysis. Belong to them:

- design and implementation of individual adaptive processes.

– designing and managing intelligence and profile-oriented courses;
– estimation of workload and costs also in the long term.

Some discussed issues are also in other cited articles. Here are presented in a shortened version. The original model is often incomprehensible, so further considerations will concern the model taking into account the organizational context.

2   The Theoretical Backgrounds

The main and constant goal of ADDIE is effective teaching. However, increasingly complex requirements, new instructional technologies, emerging automated instructional development tools, and other developments have extended the capabilities of the process ADDIE [45]. The key issue is building quality in instructional systems. The processes of the ADDIE model have evolved over the last decades. ADDIE is more than just a tool for applying behaviour-oriented learning principles. This progress has been achieved through step-by-step procedures designed to enable anyone to develop advanced instruction related to complex technological and cognitive issues, which would normally require experienced instructional design experts [16, 39].

One of the biggest challenges remains the low level of training knowledge required of teachers/trainers. These professionals need to understand the model and how to use it. Therefore, the original *Air Force One* ADDIE model has been improved by presenting it in a simpler and more intuitive way so that developers of instructional systems with different levels of knowledge can more easily understand and use it [11, 17, 27] - Fig. 1 [1, p.438] showing organizational functions, quality improvement, and author's modifications (dashed lines).

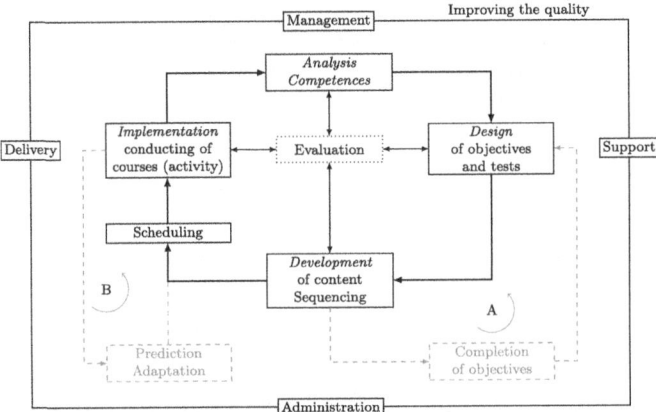

Fig. 1. Improved ADDIE model [1, p.438] with author's modifications (dashed lines).

The nature and extent of the development, update or revision activity determines entry or re-entry into a particular stage of the process and presents it in the context of organizational functions (top-level functions for training activities - *management, support, administration, delivery* and *evaluation*). The main parts of the process are *analysis, design, development, implementation* and *evaluation*. *Evaluation* is the central part of the whole model.

Organizational features built into the *quality improvement process* (QI) [1, p.438] have also been added. The *management* function involves directing or controlling the development and operation of the learning system. The *support* function maintains all parts of the system. The *administrative* function includes day-to-day processing and record keeping. The *delivery* feature provides instructions to students. Finally, the *evaluation* function collects feedback through formative, summative, and operational evaluations to assess the system and student performance.

2.1 Extended ADDIE Model

The design process proposed in the paper uses activity sequencing. The model of designing and implementing the didactic process presented in Fig. 1 works on activities (didactic events). Modifications are marked with dashed lines.

The model shows the base stages of a project (analysis, design, batching and sequencing, activity planning, and class execution) and two additional and loops added to the model (dashed lines):

A related to the design of objectives and activities,
B related to the prediction of effects and the adaptive distribution of activity over time.

The proposed approach requires the decomposition of the didactic process up to the level of activity.

2.2 Decomposition of the Didactic Process

Attention to competence has become a key focus of international debates on learning, curricula, and assessment within general education [6]. Developing competence has value not only for learners but also for the economy and society [10]. Competence is closely related to the notion of proficiency and mastery used in fields such as mathematics and language [43].

We will start the decomposition of the didactic process by defining and designing the required competences. The concept of competence can be defined in various ways. According to [47], *competence is the ability to integrate and apply contextually-appropriate knowledge, skills and psychosocial factors (e.g., beliefs, attitudes, values and motivations) to consistently perform successfully within a specified domain.*

Let's apply the decomposition of competences and arrange them as a certain sequence - Fig. 2.

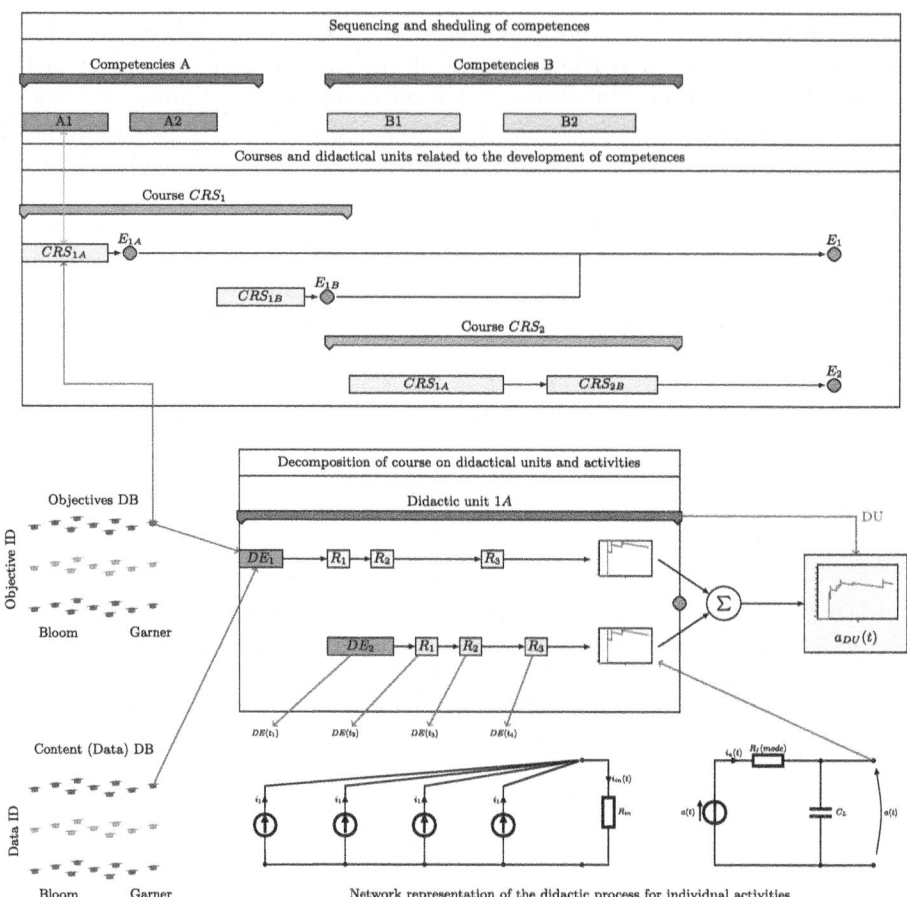

Fig. 2. Decomposition of the competency-oriented didactic process used in the implementation process.

In the didactic process, competences refer to, among others, knowledge and skills. They have partly developed during the didactical process. Therefore, competence needs to decompose into lower levels. Courses, classes and activities are both sequenced and planned. As part of the course, it is necessary to define learning objectives and design related activities and evaluation rules. Here we focus on competences in the cognitive sphere. At the lowest level, competences are decomposed into activities and linked to objectives and appropriate instructional materials.

Activity analysis, including the use of instructional materials, is possible by modeling the flow and collection of information in the process of learning and forgetting. The didactic process can therefore be presented in the form of a network (Subsect. 3.1) modeling the flow of information with a structure corresponding to the structure of activity. The parameters of the model elements correspond to the speed of learning and forgetting. Based on the network simulation results, an adaptive learning process can be designed and implemented, which will be discussed later.

2.3   Bloom's Taxonomy of Didactic Objectives

Bloom's taxonomy [5] classifies cognitive goals according to the cognitive level at which they are implemented. It is good to be aware of the cognitive level at which the classes are conducted. The well-known Bloom's taxonomy with minor modifications [3] is still used worldwide. It consists of six levels [5].

- The *knowledge* level includes the development of objectives involving the memorization, recognition and recall of basic definitions, rules, algorithms and procedures in the categories of detailed, procedural and abstract knowledge;
- The *comprehension* level includes learning objectives in the categories of translation, interpretation, extrapolation;
- The *application* level involves developing the ability to apply knowledge in practical situations as an application of terms, methods, algorithms and theories.
- The *analysis* level contains learning objectives in the categories of element, relationship and sequencing analysis;
- The *synthesis* level consists of the synthesis of ideas, procedures and structures.
- The *evaluation* level includes the categories of assessment of skills that relate to the internal system of knowledge and beliefs and related to the application of external criteria.

Virtually any level can be used as a means of enrichment depending on the cognitive level of the student. As the level increases, the workload increases (higher level of cognitive activity) due to the higher abstraction and high complexity. The expenditure on the implementation of activities at the high level is associated with a large amount of work, as it requires abstract thinking, which is characteristic of advanced students [23].

   Achievements (A_{B_K}, ..., A_{B_E}) corresponding to individual levels of Bloom's taxonomy can be presented in bar form (Fig. 3), creating a certain profile of the didactic process (the profile should be related to the type of Gardner's intelligence we describe further). The participation of individual activities in the didactic process is visible. You can see what the emphasis is on. The individual levels refer to verbs [2] representing the intellectual activity at each level. The verbs can help to formulate corresponding objectives. The reverse process is also possible. Analyzing the verbs, it can be seen that the implementation of learning objectives at the level of knowledge requires the least amount of work spent, where it is required to master *encyclopedic* knowledge. Individual activities along with their cognitive level can be associated with the appropriate type of intelligence [42].

2.4 Design Based on Theory of Multiple Intelligences and Bloom's Taxonomy

Bloom's taxonomy of cognitive levels discussed above [5] can be integrated with types/kinds of intelligence (Theory of Multiple Intelligence) [18–20] to provide a framework for an individualized instructional process [42] - Fig. 3.

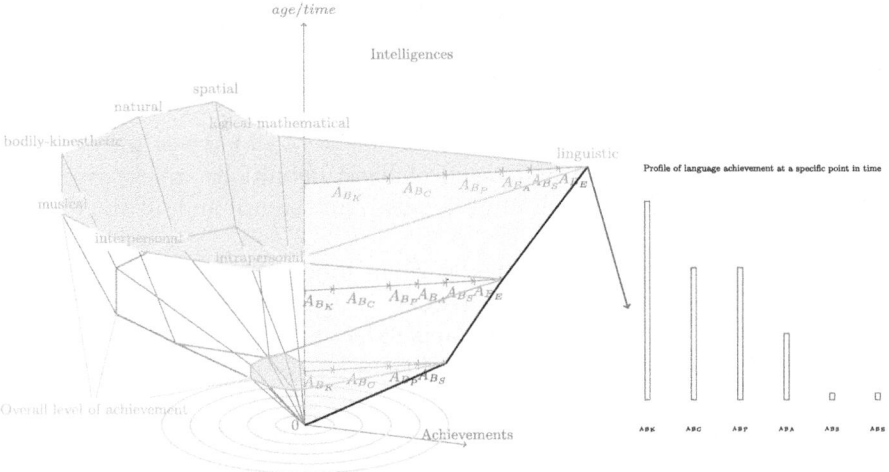

Fig. 3. Multiple intelligences and achievements in terms of Bloom's learning objectives.

Students can work on the intelligence where they feel strong and develop it, or they can choose to work on that intelligence where they do not feel strong and consciously work those areas of their intelligence.

The previously mentioned profiles of learning achievements (Fig. 3 - bars on the right) affect designing activities and instructional materials and corresponds to learning objectives. The bars show the levels of achievement (A_{B_K}, ..., A_{B_E}) corresponding to each level of Bloom's taxonomy. The profile enables the assessment of the didactic process over time. Determining the profile of achievements at the design stage makes it possible to design goals and instructional materials needed to achieve them at a specific point in time.

This observation leads to the assumption that didactic materials should be designed and include activities enabling the achievement of the assumed educational objectives. Ideally, a single didactic material serves specific goals according to a given profile for a specific type of Gardner's intelligence - Fig. 2 - *Content DB*. Instructional materials can be treated as *data* (content) - . This approach enables practical implementation described below.

3 Practical Implementation of the System

The practical implementation of the system requires the integration of the issues discussed above. The DIKW (Data, Information, Knowledge, Wisdom) model [1,

26]a The DIKW model enables the assignment of individual parts of the didactic process and the implementation of the system - Fig. 4. The DIKW model can also be presented as a flow diagram [26].

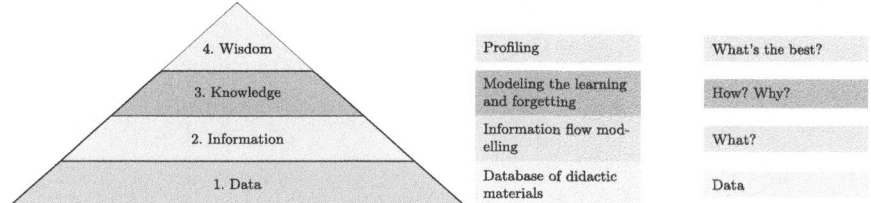

Fig. 4. DIKW model with related process elements and questions to help you understand the connections.

The DIKW models actual processes. The source of *data* may be, for example, scientific research, creative works, tasks to be performed, examples .... It represents structure content. *Information* is processed data of a static nature, presented in such a form that it is easy to conclude (e.g. in visual form). It represent communication channel. *Knowledge* makes it possible to observe phenomena from different perspectives, associated with the learner. It is constantly changing due to experience, and is difficult (in fact impossible) to transfer. At this level, learning and forgetting processes are modeled. Individual levels of achievement are determined. *Wisdom* is accumulated knowledge associated with a concrete person. Wisdom is based on experience and knowledge gained in the past. It allows the transfer of ideas from one domain to another. It can also be transferred between individuals. At this level, the designated achievement profiles for each type of intelligence are taken into account. This applies to both the design stage and the implementation of the didactic process in order to achieve a specific shape of the profile. Having associated learning objectives based on Bloom's taxonomy with activities and corresponding instructional materials, you can model and simulate didactical process.

3.1   Modeling of Activities

The main reason for dealing with the subject is the apparent lack of use of modern methods of description and simulation in the didactics [38]. However, the mathematical models using direct mathematical formulas were previously created [29]. The most important of them, from the point of view of the article, are cited. Detailed issues discussed in the article can also be found in [31–33, 35, 36].

The analysis of the learning and forgetting process is based on forgetting curves represented by direct formulas [14, 28, 50]. The didactic process can be described by the differential equations and represented in the intuitive form of a network of connected elements - Fig. 2 - *Network* .... Such a network can be shown in the form of a schematic (here electrical-like) as shown in other articles [31–33, 35, 36]. The use of the microsystems simulator [32] required defining both the network variables (Table 1) and equations by using *analogy* [7].

Table 1. Generalized variables for the electrical and educational environment.

Generalized variables		Electrical environment		Educational environment [34]	
e	*effort*	v	*voltage*	a	*achievements*
f	*flow*	i	*current*	i	*information flow*
p	*state*	q	*charge*	q	*information*
E	*energy*	E	*energy*	E	*workload*
W	*work*	W	*work*	W	*work*

The equations are represented by elements linked with their mathematical models [37]. Network variables can be represented as vector (1).

$$x = [a, i, q]^T \tag{1}$$

where: a - variables related to *achievements*, i - *information flows*, q - variables describing *unit information*.

Network Elements The representation of individual activities and their repetitions and a simplified model of the learning and forgetting process in the form of a network is presented in Fig. 2 - *Network* .... Interpretation of the basic elements is as follows [35]. The R_a (information resistance) models loss of information between nodes m_a and n_a. The C_a (information capacity) is an element that gathers information. The $a_a = f(t)$ represents the knowledge source. The $i_a = f(t)$ represents the information source. The L_a element represents the self-learning ability. Descriptions of other elements can be found in i.e. [37]. More complex models of learning and forgetting can be found in [35]. In the case of the decomposition of the didactic process in terms of Bloom's cognitive level [5] and Gardner's type of intelligence [18–20] each activity should be modelled separately. This causes a significant complexity of the description and a significant increase in the number of variables. If the activities are to cover all Bloom's cognitive levels ($b = 6$) and basic types of intelligence ($g = 8$), then the number of all learning goals $c = b \times g = 6 \times 8 = 48$. The real number of variables will be even greater due to the modelling of learning and forgetting processes. It depends on the complexity of the learning and forgetting model. In the simplest model [35] (Fig. 2 - *Network* ...), the number of additional variables per activity is 3. So the total number of variables (equations to solve) equals $192 = 48 + 48 \times 3$. This specialized software has implemented techniques for limiting the number of mathematical operations and memory occupancy (the sparse matrix technique). The Dero [32] simulator was used here to simulate the networks.

4   Results and Discussion

Based on the previously discussed issues, we will simulate selected processes. We will show how the simulation results can be used for decomposed didactic processes in the design and implementation of adaptation processes:

- for sequencing the didactic process (class schedule) in terms of the achieved effects,
- the impact of introducing short breaks on workload and learning outcomes,
- workload estimation,
- estimating learning outcomes based on activity.

4.1   Workload and Effects

The workload is closely related to the results but is not most important. The number of repetitions and their distribution is important because repetitions lead to the consolidation of memory [12,13].

The example uses the same real students' activity data with IDs GM, JM, and SA as presented in [34], where other aspects of didactic process are studied. Students represent different levels of activity. Simulation of individual didactical processes were performed. In the example, the estimated total student workload should be 30 h. Figure 5 shows the results of simulations of achievement and workload distribution.

The large number of activities and repetitions affect the course of the forgetting curve. The greater the number of repetitions, the smaller the slope of the performance curves. It also has a significant impact on the workload. The GM student is more involved than others. The workload is twice that of other students. The level of achievement is higher and the slope the achievement curve is smaller. The changes seem small, but they have a big impact on long-term results.

4.2   Predicting the Next Didactic Unit or Activity

Individual activities are most often associated with didactic units. While the sequencing of didactic units is not a problem, optimal scheduling is. The time gaps between individual classes/activities can be incorporated into the design process. The individual student can adapt and modify the schedule. If the interval between didactic units or activities is too large, the entry-level at the beginning of the next classes or course may be too low - Fig. 5. The beginning of the next didactical unit may be beyond the limit of oblivion. The following classes should start in a time window. The width of the time window differs between students. It depends on individual characteristics based on initial knowledge and the number of repetitions of the material - Fig. 5.

The minimal normalized level of achievements was set to 0.51. JM and SA students didn't even reach the required level. The workloads are relatively low. However, JM student spent more time and was more active. The slope of the achievement curve is slightly lower for the JM (zoomed area). GM student has

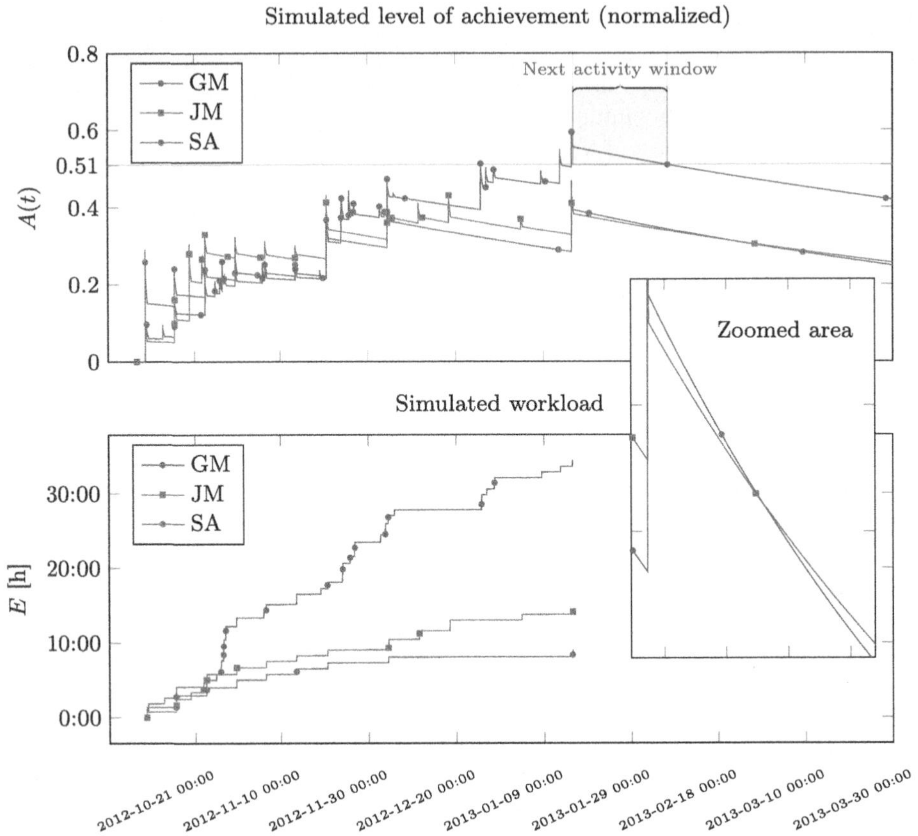

Simulated level of achievement (normalized)

Fig. 5. Simulation of achievements and workload.

exceeded the minimum achievement level at the very end of the course. The workload exceeded 30 h. The students meet requirements over some time. The *Next Activity Window* shows the amount of time since completing the course where the level of achievement is higher than expected. The next course should start within the time window. *JM* and *SA* students should not start the next course because the expected achievement level at the start of the next course is below the required level (0.51).

4.3   Time Gap and Workload

The introduction of small breaks is an effect known from the literature as a way to increase efficiency. Research [9, 25, 40] shows that students learn better when the learning process is divided into several shorter practice sessions, rather than focusing them on one longer [4] session. The practice of splitting and spacing results in better material retention than *cramming* [9, 24, 25, 40]. The *gap effect* [44] increases if the student is engaged in a distributed practice that focuses on a specific goal [48,49]. Objective-oriented practice backed by time-based feedback results in greater learning gains [2,15,41].

Forgetting that occurs during breaks is a general process that promotes cognitive development by supporting the processes of assimilation and generalization of knowledge [8,48,49]. When designing a course, you should prepare a schedule of quizzes, tests and exams. Students benefit more from repetition when exams are expected than when exams are unexpected [41]. Research shows how and when evaluation improves knowledge retention as a *major learning event* that can also contribute to faster forgetting![21].

The example shows the effect of introducing short breaks during classes and the effect of reducing breaks between individual classes (class schedule compression). Figure 6 presents the results of the simulation of achievements and workload for various variants of the didactic process. This project can be a reference on how to learn and can motivate students to manage their learning process.

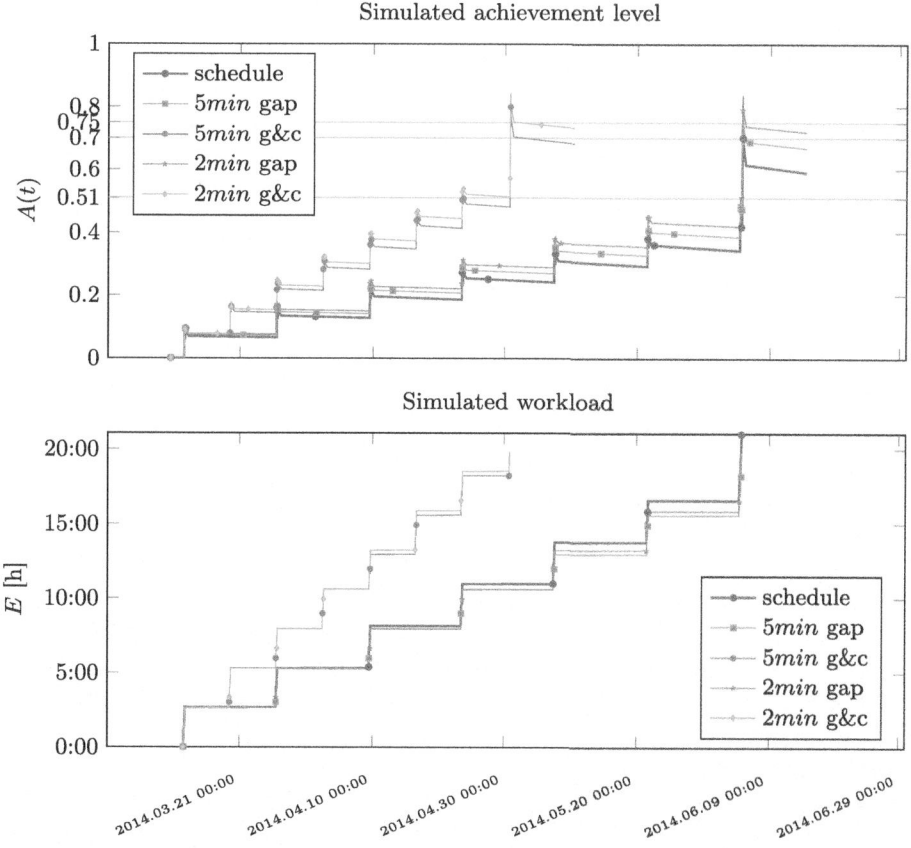

Fig. 6. Simulation of achievements and workload in the didactic process with short breaks during classes (gap) and compressed schedule of classes and short breaks during classes (g&c).

The simulation results show the most effective variants. Scheduling breaks during classes (gap) automatically shortens the time of effective learning, i.e. slightly reduces the workload. Although the workload is lower, the efficiency of the process is even higher by about 15%. The largest amount of work is incurred for the project without interruptions, which is characterized by the lowest efficiency. Interrupting the learning process and then returning to it generates a repetition effect. The repetition effect is limited if a disruptive factor occurs during the break [30]. Shorter breaks are more effective. A similar but more subtle effect occurs when compressing the schedule over time (g&c). Efficiency is marginally higher, but the slope of the forgetting curves is lower. In the long run, this leads to better retention of knowledge.

5   Conclusions and Prospects for Further Research

The article presents a modified approach to designing the didactic process based on competences and their composition up to the level of activity. The extension of the ADDIE model was shown, enabling a more effective design of the didactic process and the implementation of the adaptation process. Designing for Gardner's types of intelligence and a specific cognitive level according to Bloom's taxonomy was shown. In the process of design and implementation, the developed techniques of simulation of the didactic process with the use of a microsystems simulator were used. Simulation results of selected modified processes are shown. The results of the simulation of the impact of breaks during classes on achievements and workload as a technique for increasing efficiency were shown. It also shows the implementation of the adaptive process by simulating the results for the required level of achievement. This enables the implementation of an individualized adaptation process by designing the order and starting the time of subsequent classes/activities. The paper presents only selected examples of the use of the developed modelling and simulation techniques. Some topics can be found in previous articles. Other interesting topics not discussed here include: analyzing the relationship between parts of the teaching process by analyzing the network structure and/or equation structure, automatically creating network diagrams, detecting repetitions, and designing thematically related activities. The papers demonstrate the usefulness of the developed simulation techniques. If we have appropriate models of real processes, they give insight into the phenomena that occur during didactic processes. Understanding these phenomena and using the developed techniques can help in designing more ergonomic teaching processes. It also enables the automatic design and implementation of adaptation processes.

References

1. Allen, W.C.: Overview and evolution of the ADDIE training system. Adv. Dev. Hum. Resour. **8**(4), 430–441 (2006)
2. Ambrose, S.A., Bridges, M.W., DiPietro, M., Lovett, M.C., Norman, M.K.: How Learning Works: Seven Research-Based Principles for Smart Teaching. Wiley, New York (2010)

3. Anderson, L.W., et al.: A Taxonomy for Learning, Teaching, and Assessing: A Revision of Bloom's Taxonomy of Educational Objectives, Abridged Edition, vol. 5(1). Longman, White Plains (2001)
4. Bjork, R.A., Allen, T.W.: The spacing effect: consolidation or differential encoding? J. Verbal Learn. Verbal Behav. 9(5), 567–572 (1970)
5. Bloom, B.: Taxonomy of Educational Objectives: The Classification of Educational Goals: Handbook I: Cognitive Domain. Longmans, Green (1956)
6. Caena, F., Punie, Y.: Developing a European framework for the personal, social & learning to learn key competence (LifEComp). In: Punie, Y. (ed.) Literature Review & Analysis of Frameworks. Publications Office of the European Union, Luxembourg (2019)
7. Care, C.: Technology for Modelling: Electrical Analogies, Engineering Practice, and the Development of Analogue Computing. History of Computing. Springer, London (2010). https://books.google.pl/books?id=bkbBZcR7DG4C
8. Carpenter, S.K., Cepeda, N.J., Rohrer, D., Kang, S.H., Pashler, H.: Using spacing to enhance diverse forms of learning: review of recent research and implications for instruction. Educ. Psychol. Rev. 24(3), 369–378 (2012)
9. Cepeda, N.J., Vul, E., Rohrer, D., Wixted, J.T., Pashler, H.: Spacing effects in learning: a temporal ridgeline of optimal retention. Psychol. Sci. 19(11), 1095–1102 (2008)
10. Child, S.F., Shaw, S.D.: A purpose-led approach towards the development of competency frameworks. J. Furth. High. Educ. 44(8), 1143–1156 (2020)
11. Dick, W., Carey, L., Carey, J.O.: The Systematic Design of Instruction. Pearson, London (2005)
12. Dudai, Y.: The neurobiology of consolidations, or how stable is the engram? Annu. Rev. Psychol. 55, 51–86 (2004)
13. Dudai, Y.: Reconsolidation: the advantage of being refocused. Curr. Opin. Neurobiolog. 16, 174–178 (2006)
14. Ebbinghaus, H.: Memory: a contribution to experimental psychology (1913). https://web.archive.org/web/20051218083239/. http://psy.ed.asu.edu:80/~classics/Ebbinghaus/index.htm. Original work published in 1885
15. Ericsson, K.A., Krampe, R.T., Tesch-Römer, C.: The role of deliberate practice in the acquisition of expert performance. Psychol. Rev. 100(3), 363 (1993)
16. Gagné, R.M., Wager, W.W., Golas, K.C., Keller, J.M., Russell, J.D.: Principles of Instructional Design (2005)
17. Garcia, K.L., Ozogul, G.: The evolution of the instructional system development model in the United States Air Force. TechTrends 1–11 (2023)
18. Gardner, H.: Frames of Mind. Basic Books, New York (1983)
19. Gardner, H.: Multiple Intelligences: The Theory in Practice. A Reader (1993)
20. Gardner, H.E.: Frames of Mind: The Theory of Multiple Intelligences. Basic Books (2011)
21. Halamish, V., Bjork, R.A.: When does testing enhance retention? A distribution-based interpretation of retrieval as a memory modifier. J. Exp. Psychol. Learn. Mem. Cogn. 37(4), 801 (2011)
22. Holton, E.F., III, Swanson, R.A.: Foundations of Human Resource Development. ReadHowYouWant.com (2011)
23. Kafai, Y.: Constructionism. In: Sawyer, K. (ed.) The Cambridge Handbook of the Learning Sciences. Cambridge University Press, Washington University (2006)
24. Kang, S.H.: Spaced repetition promotes efficient and effective learning: policy implications for instruction. Policy Insights Behav. Brain Sci. 3(1), 12–19 (2016)

25. Kornell, N.: Optimising learning using flashcards: spacing is more effective than cramming. Appl. Cogn. Psychol. Off. J. Soc. Appl. Res. Memory Cogn. **23**(9), 1297–1317 (2009)
26. Liew, A.: Understanding data, information, knowledge and their inter-relationships. J. Knowl. Manag. Pract. **8**(2), 1–16 (2007)
27. Molenda, M., Pershing, J., Reigeluth, C.: Designing instructional systems. In: Craig, R.L. (ed.) The ASTD Training and Development Handbook: a Guide to Human Resource Development (1996)
28. Murre, J., Dros, J.: Replication and analysis of Ebbinghaus' forgetting curve. PLOS ONE **10**(7), 1–23 (2015). https://doi.org/10.1371/journal.pone.0120644
29. Panadero, M., Pardo, A., Panadero, J., Andreas, M.: A mathematical model for reusing student learning skills across didactical units. In: 32nd ASEE/IEEE Frontiers in Education Conference, November 2002
30. Peltokorpi, J., Jaber, M.Y.: Interference-adjusted power learning curve model with forgetting. Int. J. Ind. Ergon. **88**, 103257 (2022)
31. Plaskura, P.: Assessing the quality of the didactic process on the base of its monitoring with the use of ICT. Педагогічні науки: теорія, історія, інноваційні технології (Pedagog. Sci. Theory Hist. Innov. Technol.) **76**(2), 185–196 (2018). https://doi.org/10.24139/2312-5993/2018.02/185-196
32. Plaskura, P.: Dero 4 simulator as a didactical tool. Aparatura Badawcza i Dydaktyczna **23**(1), 44–51 (2018). http://abid.cobrabid.pl
33. Plaskura, P.: The use of ICT in improving the effectiveness of the didactical process. Педагогічні науки: теорія, історія, інноваційні технології (Pedagog. Sci. Theory Hist. Innov. Technol.) **17**, 152–159 (2018). http://dspace.pnpu.edu.ua/handle/123456789/9739
34. Plaskura, P.: Wykorzystanie technologii informacyjnych do modelowania i monitorowania jakości procesu dydaktycznego (The use of information technology for modelling and monitoring the quality of the didactical process). Wydawnictwo Uniwersytetu Jana Kochanowskiego, Piotrków Trybunalski, December 2018
35. Plaskura, P.: Modelling of forgetting curves in educational e-environment. Inf. Technol. Learn. Tools **71**(3), 1–11 (2019)
36. Plaskura, P.: Monitorowanie jakości procesu dydaktycznego z wykorzystaniem ICT (Monitoring the quality of the didactical process with the use of ICT). In: Leshchenko, M., Zamecka-Zalas, O., Kiełtyk-Zaborowska, I., Jarocka-Piesik, J. (eds.) Globalne i regionalne konteksty w edukacji wczesnoszkolnej, pp. 151–164. Wydawnictwo Uniwersytetu Jana Kochanowskiego w Kielcach Filia w Piotrkowie Trybunalskim (2019)
37. Plaskura, P.: The use of analogy to simplify the mathematical description of the didactical process. In: Ermolayev, V., Mallet, F., Yakovyna, V., Mayr, H.C., Spivakovsky, A. (eds.) ICTERI 2019. CCIS, vol. 1175, pp. 136–160. Springer, Cham (2020). https://doi.org/10.1007/978-3-030-39459-2_7
38. Rao, K., Edelen-Smith, P., Wailehua, C.U.: Universal design for online courses: applying principles to pedagogy. Open Learn. J. Open Dist. e-Learn. **30**(1), 35–52 (2015). https://doi.org/10.1080/02680513.2014.991300
39. Reigeluth, C.M.: Instructional Design Theories and Models: An Overview of Their Current Status. Routledge, Amsterdam (1983)
40. Rohrer, D., Taylor, K.: The effects of overlearning and distributed practice on the retention of mathematics knowledge. Appl. Cogn. Psychol. Off. J. Soc. Appl. Res. Memory Cogn. **20**(9), 1209–1224 (2006)

41. Rothkopf, E.Z., Billington, M.J.: Goal-guided learning from text: inferring a descriptive processing model from inspection times and eye movements. J. Educ. Psychol. **71**(3), 310 (1979)
42. Rule, A.C., Lord, L.H.: Activities for differentiated instruction addressing all levels of bloom's taxonomy and eight multiple intelligences (2003)
43. Rycroft-Smith, L., Boylan, M.: Summary of evidence for elements of teaching related to mastery in mathematics. Expresso 16 (2019). https://www.cambridgemaths.org/Images/espresso_16_mastery_in_mathematics.pdf
44. Sisti, H.M., Glass, A.L., Shors, T.J.: Neurogenesis and the spacing effect: learning over time enhances memory and the survival of new neurons. Learn. Memory **14**(5), 368–375 (2007)
45. Smith, P.L., Ragan, T.J.: Instructional Design. Wiley, New York (2004)
46. Tomczyk, L.: ICT in schools and non-formal education in Poland. Challenges of digital literacy development, modernisation of education system and digital inclusion through new media from the perspective of experts from business, education and NGO sectors. Memorias y Boletines de la Universidad del Azuay, 116–145 (2020)
47. Vitello, S., Greatorex, J., Shaw, S.: What is competence? Learning and assessment. Research report, Cambridge University Press & Assessment, A Shared Interpretation of Competence to Support Teaching (2021)
48. Vlach, H.A.: The spacing effect in children's generalization of knowledge: allowing children time to forget promotes their ability to learn. Child Dev. Perspect. **8**(3), 163–168 (2014)
49. Vlach, H.A., Sandhofer, C.M.: Distributing learning over time: the spacing effect in children's acquisition and generalization of science concepts. Child Dev. **83**(4), 1137–1144 (2012)
50. Wickelgren, W.: Single-trace fragility theory of memory dynamics. Mem. Cognit. **2**, 775–780 (1974)

Theoretical Principles of Measuring and Interpreting Levels of Attention, Involvement and Organizing Feedback of Students to the Educational Process Using Automated Software Products

Aleksander Spivakovsky[iD], Lyubov Petukhova[iD], Maksym Poltoratskyi[iD], Oleksandr Lemeshchuk[iD], Anastasiia Volianiuk[✉][iD], Olena Kazannikova[iD], Nataliia Voropay[iD], and Svitlana Chepurna[iD]

Kherson State University, Shevchenka St., 14, Ivano-Frankivsk 76018, Ukraine
{spivakovsky,petuhova,mpoltoratskyi,olemeshchuk,avolianiuk, okazannikova,nvoropay,smatvienko}@ksu.ks.ua

Abstract. The article highlights the theoretical foundations of measuring and interpreting student attention levels, engagement, and feedback, an attempt is made to characterize the possibilities of using automated software products for this purpose. The article identifies the psychological characteristics of students of the Zoomers generation, which affect the educational process and cause the revision of educational technologies. A comparative analysis of traditional and distance learning was carried out. Based on the characteristics and interpretation of visual markers for measuring the attention and involvement of the acquirers, the requirements for the software product for determining the degree of involvement and the degree of information perception are described. The article discusses the main AI-based tools that allow you to assess the level of audience attention in real time during online conferences (EmotionCues, HEADROOM and MeetingPulse) from the standpoint of the effectiveness of the analysis of the emotional component and the possibility of integration with online conference services. The theoretical analysis is the basis for further research on the measurement and interpretation of students' attention, engagement and feedback.

Keywords: Educational Process · Attention · Involvement · Organizing Feedback · Engagement · Automated Software Products · Machine Learning Technology

1 Introduction

The disruption of the digital infrastructures of different generations has led to the catalysis of mass processes, particularly in the field of education. We started work on the creation of this article during the global pandemic, quarantine measures and the introduction of distance learning into all links of the educational system. We continued the research in the conditions of martial law, when online education is the only option for obtaining education in many regions of Ukraine.

G. Antoniou et al. (Eds.): ICTERI 2023, CCIS 1980, pp. 144–159, 2023.
https://doi.org/10.1007/978-3-031-48325-7_11

The gap in communication in the remote format causes a decrease in the quality of education, but the gap from traditional learning is so great that a new look at the interaction between a teacher and a student is needed. The role of a teacher today is rapidly transforming into the role of an educational blogger, and the role of a student is transforming into a subscriber interested in content. Free access to absolutely unlimited information opens up opportunities for students to independently choose sources of knowledge, therefore, the trend of educational blogging becomes a new reality after the final loss of the monopoly on knowledge by teachers. Using the achievements of the information age opens up unlimited opportunities for the young generation both for independent learning of educational material and for improving their professional qualities and skills. At the same time, the teacher remains a trainer, coordinator, expert in his field, but his role changes to a blogger who creates and promotes educational content, is an opinion leader and an authority among student subscribers.

Note that the problem of the gap between the generation that teaches and the generation that is taught is important today. This gap is caused, in particular, by the technology of different generations, their level of digital skills and, in general, their interaction with information technologies.

The modern educational process, in the era of social networks and rapid development of technologies, differs from what it was 5–10 years ago, and is characterized by the active implementation of information and communication technologies, a combination of traditional and innovative teaching methods, and the combined use of traditional and remote forms of education. In the conditions of the informatization of the educational environment, teachers face the main tasks:

- to form the student's internal motivation to acquire knowledge and future professional activity;
- improve the skills of information culture and increase the level of digital competence;
- develop students' critical thinking;
- teach how to quickly navigate in a wide flow of information and determine its truth;
- select effective information and communication methods and teaching tools.

We consider it expedient to develop priority directions for solving these problems through the prism of the theory of generations. This research approach gained its popularity in the early 1990s, the authors of which are Neil Howe and William Strauss [19]. Researchers have argued that historical context determines human behavior: a group born in one period of time has similar personality traits common to an entire generation. According to this theory, the following generational cycles are familiar to modern society: "Baby Boomers", "Generation X", "Generation Y" (millennials), "Generation Z" (digital natives) and "Generation Alpha". Sociological studies claim that modern educational institutions educate representatives of "Generation Z".

The pedagogical community is actively researching and discussing how to teach today's youth, what forms and methods of organizing the educational environment to use, how to build an educational environment, and what information and communication means of learning will be effective. After all, the modern challenge of the forced transition to distance learning, caused by the rapid spread of the COVID-19 corona virus infection throughout the world, vividly actualized the problem of training digital natives and finding an effective educational model. We have identified contradictions, the solution of which will contribute to increasing the level of efficiency of the use of technologies in the educational process (in particular, in distance and mixed learning formats):

- contradiction between those who teach and those who are taught;
- the contradiction between the pedagogical possibilities of information and communication technologies and the low level of effectiveness of their use in the educational process;
- contradiction between the psychological features of information perception by students who constantly interact with information technologies (generation z) and traditional content and methods of learning;
- the problem of focusing attention during classes in a distance/mixed format and the teacher's inability to monitor the level and dynamics of the learner's inclusion in the educational process;
- lack of motivation and low level of self-organization of the student in the process of distance learning and the impossibility of constant control by the teacher and the need to conduct it.

Based on the above contradictions, we see the need to study the cognitive features of modern students of higher education (zoomers generation), traditional and modern learning models, classic and new views on the student's learning process, analyze the theoretical foundations of the development of tools to control the student's level of involvement and motivation.

2 Peculiarities of Studying Students of the Zoomers Generation

According to the provisions of the Theory of Generations, the majority of modern students are the generation of people born from 1997 to 2012. The most common term used in the world to refer to them is Generation Z. However, there are other less common terms - Homelanders, Homeland Generation, Zoomers, New Silent Generation [9, p. 8]. Avoiding unpleasant associations with the letter Z, caused by the war of the Russian Federation on the territory of Ukraine, in the article we will call representatives of this generation Zoomers.

The generation of Zoomers replaced the millennials born in the 80s and early 90s. Its representatives grew and matured along with the development of technology. Zoomers cannot imagine their everyday life without gadgets and quickly master any new technology. They easily use the Internet for entertainment, communication and work. A distinctive feature of Zoomers is the fact that technology is part of their everyday life. IT has significantly influenced the way of thinking, habits and aspirations of young people. The theory of generations calls this generation "digital children" [5]. Due to the abuse of smartphones, mobile Internet and social media, the "digital children" experience many problems such as technology addiction, offline phobia and problematic use of social networks. From the point of view of socialization, it can be seen that higher education students are more influenced by psychological factors related to technology, as they use it extensively in their academic studies and access to information [10]. However, it is advisable to use these features to improve their effectiveness in online learning.

As a result of the analysis of the concepts regarding the peculiarities of the education of Zoomers students, we obtain a set of the following psychological characteristics of this generation:

- difficult to recognize authorities, especially do not recognize public opinion, do not listen to the advice of adults;
- gain experience through their own victories and failures;
- freely express their opinion and dissatisfaction;
- they value their individuality and are afraid to express themselves in different forms;
- they are "looking for themselves" and their vocation for a long time, so they are in no hurry to get a job and start a family;
- picky about the conditions of their own comfort;
- open to new impressions and emotions;
- they are energetic, restless and inquisitive;
- it is difficult for them to work for the interests of the team, they value their own benefits more;
- they are multitasking, but at the same time a bit apathetic and passive;
- know how to quickly switch and grab information "on the fly";
- they strive for career and personal growth, financial well-being.

We should also note that memorizing information is not a priority for Zoomers, because they are used to everything ready on Internet pages. And the next generation after zoomers does not know life without gadgets at all. They perceive the material clip-wise, that is, by the example of changing pictures on the Internet.

2.1 A Theoretical Overview of the Traditional Learning Model and Theories of Cognition

Today, information and communication technologies are increasingly penetrating various spheres of education. This is facilitated by both external factors related to the general computerization of society and the need for appropriate training of specialists, as well as internal factors related to the spread of modern digital technologies and software in schools, the adoption of governmental and intergovernmental programs for the computerization of education, the emergence of relevant digital experience of a large number of teachers. In most cases, the use of digital tools has a real positive impact on the activation of teachers' work and on the effectiveness of student training [15]. Digital education facing COVID - 19 pandemic). The level of technological comfort, innovativeness affect the effectiveness of using digital products for online learning [12] (Table 1).

A comparative analysis of traditional and distance learning models, which coexist in parallel in the educational process of modern higher education institutions, makes it possible not only to substantiate the feasibility of using computers in the learning process, but also to make an attempt to determine the criteria requirements for the creation of educational software products in order to measure the levels of attention and involvement of students to the educational process.

Thus, training according to the traditional model takes place under the coordination of the teacher, who directs the educational process depending on the specific situation. Within the implementation of the traditional model, learning depends on the teacher's

Table 1. Comparative analysis of traditional and distance learning

Traditional education	Distance Learning
During the evaluation of the student's work, the level of knowledge, skills and abilities is taken into account. But at the same time, there is a significant part of subjectivism	The assessment of the applicant is determined only by the level of knowledge, skills and abilities. Other factors are not taken into account
Visual perception is used much less often than hearing information	The visual channel of information transmission is involved, although a significant share of audio information (audio recordings, videos, etc.) is not excluded
The main information carriers are books, abstracts. Information is presented in a form familiar to humans, easy to read, but poorly structured	The perception of ordinary text information from the screen is much worse than that of printed or handwritten information. Factors compensating for these shortcomings are a clear structure of information, a developed system of references, a reasonable combination of visual and substantive information
During the learning process, the teacher performs a variety of functions: checking the attendance of the class, presenting the material, controlling the quality of learning, establishing interdisciplinary connections	The main workload of the teacher is transferred from the stage of conducting training sessions to the stage of preparing material for filling educational software. The teacher acts as a consultant

personality: the ability to arouse interest in the topic, the ability to conduct a casual conversation. The teacher must have extensive and in-depth knowledge of various fields, skillfully use them during the teaching of the educational component. Usually, an experienced teacher, conducting a classroom lesson, can easily determine the level of attention and involvement of the learner in the material being studied, based on the physiognomy of the listeners (characteristic features and facial expressions of a person). The external manifestation of attention, of course, requires the mobilization of all the senses, but it is most dramatically manifested in the expression of the eyes. It determines the level of emotional adjustment and what is happening in the audience. Attention and concentration are unmistakably defined in a subject whose gaze is fixed, facial muscles tense, eyebrows moved to the bridge of the nose.

At the current stage, the use of the method of visual biological feedback is reflected in traditional models of learning, which involves the transfer of knowledge face to face, and the teacher can visually cover the audience during the lesson. However, during training in the distance model, which involves the relationship of the subjects of the online educational process, it is difficult, and often impossible, for a teacher, even an experienced one, to assess the level of attention and involvement in the perception of the material based on the physiognomy of the listener on the other side of the screen. Therefore, the problem of measuring the attention and involvement of students in the educational process through the use of automated software products is actualized.

In our opinion, educational software products should meet the following requirements:

- have a developed system of help and tips for both the teacher and the student;
- provide maximum informativeness with minimal user fatigue;
- present the material in the form of separate visual modules;
- information must be clearly structured;
- to ensure that the visual presentation of information matches its content.

2.2 Theoretical Review of the Modern Model of Learning and Theories of Cognition

Early attempts to operationalize the concept of "student engagement in learning" were associated with measuring the amount of time a student spends on tasks and learning in general. Emphasis on temporary indicators was largely due to the belief that learning outcomes can be judged by the time spent [6, 11].

In the process of mastering the topic, the concept of the involvement of acquirers began to be supplemented with other characteristics, as a result of which the initial construct became more complicated. For example, scientists began to talk about the fact that student involvement is expressed not only in the time spent, but also in the efforts spent. It also began to be attributed to indicators that relate to learning in an indirect way (for example, funding, pragmatics of learning, extracurricular university activity, loyalty to the university, feeling like a part of the university, etc.) [3].

In some works, the involvement of the students is interpreted as the "energy" that the student invests in his studies. The most famous definition in this sense was proposed by A. Astin: "Student involvement is a combination of physical and mental energy spent to acquire academic experience" [1]. According to the scientist, the concept of "involvement" is related to Z. Freud's concept of "cathexis", which means investing energy in objects that are outside the subject.

The topic of measuring and analyzing student involvement in the educational process has been actively developed since the 1980s - primarily in connection with research into the possibility of reducing the number of students who drop out (surveys show that from 25 to 60% of students are constantly bored in classes and distracted from educational process). The problem of monitoring student engagement is relevant today both for the traditional classroom learning process and for mass open online courses, educational games, simulators and simulators, intelligent learning systems, etc. [18].

The following methods of measuring student involvement are most common: self-assessment of involvement by students themselves; external observation using control cards and subsequent rating; automatic measurement of the level of engagement using technical means. In particular, the research is dominated by the self-assessment method. At the same time, information systems for automatic assessment of involvement have been used for quite some time. A significant part of them is based on the analysis of the speed and accuracy of students' performance of control tasks. For example, indicators of low engagement may be random answers to easy questions or very short task completion times. There are attempts to track student behavior in five states: active, transcribing, absent, distracted, and transitioning to another activity [4]. In the era of intelligent machines that sense, control and monitor human feelings, emotions and feelings, there is a need to develop automatic mechanisms for measuring student engagement [8]. There is a class of popular techniques for automatic estimation of the level of involvement, based on the processing of data from various electro- and neurophysiological sensors. For example, capturing brain signals to extract features of brain wave signals using an encephalogram while students are engaged [17], while viewing recorded lecture materials [7] or in the process of synchronous online learning using special brain headphones for encephalogram [13]. For the same purpose, a mechanism for monitoring attention and anxiety based on brain wave signals has been developed [2]. However, it is clear that large-scale application of such techniques is impossible.

3 Overview of Modern Tools for Determining the Levels of Student Involvement in the Educational Process

The development of information technologies and today's realities dictate to society new conditions for the organization of not only the educational process, but also generally make corrections in everyday work. It is not surprising that the use of such technologies as Skype, Zoom, Google Meet, WebEx, Microsoft Teams in the organization of various online conferences, hybrid, semi-hybrid events and in the educational process has now become widely popular.

Each of the mentioned software products can definitely be effectively used as a tool for organizing and conducting online events, but they do not provide an opportunity to analyze the degree of involvement and perception of information by users.

Currently, AI technologies are gaining wide popularity in the organization of hybrid and semi-hybrid online events. Since the use of this kind of algorithms makes it possible to analyze the emotional state, the level of perception of information in real time. Currently, quite a few systems have in their arsenal the possibility of automated shorthand, such a functional feature can be effectively applied at proto-level meetings, where, according to internal rules, the event must be accompanied by a shorthand. Such systems (systems with AI elements) include: • Sembly, • Fathom, • Notiv, • Hendrix.

The use of this kind of technology is also relevant in the organization of the educational process in the conditions of distance learning, since it does not allow to visually capture the attention and involvement of students in the educational process. Therefore, there is currently a need to develop unique software tools that will allow evaluating the degree of involvement, perception of information by the audience in the process of training or conducting any type of online event.

Next, we will consider several software tools that partially outline the problem of audience attention analysis.

3.1 EmotionCues

In the article [8], the authors describe a software product that allows assessing the emotional component of students. The approach described in the article makes it possible to evaluate the involvement of not only one specific student or student, but also to analyze the involvement of the entire audience as a whole.

It should be noted that previous research by the authors showed that emotions can be indicators of attention, involvement and behavior of students in general [2–6]. Based on previous research, the software product EmotionCues was developed. It will be useful not only to teachers who analyze the educational process, but also to parents who can analyze the degree of interest of their children in the educational process. The authors of this software product highlight the following system of requirements:

- obtaining the emotional status of all people in the video stream.
- obtaining the emotional status of an individual in a video stream. The possibility of getting to know the emotional status of the chosen person in more detail.
- the ability to create emotional portraits of different people.
- the ability to obtain results with further aggregation for deeper analysis.

Figure 1 shows the main dashboard of the EmotionCues software product.

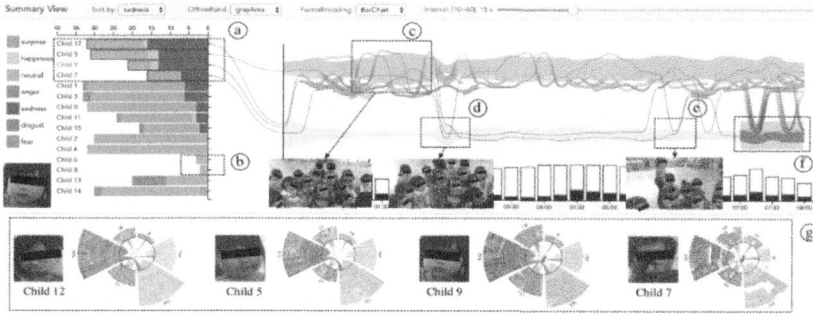

Fig. 1. EmotionCues control panel

At the first stage, a set of videos is processed and emotions are extracted using computer vision algorithms. The next stage is the preparation of interactive visualization, which can support visual analysis of the video at two different levels of detail:

- general analysis of the emotional component of the entire audience.
- analysis of the emotional component of a specifically chosen person.

Using this approach makes it possible to analyze the emotional component of not only one selected participant, but the entire audience in sufficient detail and effectively. With the help of the color system, the control panel of the EmotionCues software tool displays various emotional states of the participants of the online event, namely:

- green is used to show surprise;
- yellow determines the level of happiness;
- gray for a neutral state;
- purple - a state of anger;
- blue represents sadness;
- purple state of disgust;
- blue defines fear.

Thus, the EmotionCues system is a comprehensive solution for analyzing the level of audience engagement using emotional components. Each video is about 10 min long (1.26 GB) with a resolution of 1920 × 1080 and 30 frames per second (FPS). That is, each video for analysis consists of almost 18,000 frames in high resolution.

It is worth noting that the EmotionCues software tool is one of the most powerful tools that allows you to assess the level of audience interest. In addition, it is important to note that this software product provides an opportunity to build an emotional picture and level of interest both for the entire audience and separately for certain selected participants. But EmotionCues does not allow for real-time analysis and cannot yet be integrated with popular services: Webex, Microsoft Teams, Zoom, Vimeo.

3.2 Headroom

Headroom is another software product for remote as well as hybrid meetings. Functionality will help to balance online meetings thanks to the evaluation of the involvement of the team of participants of the online event as a whole. The main task of this software product is to improve communication at online events by replacing routine manual actions with artificial intelligence algorithms.

This software product can be effectively used both at various online events and at hybrid ones. The functionality of Headroom allows it to be used at online events of an educational nature. In addition, Headroom's internal algorithms will allow you to receive a transcript of your online event, participant notes. This kind of functionality is very similar to various minutes of meetings, which according to the rules are accompanied by a transcript.

It should also be noted that this kind of functionality will make it possible to resolve some misunderstandings that may arise in the meeting mode, to raise reasons and problems. The Headroom functionality allows you to recognize emoticons and gestures in real time. After the session is over, each participant of the online event can watch an accelerated version of the online event, which will allow you to assess how the event went, see the level of involvement of each event participant, and assess the balance of the event. Automatic compilation makes it possible to generate a video with the most vivid moments of the meeting.

Also, this system allows you to analyze: • facial expressions; • gestures; • audio accompanying the online event; • text messages accompanying the online event.

As a result of the algorithm, you will receive the aggregated results of the meeting. Also, the Headroom functionality allows you to highlight key information that was presented at the online event. It should also be noted that the information from the results of the analysis of the video stream, the video recording of the online seminar is confidential and protected information. Thus, the Headroom functionality provides the following advantages:

- audience engagement analysis, built-in algorithms for aggregating the obtained results;
- high quality audio and video recordings of the online event;
- group chat;
- this software tool is safe, all information and analysis results are confidential information;

- meeting tools that help you conduct a workshop with elements of inclusion;
- built-in automatic transcription in real time;
- built-in automatic saving of notes for the team;
- built-in gesture recognition;
- searchable meeting knowledge base (video, notes, transcript, etc.)
- custom video summaries of each meeting
- generation of a transcript based on the results of the meeting.

It should be noted that Headroom is optimized for use with Google Chrome. The team is currently working on adding full support for other popular browsers, as well as providing support for the app on mobile devices.

Since the results of the meeting (videos, audio recordings, transcripts, survey results) will be stored in the cloud, all information is confidential and only seminar participants and invited people have access to this information.

The software tool Headroom is a powerful tool for more effective holding of online meetings, seminars and conferences, the level of involvement can be assessed at the level of passing a survey with further processing.

3.3 MeetingPulse

Another representative of this kind of software is MeetingPulse. This software tool is used to hold a virtual or hybrid conference. MeetingPulse has many additional tools that event organizers need for a successful conference—from synchronous polling to sophisticated software to support online event management and delivery. This software can be integrated with the following services: • Webex, • Microsoft Teams, • Zoom, • Vimeo, • Youtube.

It should be noted that for the interaction of third-party services with MeetingPulse, the developers have previously developed an API [9] with quite detailed documentation. MeetingPulse software consists of 12 modules, each module has a certain functionality. During the online seminar, you can create a survey that can be available both immediately and at a certain time. In this way, the presenter/speaker can follow the results of the survey in synchronous mode with the conference. Such tools make it possible to better understand the state of the audience, the extent to which it perceives information.

In addition to the synchronous polling mode, there is also the option to create polls before and after the conference. This kind of survey can be effectively used to assess the residual level of information after the conference.

It should also be noted that the available voting mechanism in conference mode will allow you to shape the agenda in advance, identifying the most popular topics. Poll and voting results can be exported to a CSV file for further processing and representation of the results. During the presentation, participants or listeners can react in real time to what was said, in addition, the speaker can watch the "pulse" of the room, such auxiliary synchronous mechanisms allow the speaker to analyze the state of the audience and its ability to perceive information in real time.

At the beginning or during the online seminar, you can create questionnaires with questions for audience segmentation. This approach will make it possible to more effectively process the received information through clustering. In addition, you should note the possibility of conducting brainstorming sessions, creating and sending invitations to your planned events.

Thus, it should be noted that MeetingPulse is a comprehensive solution for the effective conduct of online seminars with a fairly large number of tools, which allows you to assess interest in real time using the results of the participants' completion of various survey plans. But it is quite difficult to fully assess the level of audience engagement based on the emotional component using the MeetingPulse software tool.

4   A Software Tool for Analyzing the Degree of Involvement and the Degree of Information Perception

4.1   Requirements for the Software Product

In the course of the literature analysis, we highlight the following register of priority requirements. The main task of our software product is the ability to analyze the state of attention, interest and subsequent perception of information by the audience. Thus, we distinguish the following requirements:

- integration with the main systems for conducting virtual or hybrid conferences: Webex, Microsoft Teams, Zoom, Vimeo;
- the possibility of building an emotional portrait and the level of interest of an individual participant of a virtual conference from the corresponding video stream in real time;
- building an emotional portrait and the level of interest of the virtual conference audience from the corresponding video stream in real time;
- possibility of comparative analysis of received analytical information;
- possibility to cluster information by gender and age;
- creation of appropriate UI for aggregation and representation of relevant analytical information;
- conference participant registration;
- the ability to create surveys:

  - display polls before the start of the video conference,
  - displaying polls synchronously with the conference,
  - display of polls after the conference;

- access to the chat in parallel with the conference;
- export of results of polls and votes in CCV format;
- sending invitations to the relevant online event;
- creation of an interactive board for interaction of participants;
- creation of an appropriate API for further interaction of third-party services.

Each of the components of the register of requirements is an important component of the analysis and provides the teacher with information about the state of preparation and student engagement. For example, the possibility of creating a survey in parallel with the conference is useful at various stages: at the beginning of the lesson, it allows you to update basic knowledge, identify gaps on a specific topic (which helps the teacher understand the level of preparedness of the audience, make the right emphasis in teaching the material, recommend material for repetition); in the middle of the lesson, the survey helps to maintain attention, to examine the level of perception; the survey at the end of the conference gives an opportunity to assess the remaining level of information, to repeat the main points. In general, the tool for creating surveys allows you to more qualitatively assess how the audience perceives information. The possibility of exporting to CSV format will enable better processing and analysis of the received information. Access to the chat during the conference performs several useful functions: communication (in the chat you can send text messages, various files with tasks, etc.), interaction with the audience, interactivity of the lesson (using the chat for students' answers to the questions, using quick reactions or short answers), feedback function (a student can write a comment or ask a question). The use of the chat allows you to partially check the level of attention during the lesson, as it requires additional actions from the student in accordance with the teacher's task/question (for example, typing a message or putting a mark in the chat).

Using an interactive whiteboard allows you to actively involve students in the lesson and motivate them to study, make the lesson dynamic, interactive and multimedia. The interactive whiteboard in online education is a useful tool for visualization and collaboration in creating and editing documents and images in real time, therefore it also allows analyzing the state of students' attention, their interest and involvement in the educational process.

For now, we would like to emphasize that the requirements described above are the most prioritized and those that significantly affect the ability to analyze the degree of interest and the ability to perceive information. Next, we will consider the system of external signs and visual markers that can be used in the analysis of the corresponding video stream of the virtual conference.

4.2 Characterization and Interpretation of Visual Markers for Measuring Student Attention and Engagement

So, attention is defined as a special form of mental activity, which manifests itself in the orientation and concentration of consciousness on objects significant for the individual, phenomena of the surrounding reality, on one's own experiences [16, p. 71].

The content of mental activity turns out to be quite dynamic, that is, it constantly changes with a relative delay on certain objects or actions. This fascination of a person with something, orientation and concentration is called attention. One of the characteristic features of attention is that a person subjectively feels himself where attention is concentrated. Attention is a deeply personal phenomenon. Behind it are always the subject's needs, motives, goals, and attitudes. A person's attitude towards the world and other people is manifested in emotions, feelings, and desires. This attitude is also reflected in attention. The surrounding reality, the phenomena in it, which correspond to the needs and interests of the subject, determine his attention. A change in the attitude of the subject to the object causes a change in attention. This is expressed in a change in clarity (clarity, expressiveness) of the content on which the subject's consciousness is focused.

From the numerous signals of the environment, a person selects what is necessary for purposeful activity at the moment, depending on this, assigning a certain meaning to each object. In the interaction with the environment, a selective display of objects and phenomena by consciousness is formed, which is provided by attention. Selectivity, which regulates cognitive and productive activity, is determined by the capabilities of the individual, his orientation, the purpose of his activity. Processes of voluntary regulation of activity (will) serve as mechanisms of selective attention.

Attention is not an independent reflection, not an independent mental process, attention does not have its object of knowledge. Attention to monitoring the dynamics of mental processes is a condition for reflective activity at different levels of consciousness [15, p. 247].

Like any mental phenomenon, attention cannot be studied directly, because mental phenomena are subject to direct observation only by the person who experiences them. Mediated markers are used to assess attention, such as non-verbal means of mental activity, namely non-verbal sign systems. Among such systems, the following are distinguished: optical-kinetic, para- and extralinguistic, proxemics, visual contact. Nonverbal systems are a situational reflection of a person's personal characteristics, namely his mood, emotional background, and attitude. Nonverbal manifestations are involuntary and spontaneous (nonverbal markers carry 60–80% of information, respectively, 20–40% is verbal communication).

In terms of research, the optical-kinetic system and visual contact are important. The optical-kinetic system of signs includes gestures, facial expressions, pantomime, which are presented as a property of perceived general motility. Various manifestations of facial expressions, gestures and pantomime signal the subject's emotional state. Visual contact is related to visual perception (eye movement). Visual contact helps to demonstrate attention and interest. Means of non-verbal interaction are natural and unconscious, which can be an argument for the objectivity of the non-verbal mechanism.

The face is the main source of information about a person's condition. The act of looking is a signal of interest, therefore, the most accurate information is conveyed by a person's gaze, since the expansion or contraction of the pupils is not controlled by a person. When a person is interested, his pupils are dilated (by 4 times), when negative emotions predominate, the pupils narrow (we are talking about expressed emotional states).

Based on the analysis of the literature on the researched problem, it was determined that external signs of attention include:

- a fixed, interested look (every 10 s);
- steady gaze in the direction of the speaker;
- looking into the speaker's eyes (a sign of respect, readiness for contact);
- facial muscle tension (optional);
- slightly shifted eyebrows to the bridge of the nose; • slightly squinted eyes;
- eyes are dilated, seem larger than usual;
- vertical wrinkles on the forehead (concentration); • head tilted to the side;
- nodding the head (an act of approval); • slightly squinted eyes;
- touching the chin; • the hand touches the cheek;
- the index finger is vertical to the temple, the thumb supports the chin (critical evaluation attitude);

- rubbing the bridge of the nose, manipulating glasses (concentration);
- slightly raised eyebrows, wide-open eyes (surprise); • the body is bent forward;

We draw your attention to the fact that all visual signs are arranged in order from the most significant to additional, auxiliary ones.

The opposite concept of attention, which negatively affects the student's involvement in the educational process and the quality of material perception, is inattention, which is manifested by the following visual indicators:

- rubbing the cheek with the hand (it became boring);
- wandering gaze; • looking to the side; • head rubbing;
- touching the ear (not interesting to listen to);
- supporting the chin with the palm of the hand (uninteresting, fighting the desire to fall asleep);
- lowers eyelids or takes off glasses (gets fed up when it's over);
- tapping with a finger or pen (uninteresting);
- plays with objects with fingers (careless); • supporting the cheek with the hand;
- covering the mouth with the palm of the hand;
- hands crossed on the chest (indifference);
- the body is thrown back (not interesting); • looking at the clock;

Taking into account the priority and place in the hierarchy of visual indicators, as well as the degree of complexity of their implementation in software, at this stage of the research we selected the following main visual markers of attention and involvement of students in the educational process as a basis for the development of the software product: a fixed, interested look (every 10 s), steady gaze in the direction of the speaker; eyebrows slightly shifted to the bridge of the nose; facial muscle tension (optional); head tilted to the side; nodding the head (an act of approval).

Among the visual signs of inattention, the following indicators were chosen within the scope of the study: wandering gaze; look away; covering the mouth with the palm of the hand.

4.3 Use of Machine Learning Technology

The main driving force behind video stream analysis is the use of machine learning methods and algorithms. We distinguish two main parts:

- algorithms that identify formalized visual markers from a video stream.
- algorithms that determine the level of interest of the audience and the degree of perception of information.

It can be represented schematically as follows (Fig. 2).

The use of machine learning methods will make it possible to automatically analyze and identify the visual and behavioral markers described above from the video stream. The specified visual markers must be formalized in a certain way, convenient for use and further processing by ML algorithms. With the help of built-in algorithms, the machine learning system can distinguish certain visual factors that, in our opinion, can help analyze the level of interest, as well as the degree of perception of information.

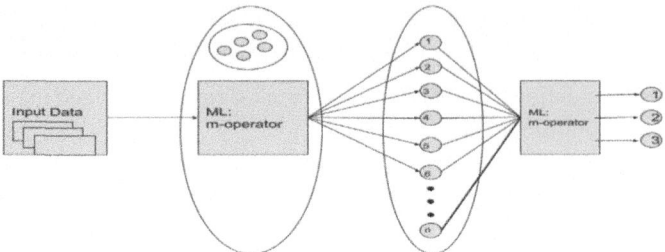

Fig. 2. Scheme of the use of ML technology in the video stream analysis system of the conference

The next step is the need to aggregate and calculate the degree of attention. This task can also be solved using machine learning methods. After extracting visual markers from the video stream, we evaluate them as indicators of the degree of interest for each specific participant. Based on the collected information, we have the opportunity to aggregate, cluster the received information and build an emotional portrait of the entire audience.

Thus, thanks to machine learning methods, we solve two main problems:

– automated selection of visual factors from the video stream.
– analysis of the influence of the received factors on the attention of the conference participants
– aggregation and further interpretation of further results.

More specifics and details of the use of machine learning methods will be described in our next scientific studies.

5   Conclusions

We reviewed the main AI-based tools that allow you to assess the audience's attention level in real time during online conferences. Among them, EmotionCues, HEADROOM and MeetingPulse. These tools allow you to analyze the emotional component of the audience in sufficient detail and effectively, but not all of them can be integrated with popular online conference services, such as Microsoft Teams or Zoom, and they also do not allow you to analyze the video stream in real time. That is why there is a need to develop a specialized software product that will allow to fully assess the level of audience engagement based on the emotional component, and can also be integrated with services for conducting online meetings.

We identified the main requirements for the software product, described the visual markers for measuring students' attention and engagement. Also, we came to the conclusion that the main technology for the development of this product is the methods and algorithms of machine learning.

References

1. Astin, A.: Student involvement: a developmental theory for higher education. J. Coll. Stud. Pers. **25**(4), 297–308 (1984)

2. Chen, C., Wang, J.: Effects of online synchronous instruction with an attention monitoring and alarm mechanism on sustained attention and learning performance. Interact. Learn. Environ. **26**(4), 427–443 (2018). https://doi.org/10.1080/10494820.2017.1341938

3. Coates, H.: Student Engagement in Campus Based and Online Education: University Connections. Taylor and Francis, London (2006)

4. Dinesh, D., Athi Narayanan, S., Bijlani, K.: Student analytics for productive teaching/learning. Paper Presented at the Proceedings - 2016 International Conference on Information Science, ICIS 2016, pp. 97–102 (2017). https://doi.org/10.1109/INFOSCI.2016.7845308

5. Features of generation Z. https://buki.com.ua/news/pokolinnya-z/. Accessed 15 Dec 2022

6. Fredrick, W.C., Walberg, H.J.: Learning as a function of time. J. Educ. Res. **73**(4), 183–194 (1980)

7. Gao, Q., Tan, Y.: Impact of different styles of online course videos on students' attention during the COVID-19 pandemic. Front. Public Health **10** (2022). https://doi.org/10.3389/fpubh.2022.858780

8. Ho, M., Mantello, P., Ghotbi, N., Nguyen, M., Nguyen, H., Vuong, Q.: Rethinking technological acceptance in the age of emotional AI: Surveying gen Z (zoomer) attitudes toward non-conscious data collection. Technol. Soc. **70** (2022). https://doi.org/10.1016/j.techsoc.2022.102011

9. Howe, N.: How the Millennial generation is transforming employee benefits. Benefits Q. **30**(2), 8–14 (2014)

10. Kurmanova, A., Kozhayeva, S., Ayupova, G., Aurenova, M., Baizhumanova, B., Aubakirova, Z.: University students' relationship with technology: psychological effects on students. World J. Educ. Technol. Curr. Issues **14**(4), 1225–1233 (2022). https://doi.org/10.18844/wjet.v14i4.7743

11. McIntyre, D.J., Copenhaver, R.W., Byrd, D.M., Norris, W.R.: A study of engaged student behaviour within classroom activities during mathematics class. J. Educ. Res. **77**(1), 55–59 (1983)

12. Negm, E.: Intention to use internet of things (IoT) in higher education online learning – the effect of technology readiness. High. Educ. Skills Work-Based Learn. (2022). https://doi.org/10.1108/HESWBL-05-2022-0121

13. Peng, S., Chen, L., Gao, C., Tong, R.J.: Predicting students' attention level with interpretable facial and head dynamic features in an online tutoring system. Paper Presented at the AAAI 2020 - 34th AAAI Conference on Artificial Intelligence, pp. 13895–13896 (2020)

14. Pobegaylov, O.: Digital education facing COVID - 19 pandemic: technological university experience. Paper Presented at the E3S Web of Conferences, p. 273 (2021). https://doi.org/10.1051/e3sconf/202127308090 (2021)

15. Psychology: Textbook, Edited by Trofimov, Y. 4th edn., 560 p., Stereotype. Kyiv, Lybid (2003)

16. Serdyuk, L.: Psychology: study guide. For distance learning, 233 p. "Ukraine" University, Kyiv (2005)

17. Shaw, R., Patra, B.K., Pradhan, A., Mishra, S.P. Attention classification and lecture video recommendation based on captured EEG signal in flipped learning pedagogy. Int. J. Hum.-Comput. Interact. (2022). https://doi.org/10.1080/10447318.2022.2091561

18. Solovyov, V., Kuklina, D., Slavgorodskyi, A., Pukhov, I., Tytko, M. Monitoring student involvement in the educational process. https://www.osp.ru/os/2018/2/13054177. Accessed 14 Oct 2022

19. Strauss, W., Howe, N.: Generations: the history of America's future, 1584 to 2069. Harper Perennial, 538 p. (1991). https://archive.org/details/GenerationsTheHistoryOfAmericasFuture1584To2069ByWilliamStraussNeilHowe/page/n3/mode/2up

Self-directed Learning in Chemistry Laboratory via Simulations

Fatma Alkan[✉]

Hacettepe University, Beytepe, 06800 Ankara, Turkey
alkanf@hacettepe.edu.tr

Abstract. Self-directed learning is seen as a key competence for survival in the twenty-first century. Self-directed learning is not about learning; instead, it is a meta-theory about learning how to learn. During the pandemic process, many theoretical and applied courses were conducted via distance education. One of these courses is the general chemistry laboratory. While the laboratory course was conducted with distance education, simulation applications were used. For the general chemistry laboratory experiments, first of all, theoretical lectures were made over zoom. Afterwards, the students performed experiments using simulation. The aim of this research is to determine the effect of the general chemistry laboratory conducted with distance education and simulations on the chemistry laboratory self-efficacy. The research was designed in a quasi-experimental design model. The sample group of the research consists of 25 pre-service chemistry teachers studying at a state university. Data were collected with the chemistry laboratory self-efficacy scale. The data obtained from the chemistry laboratory self-efficacy scale were analyzed. As a result of the research, it was determined that the simulation-supported laboratory application had a significant effect on the psychomotor self-efficacy and cognitive self-efficacy of the pre-service teachers. According to the results of the research, pre-service teachers are worried and afraid that they will not be able to do the experiment in the laboratory, so simulations are effective in increasing psychomotor and cognitive self-efficacy as they are very useful for preparation for the experiment.

Keywords: General chemistry laboratory · chemistry laboratory self-efficacy · pre-service chemistry teachers · Covid-19 Pandemic

1 Introduction

Students see chemistry as quite boring, difficult and challenging. Some subjects of chemistry, such as the structure of the atom and visualizing how chemical bonds occur, are thought to be difficult to understand due to their nature [1]. It is stated that chemistry is a difficult science and it is difficult to make a career in science [2]. Students generally do not want to study science courses except compulsory courses. In high school and university, while students prefer biology-related courses, they prefer chemistry and physics-related courses less [3]. Students are more interested in science during secondary school and can even acquire science knowledge. It is very important to ensure that this

© The Author(s), under exclusive license to Springer Nature Switzerland AG 2023
G. Antoniou et al. (Eds.): ICTERI 2023, CCIS 1980, pp. 160–172, 2023.
https://doi.org/10.1007/978-3-031-48325-7_12

is also gained by students at other levels [4]. Many of the events in our daily life are related to chemistry [5, 6]. Students are quite aware that chemistry is related to various areas of our lives. In fact, these students are aware that problems will arise when they think that their chemistry knowledge is insufficient to explain the relationship in these areas [1].

Chemistry is an experimental science. In order to be an expert in this field, designing an experiment is the most important requirement. For this reason, it is unacceptable that traditional laboratory programs have content that provides little training for the development of this skill [7]. Students do not forget very little of what they have learned and they have difficulty in applying this information they have not forgotten [8]. In order to make chemistry more meaningful and interesting, more laboratory activities and practical work should be done, and it is emphasized that it may be useful to associate chemistry with daily life [9]. Practical activities related to chemistry will enable students to understand that chemistry will be understandable and applicable [2]. Laboratory work in chemistry should be an important component of the assessment of what is learned in the chemistry course. Otherwise, failure will occur [10].

Educational institutions aim at the optimum use of technology in order to equip their students with theoretical and practical knowledge so that they can use them in their business life. It is estimated that the impact of the development of technology on higher education will increase in the coming years [11, 12]. It is the responsibility of higher education institutions to train the teachers of the future. In this context, education programs should take necessary initiatives for future teachers to be aware of and learn how to use educational technologies effectively in their teaching practices [13]. It is expected that pre-service teachers have important technology skills and move from being computer literate to technology competence [14]. According to the studies, it has been noticed that pre-service teachers are willing to use technology in their future course applications [15]. The Covid-19 pandemic process necessitated the integration of technology and web tools into education. In this process, all theoretical and practical courses were carried out with distance education. The biggest savior of practical lessons, namely laboratory lessons, has been simulations. Simulations are explained as educational software that allows learners to observe representations of processes that they cannot actually experience, and to create interaction within the program. Natural events at various scales are modeled by scientists, simulations are developed and applied in order to understand natural events [16].

With the development of technology, many experiments aimed at reaching large audiences, especially expensive, dangerous and hard-to-reach experiments, are offered to people in virtual laboratories with interactive simulations. In this way, students can see the work done in the laboratory in the form of video in the computer environment for educational purposes and experience it in the virtual environment [17, 18]. During the pandemic process, many theoretical and applied courses were conducted via distance education. One of these courses is the general chemistry laboratory. While the laboratory course was conducted with distance education, simulation applications were used. It is the best example of the application of the self-directed learning method with simulation. In self-directed learning, the student experiences how to learn a subject better. Self-directed learning provides each individual with the opportunity to learn at their own pace.

The aim of this research is to determine the effect of the general chemistry laboratory conducted with distance education and simulations on the chemistry laboratory self-efficacy perception of pre-service teachers.

2 Method

Quasi experimental design was used in the research. The single-group pretest-posttest quasi-experimental design is a type of research design often used by behavioral researchers to determine the effect of a treatment or intervention on a given sample. This research design is characterized by two features. The first feature is the use of a single group of participants (i.e., single group design). This property indicates that all participants are part of a single condition; that is, all participants are given the same application and assessments. The second feature is a linear sequence (i.e., a pretest-posttest design) that requires evaluation of a dependent variable before and after an implementation. In the pretest-posttest research designs, the effect of the application is determined by calculating the difference between the first evaluation and the final [19]. Before starting the research, a pre-test is administered to the participants. The teaching method whose effectiveness will be examined is carried out by the teacher, followed by a post-test. Then, the change in pre-test and post-test scores is examined [20].

2.1 Sample

While determining the sample of the research, the purposeful sampling method, which is one of the non-random sampling methods, was used. Purposeful sampling is the selection of a sample using the personal judgments of researchers to select a sample based on previous knowledge of a group and the specific purpose of the research [21]. This sampling method offers researchers opportunities to overcome time-related problems and to be quickly and easily accessible to participants [22].

The sample group of the study consists of 25 volunteer pre-service chemistry teachers who take the analytical chemistry laboratory course at Hacettepe University Faculty of Education. The research was carried out in the spring semester of 2021–2022 in the general chemistry laboratory course. The frequency and percentage values of the demographic characteristics of the sample group are given in Table 1.

2.2 Data Collection Tools

Data were collected with the chemistry laboratory self-efficacy perception scale. The chemistry laboratory self-efficacy perception scale was developed by Alkan (2016) [23]. Consisting of 14 items, the scale has a 2-factor structure. Cronbach Alpha internal consistency coefficient for the whole scale 0.885. Cronbach Alpha coefficient for sub-dimensions was psychomotor self-efficacy (PSE) 0.847, cognitive self-efficacy (CSE) 0.818.

Table 1. Demographic details and characteristics of sampling.

Categories		f	%
Gender	Female	21	84
	Male	4	16
Age	18	19	76
	19	4	16
	20	2	8
Grade	1	21	84
	2	3	12
	3	1	4
Total		25	100

2.3 Application Process

While the general chemistry laboratory course was conducted with distance education, simulation applications were used. For the general chemistry laboratory experiments, first of all, theoretical lectures were made over zoom. Afterwards, the students performed experiments using simulation. The simulation steps developed by the Royal Society of Chemistry used in the application are given below (Figs. 1, 2, 3, 4 and 5).

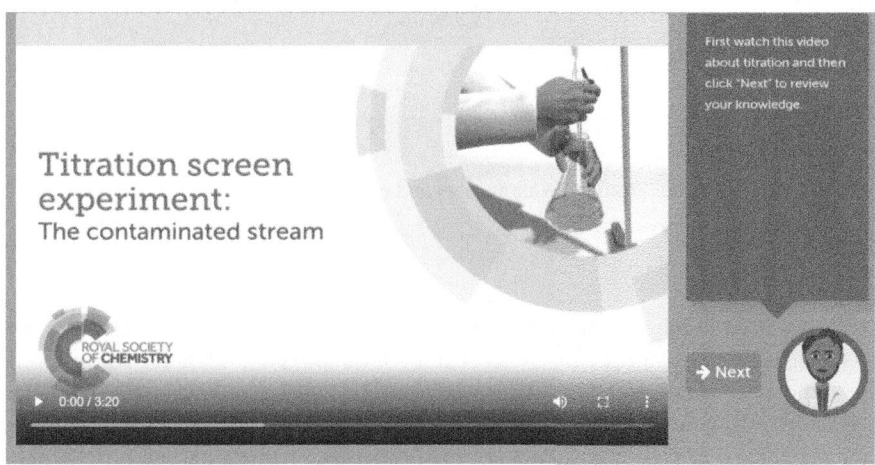

Fig. 1. Simulation steps.

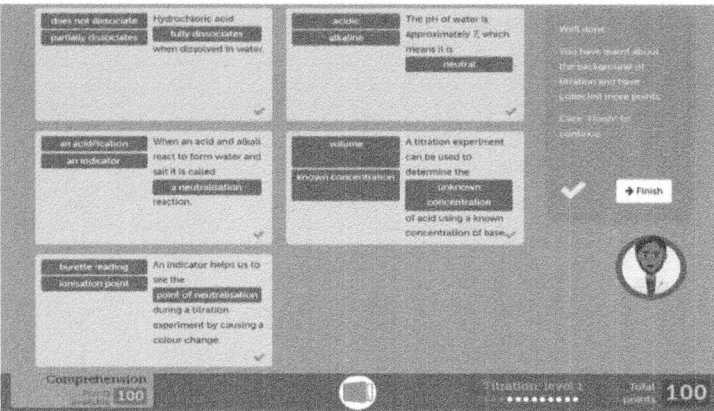

Fig. 2. Simulation steps theoretical knowledge.

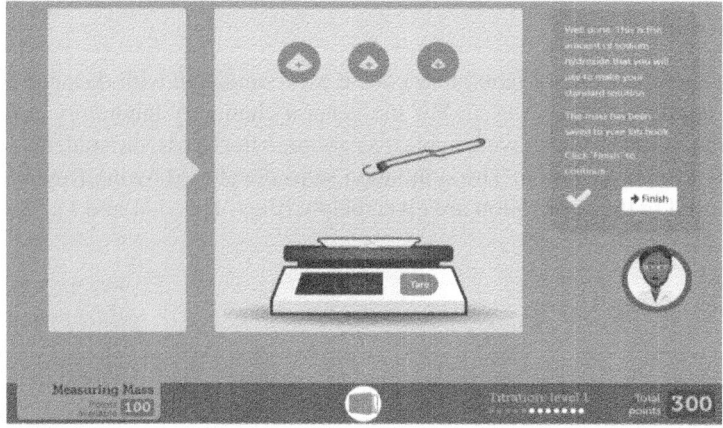

Fig. 3. Simulation steps tools and equipment knowledge.

2.4 Data Analyze

SPSS23 was used in the analysis of the quantitative data obtained from the study. After the distance education and chemistry laboratory simulation applications, the difference between the pre-test and post-test scores was examined with the Wilcoxon Signed Rank test. Before starting the analysis, the assumption of normality was examined. According to the literature, non-parametric tests are recommended if the sample size is <30 and the distribution is not normal [24]. Another literature states that the number of samples reaching 30 ensures that the distribution of the averages approaches normal [21]. In this study, the assumption of normality was tested statistically through the significance level of the Shapiro-Wilks test. According to the results of the analysis, since Shapiro-Wilks $p < .05$ normality assumption was not provided, non-parametric tests were used [25, 26].

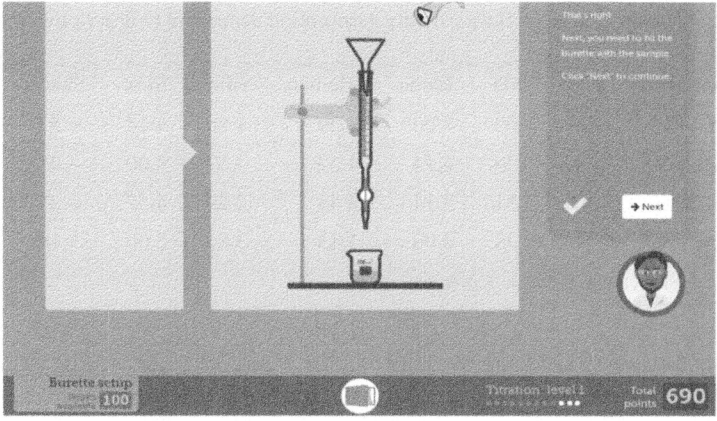

Fig. 4. Simulation steps virtually experiment.

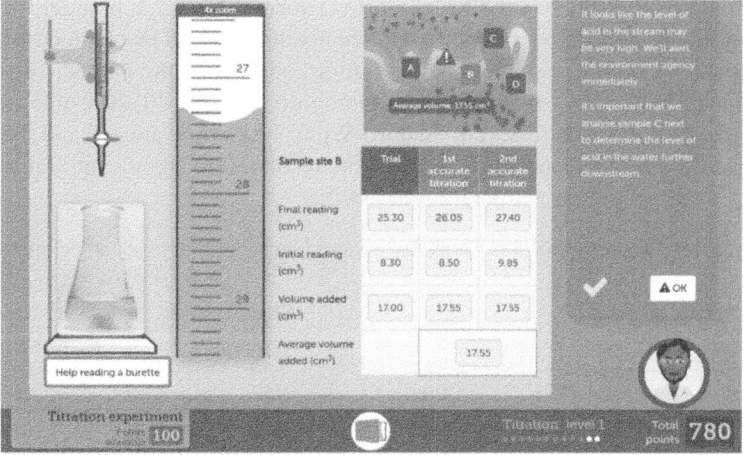

Fig. 5. Simulation steps calculation.

3   Results

Results on the normality assumption

The normality of the data obtained in a study can be determined by using descriptive, graphical and statistical methods [27]. Whether the data show a normal distribution in studies can be determined by using descriptive, graphical and statistical methods [28]. In this study, the normality assumption of the chemistry laboratory self-efficacy scale data was analyzed using descriptive methods and statistical analysis. In descriptive methods, some statistics such as arithmetic mean, standard deviation, mode, median, kurtosis and skewness coefficients were analyzed [29]. The results are given in Table 2.

When Table 2 is examined, it is seen that the mode and median values of the pretest and posttest data of the scale are different from each other. It is noteworthy that the

Table 2. Pretest-posttest scores normality assumption according to descriptive methods.

Observed variables		Mean	SD	Mode	Median	Min	Max	Skew.	Kurt.
PSE	Pre-test	3.18	.49	3.00	3.29	1.86	4.14	−.57	1.31
	Post-test	4.33	.39	4.14	4.29	3.57	5.00	−.01	−.63
CSE	Pre-test	3.36	.51	3.14	3.43	2.29	4.57	−.31	1.27
	Post-test	4.39	.38	4.00	4.43	3.86	5.00	.16	−1.47

Note: Skew. = Skewness; Kurt. = Kurtosis.

kurtosis and skewness values are not between +1.5 and −1.5. According to descriptive methods, it can be said that the distribution is not normal according to kurtosis, skewness, mode and median values.

The normality assumption of the chemistry laboratory self-efficacy scale data was examined by statistical analysis. In statistical methods, examinations are usually made with the Shapiro-Wilks Test or the Kolmogorov-Smirnov Test [27]. In order for the data set to show a normal distribution, non-significant values should be obtained from these tests. Shapiro-Wilks Test results are taken into account when the number of people in the examined group is less than 50, and Kolmogorov-Smirnov Test results when it is more than 50 [30]. For this research, Shapiro-Wilks Test was evaluated and the data set does not show a normal distribution because the results were significant (Table 3).

Table 3. Pretest-posttest scores normality assumption according to statistical methods.

Observed variables		Shapiro-Wilk		
		Statistic	df	Sig.
Psychomotor self-efficacy (PSE)	Pre-test	.940	23	.247*
	Post-test	.907	23	.496*
Cognitive self-efficacy (CSE)	Pre-test	.916	23	.051
	Post-test	.799	23	.024

* $p > 0.05$.

The sample of the study was less than 50, Shapiro-Wilks Test results were evaluated while analyzing statistical methods. Although this value is not significant in most of the sub-dimensions of the scale, it is seen in the dimensions that are significant. It was observed that the data deviated from the normal distribution and it was determined that it did not provide the statistically normality assumption.

Chemistry laboratory self-efficacy perception pre-test post-test results
Chemistry laboratory self-efficacy perception pre-post test scores should be examined from descriptive data. In the psychomotor self-efficacy (PSE) dimension, the pretest X:3.18, the posttest X:4.33, is. It is noticed that the pretest X:3.36, posttest X:4.39, in the cognitive self-efficacy (CSE) dimension.

As a result of the examinations, the data obtained from the research do not meet the normality assumption descriptively. The effect of general chemistry laboratory practices supported by simulations on pre-service teachers' chemistry laboratory self-efficacy perception was investigated. For this purpose, Wilcoxon Signed Rank test was used to examine whether there was a significant difference between pre-test findings and post-test findings in the chemistry laboratory self-efficacy perceptions of pre-service teachers. The results are shown in Table 4.

Table 4. Wilcoxon signed-rank test results chemistry laboratory self-efficacy perception scale pretest-posttest scores.

Observed variables		Mean rank	Sum of ranks	Z	p
Psychomotor self-efficacy (PSE)	Negative ranks	4.00	4.00	−4.269	.000
	Positive ranks	13.38	321.00		
Cognitive self-efficacy (CSE)	Negative ranks	5.00	5.00	−4.249	.000
	Positive ranks	13.33	320.00		

* p < 0.05
a. Based on negative ranks.

When the table is examined, it is seen that there is a statistically significant difference between the psychomotor self-efficacy (PSE) sub-dimension of the chemistry laboratory self-efficacy scale and the pre-test scores of the cognitive self-efficacy (CSE) dimension ($Z = -3.802, -3.034; p < 0.05$). It is noteworthy that the pre-application average (X: 3.18) of the pre-service teacher in the psychomotor self-efficacy dimension was after the application (X: 4.33). While it is average (X: 3.36) according to the cognitive self-efficacy dimension, it is seen that after the application (X: 4.39). Based on these results, it can be said that distance education and experiment simulation applications in the general chemistry laboratory have a significant effect on the development of pre-service teachers' chemistry laboratory self-efficacy perception.

4   Discussion and Conclusion

This research was carried out to determine the effect of distance education supported by simulation applications in the general chemistry laboratory on the perception of chemistry laboratory self-efficacy of pre-service chemistry teachers. As a result of the research, it was determined that distance education had a significant effect on developing the chemistry laboratory self-efficacy perception of pre-service chemistry teachers. The effect of distance education on chemistry laboratory self-efficacy perception is also supported by other research results. When the sub-dimensions of the chemistry laboratory self-efficacy perception scale were examined in the study, it was determined that there was a difference between the pretest/posttest scores in the psychomotor self-efficacy and cognitive self-efficacy dimensions of the scale.

The positive effect of simulation-supported applications obtained as a result of the research on affective properties is supported by the literature. Technology-assisted instruction is effective in facilitating behavioral, cognitive and emotional participation of students [31]. Simulations appear as technology-supported applications. The simulations used in the general chemistry laboratory enabled the students to better understand the steps followed in the experiment process. For example, they comprehended the pre-preparations and process steps of applications such as weighing, mixing solutions, stoichiometry, titration, the execution of the experiment and the use of the data obtained as a result of the experiment in the calculations, with the steps followed in the simulation.

As a result of the research, it was determined that the psychomotor self-efficacy and cognitive self-efficacy of the simulation-supported laboratory application pre-service teachers increased significantly. In simulation, before the student starts an experiment, prior knowledge about that experiment is checked. Concepts, units, and materials related to the experiment are subjected to a test that tests the knowledge of the material to be used in the experiment. The student who cannot get the valid score from this test cannot proceed to the next stage. At this stage, pre-service teachers developed cognitive self-efficacy. Technology-supported applications increase class participation and motivation. This is emphasized by teachers as the strength of technology [32]. There is a great deal of evidence proving that technology-assisted learning applications provide affective and cognitive benefits and create an interactive and social learning environment [33–35].

It is reported that the computer-assisted simulation program in the chemistry course contributes positively to the understanding of the concepts and principles [36]. The effect of simulations on academic success related to chemistry course was also examined. As a result of the research, it was determined that the simulation activity used in the chemistry lesson significantly increased the success of the lesson [37–39]. It was also investigated whether it caused differentiation in self-regulation skills in science lessons taught using the simulation program. As a result of the research, it was stated that the simulation activity was effective in the acquisition of intuitive knowledge, could define and interpret the simulation outputs, and as a result, self-regulation skills developed [40]. In the research on the effect of simulations on science achievement, the course was taught with the PhET simulation application. According to the results of the research, it has been reported that there is a significant difference between the lessons taught with the traditional method and the lessons taught using the simulation method in terms of science achievement [41]. Simulation laboratory activity offers students opportunities such as communicating learning objectives, giving detailed information about activities, data visualization and working with chemical solutions [42]. As a result of the research, cognitive self-efficacy of pre-service teachers is supported by the positive effect of technology-supported applications on achievement and attitude.

As a result of the research, it is another research finding that pre-service chemistry teachers developed psychomotor self-efficacy. It is surprising to achieve such a result without going into the laboratory and touching the materials one by one. This can be explained by the fact that the technology-assisted learning system increases students' learning satisfaction and motivation [43]. Educational technologies, virtual applications and mobile technologies provide access to these technologies outside the classroom. These technologies have the potential to enrich the teaching and learning experience in different subject areas [44, 45]. Technology-supported activities have many advantages in chemistry as they allow students to mix, run and react chemical solutions with a few mouse clicks [46].

Science learning includes many complex and abstract concepts that are difficult to understand [47]. Abstract concepts have a very negative effect on students' learning of chemistry. It is clear that there is disconnect between theoretical knowledge and skill practice [48]. Simulations can be used as the only tools to overcome all these negativities simulations. It is a system of multiple, complex and advanced non-stationary models. With the virtual reality created by simulations, it helps to better understand many situations that may pose a danger. In the educational environment, teachers provide students with the opportunity to create experiments without a risk environment and to observe the result [49]. Chemistry laboratory is the only places for self-directed learning. Here, the student tries to learn the subject he wants to learn by making experiments with different aspects [50]. Pre-service teachers can access simulations and repeat experiments whenever they wish. Simulations are very good resources for applying the self-directed learning method. With the simulation, the student progresses and learns by repeating each step as much as they want or by going fast. Therefore, they feel themselves developed psychomotorly.

In this study, the effect of simulation-supported general chemistry laboratory practices on pre-service teachers' perception of laboratory self-efficacy was investigated. Simulations test whether the pre-service teachers know the tools and equipment to be used in the experiment and the necessary chemicals, and check the theoretical knowledge about the subject. If the candidate's knowledge is at a sufficient level, he moves on to the next stage of the simulation. If it's not enough, it tries until it's enough. In the next step, the experiment is performed virtually. Here, the candidate has the opportunity to try the experiment as many times as he/she wants. In these trials, the experiment can reach the state it should be under the best conditions or it tests what results it should have when it should not be at all. Students are curious in the laboratory. Teachers often say, "What would happen if we exposed it to more heat?" confronted with such questions. Simulation allows the student to test all the questions he/she is curious about and see the result. Finally, the simulation shows how to use the data from the experiment to calculate the experiment result. Simulations allow each individual to progress at their own learning pace. Virtual applications cannot provide the competencies that will be gained by experimenting in the laboratory. For this reason, it is recommended for teachers to use simulations to prepare students for the experiment in the post-pandemic period, or to offer them as helpers to remember things they forgot after the experiment.

References

1. Rüschenpöhler, L., Markic, S.: Secondary school students' acquisition of science capital in the field of chemistry. Chem. Educ. Res. Pract. **21**, 220–236 (2020)
2. Mujtaba, T., Sheldrake, R., Reiss, M.J.: Chemistry for all. Reducing inequalities in chemistry aspirations and attitudes. Royal Society of Chemistry, England (2020)
3. Gatsby. Key indicators in STEM education. The Gatsby Charitable Foundation, London (2018)

4. Sheldrake, R., Mujtaba, T.: Children's aspirations towards science-related careers. Can. J. Sci. Math. Technol. Educ. **20**, 7–26 (2019)
5. Özmen, H.: Learning theories and technology supported constructivist learning in science teaching. Turk. Online J. Technol. **3**(1), 100–111 (2004)
6. Özden, M.: Qualitative and quantitative evaluation of chemistry teachers' problems encountered during chemistry teaching: samples of Adıyaman and Malatya Pamukkale university. J. Educ. **22**, 40–53 (2007)
7. Pickering, A.: Constructing Quarks. A Sociological History of Particle Physics. Edinburgh University Press, Edinburgh (1984)
8. Saint-Jean, M.: L'apprentissage par problèmes dans l'enseignement supérieur. Service d'aide à l'enseignement, Université de Montréal, Québec (1994)
9. Broman, K., Ekborg, M., Johnels, D.: Chemistry in crisis? Perspectives on teaching and learning chemistry in Swedish upper secondary schools. Nordic Stud. Sci. Educ. **7**(1), 43–60 (2011)
10. Wilson, H.: Problem-solving laboratory exercises. J. Chem. Educ. **64**(10), 895–897 (1987)
11. Becker, S.A., et al.: NMC horizon report: 2018 higher education edition (2018)
12. Becker, S. A., Cummins, M., Davis, A., Freeman, A., Giesinger, H.C., Ananthanarayanan, V.: NMC horizon report: 2017 higher education edition (2017)
13. Sat, M., Ilhan, F., Yukselturk, E.: Comparison and evaluation of augmented reality technologies for designing interactive materials. Educ. Inf. Technol. (2023)
14. Smarkola, C.: Efficacy of a planned behavior model: beliefs that contribute to computer usage intentions of student teachers and experienced teachers. Comput. Hum. Behav. **24**(3), 1196–1215 (2008)
15. Yilmaz, R.M., Baydas, O.: Pre-service teachers' behavioral intention to make educational animated movies and their experiences. Comput. Hum. Behav. **63**, 41–49 (2016)
16. Honey, M., Hilton, M. (eds.): Learning Science Through Computer Games and Simulations. National Academy Press, Washington, DC (2011)
17. Wuttke, H.D., Henke, K., Ludwig, N.: Remote labs versus virtual labs for teaching digital system design. In: International Conference on Compute Systems and Technologies-CompSysTech 2005 (2005)
18. İnce, E.Y., Kutlu, A.: Web based laboratories [Web Tabanlı Laboratuvarlar] (2016). http://ab.org.tr/ab14/bildiri/34.pdf. Accessed 24 Apr 2020
19. Cranmer, G.: One-group pretest–posttest design. In: Allen, M. (ed.) The Sage Encyclopedia of Communication Research Methods, pp. 1125–1126. SAGE Publications, Inc. (2017)
20. Mertler, C.A.: Chapter 7 quantitative research methods. In: Introduction to Educational Research, 3rd edn. SAGE Publications, Los Angeles (2022)
21. Fraenkel, J.R., Wallen, N.E., Hyun, H.H.: How to Design and Evaluate Research in Education, 8th edn. Mc Graw HIll, New York (2012)
22. Yıldırım, A., Şimşek, H.: Qualitative Research Methods in the Social Sciences, 9th edn., pp. 110–122. Seçkin Press, Ankara (2013)
23. Alkan, F.: Development of chemistry laboratory self-efficacy beliefs scale. J. Baltic Sci. Educ. **15**, 350–359 (2016)
24. Chin, R., Lee, B.Y.: Analysis of data. In: Principles and Practice of Clinical Trial Medicine, pp. 325–359. Academic Press publications, Elsevier (2008)
25. Green, S.B., Salkind, N.J.: Using SPSS for Windows and Macintosh (Analyzing and Understanding Data), 5th edn. Pearson Prentice Hall, New Jersey (2008)
26. Leech, N.L., Barrett, K.C., Morgan, G.A.: SPSS for Intermediate Statistics, Use and Interpretation, 2nd edn. Lawrence Erlbaum Associates Inc., Mahwah (2005)
27. Abbott, M.L.: Understanding Educational Statistics Using Microsoft Excel and SPSS. Wiley, USA (2011)

28. Ghasemi, A., Zahediasl, S.: Normality tests for statistical analysis: a guide for non-statisticians. Int. J. Endocrinol. Metab. **10**(2), 486–489 (2012)
29. Kirk, R.E.: Statistics an Introduction, 5th edn. Thomson Higher Education, USA (2008)
30. Büyüköztürk, Ş.: Manual of data analysis for social sciences. [Sosyal Bilimler İçin Veri Analizi El Kitabı]. PegemA Yayınları, Ankara (2006)
31. Wang, Y.: Effects of augmented reality game-based learning on students' engagement. Int. J. Sci. Educ. Part B **12**(3), 254–270 (2022)
32. Nikimaleki, M., Rahimi, M.: Effects of a collaborative AR-enhanced learning environment on learning gains and technology implementation beliefs: evidence from a graduate teacher training course. J. Comput. Assist. Learn. **38**(3), 758–769 (2022)
33. Arici, F., Yilmaz, M.: An examination of the effectiveness of problem-based learning method supported by augmented reality in science education. J. Comput. Assist. Learn. **39**, 446–476 (2022)
34. Baabdullah, A.M., Alsulaimani, A.A., Allamnakhrah, A., Alalwan, A.A., Dwivedi, Y.K., Rana, N.P.: Usage of augmented reality (AR) and development of e-learning outcomes: an empirical evaluation of students' e-learning experience. Comput. Educ. **177**, 104383 (2022)
35. Çetin, H., Ulusoy, M.: The effect of augmented reality-based reading environments on retelling skills: formative experiment. Educ. Inform. Technol. (2022). https://doi.org/10.1007/s10639-022-11415-8
36. Mihindo, W.J., Wachanga, G.B., Anditi, Z.O.: Effects of computer-based simulations teaching approach on learners' achievement in the learning of chemistry among secondary school learners in Nakuru sub county, Kenya. J. Educ. Pract. **8**(5), 65–75 (2017)
37. Alkan, F., Koçak, C.: Chemistry laboratory applications supported with simulation. Procedia Soc. Behav. Sci. **176**, 970–976 (2015)
38. Kocak, A.C., Alkan, F.: Technology-supported teaching of volumetric analysis. Cumhuriyet Int. J. Educ. **5**(1), 1–9 (2016)
39. Nkemakolam, O.E., Chinelo, O.F., Jane, M.C.: Effect of computer simulations on secondary school students' academic achievement in chemistry in Anambra state. Asian J. Educ. Train. **4**(4), 284–289 (2018)
40. Eckhardt, M., Urhahne, D., Harms, U.: Instructional support for intuitive knowledge acquisition when learning with an ecological computer simulation. Educ. Sci. **8**, 94 (2018)
41. Hannel, S.L., Cuevas, J.: A study on science achievement and motivation using computer-based simulations compared to traditional hands-on manipulation. Georgia Educ. Res. **15**(1), 40–55 (2018)
42. Barba, L.A.: Engineers Code: reusable open learning modules for engineering computations. Comput. Sci. Eng. **22**, 26–35 (2020)
43. Cascales-Martínez, A., Martínez-Segura, M.-J., Pérez-López, D., Contero, M.: Using an augmented reality enhanced tabletop system to promote learning of mathematics: a case study with students with special educational needs. Eur. J. Math. Sci. Technol. Educ. **13**(2), 355–380 (2017)
44. Crompton, H., Burke, D., Gregory, K.H., Gräbe, C.: The use of mobile learning in science: a systematic review. J. Sci. Educ. Technol. **25**(2), 149–160 (2016)
45. Lin, Y.-N., Hsia, L.-H., Hwang, G.-J.: Fostering motor skills in physical education: a mobile technology-supported ICRA flipped learning model. Comput. Educ. (2021)
46. Grazioli, G., Ingwerson, A., Santiago, D., Jr., Regan, P., Cho, H.: Foregrounding the code: computational chemistry instructional activities using a highly readable fluid simulation code. J. Chem. Educ. **100**(3), 1155–1163 (2023)
47. Gerard, L., Linn, M.C., Berkeley, U.C.: Computer-based guidance to support students' revision of their science explanations. Comput. Educ. **176**, 104351 (2021)
48. Sus, M., Hadeed, M.: Theory-infused and policy-relevant: on the usefulness of scenario analysis for international relations. Contemp. Secur. Policy **41**, 1–24 (2020)

49. Minaslı, E.: Fen ve teknoloji dersi maddenin yapısı ve özellikleri ünitesinin öğretilmesinde simülasyon ve model kullanılmasının başarıya, kavram öğrenmeye ve hatırlamaya etkisi (Tez No: 250817) [Yüksek Lisans Tezi, Marmara Üniversitesi, Eğitim Bilimleri Enstitüsü] YÖK Ulusal Tez Merkezi. (2009)
50. Alkan, F.: Self-directed learning in the analytical chemistry laboratory: examining ınquiry skills. In: Advancing Self-Directed Learning in Higher Education, pp. 43–76. IGI Global Publications (2023)

The Use of Digital Tools for Mastering Practical Disciplines in the Distance Format of Training Bachelors of Preschool Education

Larysa Harashchenko, Olena Kovalenko(ID), Liudmyla Kozak(ID), Olena Litichenko(✉) (ID), and Dana Sopova(ID)

Boris Grinchenko Kyiv University, Kyiv, Ukraine
{l.harashchenko,o.kovalenko,l.kozak,o.litichenko,
d.sopova}@kubg.edu.ua

Abstract. The article examines the peculiarities of the use of digital tools (DT) in the education of students of the specialty 012 Preschool Education in the discipline «Art needlework». The authors identified the aspects of the study of the problem of using digital tools in the professional training of future teachers, carried out by modern scientists. The relevance of the application of DT for mastering practical disciplines by bachelors in the process of distance learning is determined. It was established: the lack of development of future bachelors in the field of preschool education competence regarding the application of DT for the organization of artistic and productive activities of preschool children. A model of the use of these digital tools in the process of teaching the discipline «Art needlework» has been developed. The experience of using various digital tools of visualization, collective interaction, game services, augmented reality in mastering disciplines is described, such as: Genially, Jamboard, Conceptboard, Kahoot, H5P, Craiyon, Deepdreamgenerator, Dreamstudio, Canva, Fotor, LightShot, Fanny Pho.to, Blippbuilder. The effectiveness of the use of digital tools for mastering practical disciplines in the distance format of the training of bachelors of preschool education has been proven.

Keywords: digital tools · ICT competence · higher education · educational process · artistic construction

1 Introduction

1.1 Problem Statement

The global challenges of recent years have strengthened the modern trends in the development and implementation of distance education. Distance learning has become a forced reality for the Ukrainian education system. Due to the full-scale invasion of the Russian occupation forces in Ukraine and the potential danger for students, education is carried out in a mixed format with a significant predominance of distance education. New approaches, means, methods and tools for organizing the educational process have

G. Antoniou et al. (Eds.): ICTERI 2023, CCIS 1980, pp. 173–188, 2023.
https://doi.org/10.1007/978-3-031-48325-7_13

become extremely relevant. Digital tools have become the object of close attention from educators at all levels. The experience of their effective use has shown potential opportunities for solving the problems of youth education, training of qualified specialists, however, a large part of the issues needs further consideration.

Before the start of the coronavirus pandemic (2020), electronic training courses (ETC) were created at the Borys Grinchenko Kyiv University, and considerable experience of educational interaction using ICT in a distance format was accumulated.

ETC were developed on the Moodle platform and were filled in accordance with the educational program of the specialty 012 Preschool Education. The development of electronic training courses for the specialty 012 Preschool education was carried out taking into account the specifics of the specialty, the main difficulty of which is the social-communicative direction of the disciplines, which involves the formation of a set of student competencies related to interaction with people.

Comparing educational disciplines, it is worth noting that they have a different ratio of the amount of theoretical knowledge and practical skills that a student must master. This is important for filling electronic courses of academic disciplines and selecting digital tools for their provision. The most difficult is the development and implementation of practical direction disciplines, such as «Art needlework» (for students of the 1st year study, the first (bachelor) level of higher education of the educational and professional program 012 Preschool education), since they contain more than 80% of practical material and require the development of their own manual skills the ability of students, which is accompanied by the accumulation of methodical knowledge and skills in the use of various techniques and methods of creative activity in the educational process of a preschool education institution. The task of effective organization of practical classes in online and remote format is a real challenge for teaching practical disciplines. This is explained by the fact that practical activities in such a discipline are based on the interaction of the participants of the educational process and require, along with the acquisition of special professional competences, students to master a number of technical manual skills of making products.

1.2 Literature Review

Research related to distance learning of practical disciplines by means of ICT technologies allows us to note that the search for effective means of distance learning is particularly relevant. Many scientists and practitioners pay attention to the issue of integration of various forms of educational information, as well as combining methods of communication and interaction in university education: V. Bykov, O. Buynytska, S. Vasylenko, L. Varchenko-Trotsenko, N. Morse, O. Spirin, O. Pinchuk [4, 5, 15, 16].

In Ukrainian pedagogy, the problem of using digital tools in the process of professional training of future teachers is presented in the works of Bielienka G., Kovalenko O., Kozak, L., Nezhyva L., Miyer T., Palamar S., Ponomarenko T. [3, 10–12, 18, 22].

The publications of Ukrainian scientists consider a variety of digital tools for organizing various forms of educational interaction, in particular the use of cloud technologies (Korobeinikova T. [13], Proshkin, V. [21]), word cloud services for the students development of thinking functions (Nezhyva, L., Palamar, S., Marienko, M. [18]), AR technologies - 2D and 3D modeling in the educational process (Kozak L, Kozlitin D. [11]), creation of open resources for practice (Velychko V., Fedorenko E., Soloviev V.

Dolinska L. [6]). The question of the necessity of forming ICT competence among educators, since media literacy of preschool children is an integral component of the content of modern preschool education, was analyzed by Yankovych, O., Chaika, V., Ivanova, T., Binytska, K., Kuzma, I., Pysarchuk, O. Falfushynska, H. Arguing the need to modernize the training of young specialists Yaroshenko O. Samborska O., Kiv A. suggest using a comprehensive approach to the digital training of future teachers [13]. The experience of teaching the disciplines of the specialty 012 Preschool education in the online format is revealed in the study of Bielienka, A., Polovina, O., Kondratets, I., Shynkar, T., Brovko, K [3].

Modern foreign authors investigate various aspects of the problem of the digital tools introduction in the educational process of higher education institutions, in particular the integration of digital technologies into teaching and learning of students (Wan Ng [19]), which prove the effectiveness of using digital tools for learning in higher education, attention is focused on the importance searches for effective means of distance and online education, because, according to scientists, this is what awaits us in the future (Kim J. [9]; Pérez-Jorge D. [20]).

On the other hand, the study of the students preferences from different countries of the world testifies to the importance of educational communication in the process of studying creative disciplines (Akinci M. [2]), because students feel a strong need for live communication in order to master creative and socially oriented professional subjects (Jalongo M. [8]). The problem of conducting creative workshops in which direct physical contact with materials in higher education is important is highlighted in the work of Davidaviciene V., Zvirble V., Daveiko J. [7]. This problem is relevant for us as well. Therefore, when developing the content of the discipline, we focus on online interaction with students; this form has a number of advantages for practical educational subjects [14].

According to the professional standard of «Educator of a preschool education institution», (https://mon.gov.ua/ua/npa/pro-zatverdzhennya-profesijnogo-standartu-vihovatel-zakladu-doshkilnoyi-osviti) (the main document that regulates professional (job) duties) information and communication competence is included in the list of professional skills of an educator. It includes the ability to use ICT and electronic educational resources to organize the educational process in a preschool education institution. In accordance with these requirements, educational programs for students of the specialty 012 Preschool education are being built. Therefore, educational disciplines, along with professional tasks, include mastering ICT and creating educational content for children of early (1–2 years) and preschool (3–6 years) age.

1.3   The Aim of the Research

In view of the above, the purpose of the article is to substantiate the expediency of using DT in disciplines of a practical direction and highlight the experience of forming the professional competences of future teachers of preschool education using digital tools in the process of teaching the educational course «Art needlework» on the Moodle platform.

To achieve this goal, it is necessary to solve the following tasks: to analyze scientific sources on the problem of DT application in pedagogical education; to develop criteria, indicators and levels of formation of ICT competence in the process of learning the educational discipline «Art needlework»; to investigate the state of formation of ICT competence while studying the discipline «Art needlework»; to consider the possibilities of using DT in interaction with students in the process of studying the discipline «Art needlework».

The following theoretical and empirical methods were used in the research process: analysis of electronic resources and educational programs, methodological manuals; observation method, student questionnaires, diagnostic tasks with the aim of identifying the level of ICT competence of future specialists in the specialty 012 Preschool education.

2 Results of the Research

In order to verify the results obtained during the theoretical study, an empirical study was conducted. A group of students of the first and second year of the bachelor's degree in specialty 012 «Preschool education» (132 full-time and part-time students) of the Faculty of Pedagogical Education of the Borys Grinchenko Kyiv University participated in the research.

At the first stage of the research, we determined the level of students' ICT competence based on the «Description of Digital Competence of a Pedagogical Worker» (2019) proposed by N. Morse, O. Bazelyuk, I. Vorotnikova [17], adapted in accordance with the professional standard «Preschool Teacher» (2021) [1].

According to the «Description of Digital Competence of a Pedagogical Worker» (2019) adapted in accordance with the professional standard «Educator of a Preschool Education Institution» (2021), criteria and indicators for evaluating the level of ICT competence formation in the process of mastering the practical discipline «Art needlework» among students of specialty 012 Preschool education were determined (Table 1).

The levels of formation of ICT competence of education seekers were also determined. We set ourselves the goal of forming ICT competence at the «Beginner» and «Integrator» levels, since the discipline «Art needlework» is studied by students in the 1st year study and in the first semester of the 2nd year study. Further improvement of skills to the «Expert» level takes place in senior courses in the process of studying the disciplines of the specialty 012 Preschool Education.

To conduct the research, we added the level «Elementary» to indicate students who have a low level of digital literacy formation.

Elementary – has general ideas about the use of modern gadgets, Internet resources, knows about different ways of obtaining information, orients himself in the use of digital services, uses services of an entertainment nature, does not have experience in using and creating various services and resources for education.

Beginner - observes safety techniques in handling digital means, knows about safe behavior in the digital space, adheres to the principles of academic integrity. Consciously uses digital services for collaboration in the process of educational activity. Uses Internet resources to acquire knowledge and skills. Uses digital educational resources offered by the teacher in class. Creates text and digital presentations for educational purposes. Stores and organizes digital educational resources for personal use. Uses digital services in the organization of artistic and productive activities with preschool children (visualization,

Table 1. Criteria and indicators for assessing the level of formation of students' ICT competence in the process of mastering the practical discipline «Art needlework»

Criteria	Indicators
Digital literacy	understanding the importance of safe behavior in the digital space; understanding the role of digital tools and resources in educational activities; avoiding risks to health (physical and psychological) in the digital space; assessment of data reliability and reliability of digital sources and resources; distribution and joint use of digital educational resources using links or attachments, letters, joint documents; creation of electronic documents (text and multimedia) for training; creation and collective use of digital educational resources
The use of digital tools and applications for educational activities	use, editing, modification, combination of digital resources and services in accordance with the educational goal and task; independent monitoring and analysis of effective digital technologies that can be used for training; use of digital educational environments; creation of text and multimedia educational products
The use of digital tools for organizing interaction with children	selection of digital resources necessary for educational interaction with parents and children, taking into account the purpose, conditions, age and needs of pupils; evaluation of their effectiveness in interaction with preschool and early-age children; creation of digital educational resources (clips of cartoons, presentations, games, tasks), use of cartoons together with children; selection of educational content for children in accordance with the didactic requirements of video content, digital environment, augmented reality for the development of children's artistic and creative competence; using digital services to create interest and present interesting material to engage children in artistic and productive activities

selection and use of ready-made educational resources). Uses digital services to create simple educational products (presentation, video).

Integrator – uses digital services to jointly create new digital educational resources. Demonstrates independence in finding learning sources for the development of ICT competence. Confidently uses digital services for own educational activities. Exchanges educational materials with classmates. Methodically competently uses digital technologies in interaction with pupils and parents. Creates educational products using various digital tools and services. Effectively uses self-created digital educational resources. Combines and adapts digital services for use in joint activities with children of preschool age. The study of students' readiness to use digital tools in educational and professional activities shows that 78.78% of respondents feel the need to improve their own digital competence level, 84.84% of respondents noted that learning requires mastering (acquiring skills) of various digital services. 91% of students want to acquire the ability to use digital services to organize the artistic and productive activities of preschool children.

At this stage of the experiment, we offered students to perform practical tasks on creating educational content (cartoons, games, video tasks) using DT. Students were unable to complete the proposed tasks (100%), which indicates a lack of knowledge and skills in using DT.

The analysis of the results of the answers to the questions made it possible to determine the level of formation of the ICT competence of the education seekers at the ascertainment stage of the research. 86 respondents (65.15% of those interviewed) were assigned to the elementary level, 34.84% of the students were elementary level.

The study of students' opinions regarding the forms of studying the discipline in distance and online formats shows that 84.84% of respondents consider online learning the most effective (lectures and practical classes are carried out in real time with the presence of a teacher), 20 respondents, i.e. 15.16%, consider it the most convenient distance format is considered for studying the discipline. This distribution of opinions in favor of online learning was argued by the respondents as follows: the information provided by the lecturer is easier to understand, there is an opportunity to clarify unclear things, in the discussion process it is possible to express one's own opinion and hear the opinion of fellow students, there is an opportunity to present one's practical work, to ask for advice on its implementation in if necessary. Also, among the advantages of synchronous online learning, students mention significantly less tiredness, the ability to collectively interact, communicate, and receive emotional contact, which increases the level of productivity and interest in the process of joint activities, and thus contributes to the memorization of information. Respondents were not asked the question of the desire to master the discipline in an offline format, due to the security situation in Ukraine and the impossibility of providing 100% security guarantees for students.

The analysis of the results of the tasks and the survey allowed us to come to the conclusion that distance tasks require concentration of attention, rational distribution of time and significant self-discipline from students, therefore this format is much more difficult. According to the respondents, the use of various digital services in the organization of educational interaction facilitates the perception of information in the distance format of obtaining education.

2.1   Results of the Formative Stage

The formation of professional competences in the organization of artistic and productive activities with children of early and preschool age in the process of studying the discipline «Art needlework» was combined with the acquisition by ICT students of competences related to future professional activities and was carried out in several directions: formation of digital literacy; development of skills in the use of digital tools and applications for educational activities; using digital tools to organize interaction with preschool children.

At the formative stage of the study, the success of tasks was systematically monitored and students surveyed about the effectiveness of using various digital services in the process of studying the discipline, with the aim of preventing difficulties and optimizing educational activities (https://elibrary.kubg.edu.ua/id/eprint/27905/1/digital%20comp%20teacher%20Morze.pdf).

Based on the results of the analysis of scientific sources on the problem of the application of DT in the educational field and the results of a diagnostic study, we have developed a model of the use of digital services in educational interaction with students in the discipline «Art needlework» (Fig. 1).

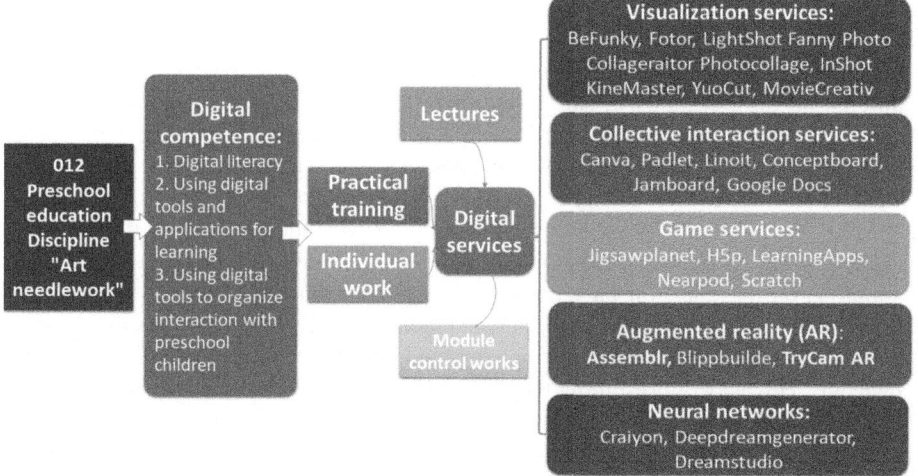

Fig. 1. The use of digital services in educational interaction with students in the discipline «Art needlework»

In our opinion, online training organized with the help of various video communication services is more rational, efficient and effective. In the course of work, the most convenient to use were Google Meet and Webex Meeting. These services were used not only for conducting lectures, but also for practical classes. The execution of products, the use of techniques was demonstrated with the help of a vertical camera attached to a tripod. This method is very convenient, because it happens in real time, there is an opportunity to work with the audience, answer questions and demonstrate work with the product.

At the first stage of mastering the discipline in the online format, considerable attention is paid to the formation of skills in using the resources of the electronic training course. Formation of skills to work synchronously in online services, completing educational tasks, to stick to the specified time, to rationally distribute responsibilities, to navigate in information resources, to communicate in a group and with a teacher, to draw up the results of completing tasks properly. For full-time students, this stage passes quickly enough, already during the second use of the service, the interaction is more harmonious and productive. Part-time students experience difficulties much longer. The results of the survey indicate that 80% of students experienced significant difficulties in using the Moodle platform and digital services. They need a detailed explanation and spend much more time during the couple on using services and working with electronic resources.

For coordinated and effective work, students are recommended to use a personal computer and a smartphone, as well as, taking into account the specifics of the discipline, a holder for a camera or phone. The use of these gadgets allows you to communicate and interact in class at the same time. Elements of gamification, augmented reality, and interactive interaction services are used in classes for the formation of relevant professional competences in the students of education, in the process of remote mastering of practical disciplines.

Interactive presentations, quizzes, quests, educational games were used in the process of studying theoretical material at online lectures. For example, at the lecture «Using natural and residual material in construction and modeling activities with preschool children», all the questions of the educational session were presented in the form of a quiz developed in the Kahoot service. When answering the questions, all students had the opportunity to express their own opinion, see the answers on the infographic, evaluate the correctness or falsity of the answer, discuss and get explanations from the lecturer on the issues under consideration. In this way, students: analyze the question, give an answer to it based on their own knowledge and experience, see the distribution of opinions on the question on the graph, hear the arguments of others, defend their own opinion, get closer to understanding the essence of the problem and its importance, understand the gaps in their own knowledge and have the opportunity to expand their knowledge. This way of presenting the material, namely, interest in the question, before its consideration and explanation, helps to deepen the interest and attention of students to the topic of the lecture. Kahoot, H5P, Poll-maker, Ranked List services were used to organize the quizzes.

Creating interactive elements of lectures with the help of Interacty and Genially services allows you to keep the attention of the audience by including them in various activities related to the subject of the lesson, reduces fatigue, maintains interest, and enriches the experience of students in using digital learning tools. The process of presenting theoretical material allows you to visualize what has been said, supplementing it with examples using various tools. For example, tools in the service Interacty Then&Now allows you to demonstrate children's products made for the first time and after 3–4 times of practice.

The Memory game offers to combine two identical pictures, helps memorization. The pictures contain images that illustrate the main positions of the practical lesson, for example, «Safety techniques in the art work lesson for preschool children». Among the problems that students have in the process of preparing for the online class in the discipline «Art needlework» is the lack of a complete list of materials and tools for the class. In the face-to-face format of the meeting, students quickly solve such problems by sharing among themselves. However, improper preparation for practical classes in synchronous online mode does not allow the student to focus on practical activities, which leads to lagging behind the pace of work in the group, inattention and difficulties in completing the educational task. Such situations arise due to the inattentiveness of the applicants in the process of preparing for the lesson, therefore, at the end of the presentation, the students are offered a list of the necessary materials, as well as a Memory game, which contains the materials needed for the next pair in the images. They have the opportunity to play it while preparing for the next match. The survey shows that all students like this game, although students don't always play it, sometimes they use a normal list of materials.

«Bank of ideas» are created during the discussion in practical classes, and Jamboard, Padlet, Linoit, and Conceptboard are used to activate learning activities. At the beginning of the pair, while discussing the main theoretical questions of the lesson, students are invited to look at thematic pictures on the Internet for a few minutes and choose the most interesting ideas, place them on the board for further analysis. For example, during a practical lesson, students are asked to search for products made of natural material, review the options offered by the Internet and choose the one that will correspond to the older, middle and younger age groups of the kindergarten in terms of material requirements, aesthetics, and level of manufacturing complexity. Students mix the selected images on Jamboard pages, creating an «Bank of ideas». The selected pictures are discussed in the group, the proposed ideas are analyzed according to the criteria, and then the products are made according to the task, taking into account the discussed features of the product (Fig. 2).

An interesting tool for creating a discussion and checking the level of students' preparation for practical classes is the use of neural networks. During practical classes, we used services that quickly generate images Craiyon, Deepdreamgenerator, Dreamstudio. Each student is invited to generate his own image based on a reference word, for example «origami» and with the addition of concepts related to the topic. The students evaluate and discuss the generated images from the point of view of the features of the product in the specified technique, share interesting images, and give suggestions for improvement.

The acquisition by future teachers of the necessary practical skills and skills of manufacturing and artistic design of products, technological methods of processing various materials, working in various techniques requires practical training. Therefore, while making products, students record each stage of work by creating a technological map, video clip or slide show with the help of various services.

Students publish finished technological maps on online boards, with the aim of sharing experiences, successes and shortcomings, as well as evaluate each other's work, commenting and giving advice. The exchange of experience in the process of collective interaction, the analysis of successes and failures is an important component of educational activity, it allows you to easily remember the main theoretical propositions, supporting them with a direct example (Fig. 3).

Fig. 2. «Bank of ideas» was created during a practical session by students using the Jamboard service

Fig. 3. Technological map of the product

To form students' ability to use educational resources, technically prepare works in various services, and collectively interact online, we used the most simple, easy-to-use and easy-to-use digital services. This is due to the fact that in one practical lesson, educational goals combine the formation of several professional competencies, and therefore the time allocated for each activity is clearly planned and executed.

Having formed the basic skills of digital literacy in students, we deepen their ability to effectively use various digital services to create new educational resources necessary for future professional activities. The creation of a digital pedagogical portfolio is an important element of students' acquisition of professional competencies as a result of studying the discipline «Art needlework». It contains all the student's work, developments, notes, technological maps, schemes, descriptions, accumulated during the study of the discipline and are ready for use by students in pedagogical practice.

A technological map is an instruction with illustrations that demonstrates all stages of the product's execution (Fig. 4.). Methodical tips are added to it, commenting on the peculiarities of the product's manufacture at each stage. Services help to quickly and efficiently create technological maps of products, collages, photo instructions are Canva, BeFunky, Fotor, LightShot Fanny Photo Collageraitor, Photocollage. Among their various functionalities, such services help to attach a photo and create a text with an inscription or a small instruction. Ready-made templates significantly speed up the production of the technological map of the product, they can be made in just a few minutes.

Fig. 4. The technological map of the product was made in the Canva service

To check the accuracy and correctness of the instructions given in the technology map, students exchange them and make the proposed product. If deficiencies are found, the technological map is corrected. Students make simple technological maps, schemes and instructions for preschool children, since among the tasks of artistic construction there is a task - forming in preschool children the ability to understand graphic instructions and make products accordingly.

When studying the discipline, we give preference to visualization services, because it is impossible to replace the practical component with digital tools. With their help, practically completed tasks are recorded, as well as notes and blanks for the portfolio. For example, the topic «Construction from decorative wire with preschool children» includes the following tasks: get acquainted with the techniques of construction from wire, make 2 products for each age group, make a blank for manufacturing the product: according to the topic of choice (for example, «Crocodile»), choose a literary work, a

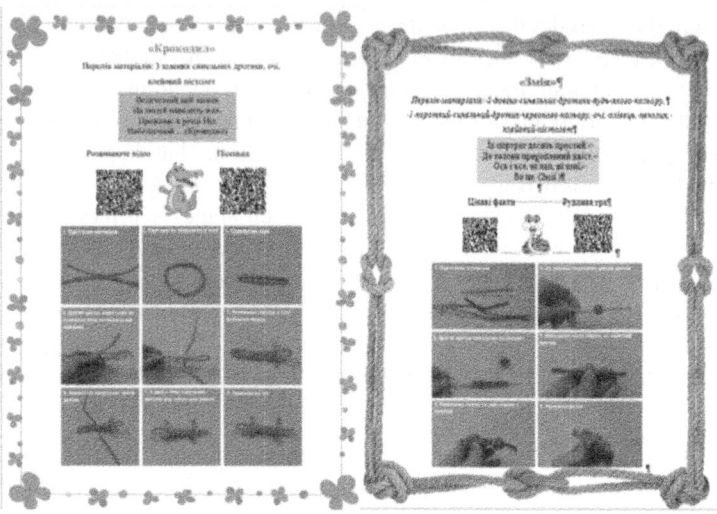

Fig. 5. Development of the manufacture of a product on the theme «Crocodile»

cartoon, a melody and place it in the form of a QR code, make a technological map of the product with instructions (service - at the student's choice), place the specified information on 1 page of A4 (Fig. 5).

The practical lesson «Design and decoration of a kindergarten group for festive events» (4 h) in the distance format consists in the execution of a subgroup project. From Internet resources, students select and model a sketch of a group room of a preschool education institution with holiday decoration elements (for example, «Spring Festival»). The sketch should contain products made by children, teachers together with children, and elements of decor, educational environment, made by teachers. In addition, students should think of forms of interaction with the parents of the pupils, with the aim of involving them in the production of interior items together with their own child. With the help of banks of images and clip-arts, students select decorative elements that should be used in the decoration of kindergarten groups, and students choose the service for making a sketch at their own discretion. The sketch is accompanied by instructions for the manufacture of each product proposed in the sketch with methodical recommendations.

Another task of forming ICT competence among future educators is the formation of the ability to create electronic content for children, as well as the use of digital services in the process of educational interaction together with preschool children. This task was implemented at the III stage of the study, as it requires the ability to freely use digital services in one's own educational activities. Having accumulated considerable experience in the use of digital tools, students are offered tasks aimed at forming their skills of educational interaction with preschool children.

Using image banks, students develop various projects, planning educational activities using digital tools. For example, creating and using interactive collages, photo layouts, schemes.

However, the use of images of one's own products is also interesting. Having processed photos of spring birds made using the origami technique and using augmented reality (AR) tools, for example, using the Blippbuilder application, the teacher can place them in a panoramic way and use them to maintain interest in the process of educational interaction with preschool children.

Module control works in the discipline «Art needlework» are creative projects that students perform independently or in a subgroup. For example, «Creating a cartoon for preschool children». There are various options for completing such a task, students independently choose the way to create a cartoon. It can be an individually completed task, a video recording of a fairy tale playing out with the help of self-made characters, a video clip from photos with sound overlay. It is interesting to implement such a project by a subgroup in the process of synchronous online work. In such a cartoon, digital images are combined, photos of manufactured products, for example, molded from clay and trimmed in Photoshop along the contour, recorded sound is superimposed. In future professional activities, students can create such a cartoon together with children and voice it.

At the stage of intermediate control to evaluate the effectiveness of the work performed, an evaluation and analysis of the effectiveness of the formative stage of the research was carried out. The evaluation included two blocks: 1) reflective analysis of impressions and achievements based on students' answers to questionnaire questions; 2) credit evaluation based on the performance of diagnostic tasks. The generalized results of the intermediate control of the study showed an increase in the level of formation of ICT competence among education seekers, 75% of respondents reached the level of integrator, 25% of respondents reached the level of beginner (Fig. 6).

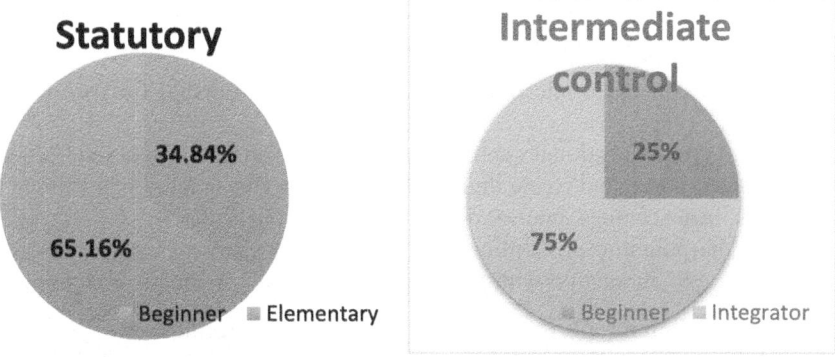

Fig. 6. Diagram of the levels of formation of ICT competence at the ascertainment stage and the stage of intermediate control of the research

3 Problems and Prospects

It is promising to further study the possibilities of using digital tools in the process of teaching students of the specialty 012 Preschool Education. Since the professional competence and ability of the teacher to organize effective educational interaction with the help of modern DT expands and diversifies the development opportunities of preschool children, since artistic activity is important for obtaining sensory experience and developing intellectual abilities of a person at the stage of preschool childhood. In the process of artistic work, children learn the properties of materials, accumulate constructive abilities and skills, develop thinking processes: analysis, generalization, isolation, combination, the ability to construct, understand and use schemes, solve various creative and logical tasks, develop creativity, the ability to construct and model products that contribute to the formation of the foundations of systemic, creative and constructive (engineering) thinking as opposed to superficial (clip) thinking. Another interesting aspect for further research is the study of the possibilities of using digital tools for the development of preschool children.

4 Conclusions

The issue of using digital tools for mastering practical disciplines in the remote format of training bachelors of preschool education is relevant and insufficiently studied. The conducted analysis of the study of the state of formation of students' ICT competence at the beginning of the course «Art needlework» revealed the problem of unformed digital competence of 1st year students (75% - elementary, 25% - beginners).

The organization of educational activities using DT in the process of studying the discipline «Art needlework» contributed to the formation of ICT competence among students. They mastered the system of knowledge, skills and abilities to:

- create, use, edit electronic documents (text and multimedia) for own educational activities;
- develop digital educational content in accordance with the educational program for preschool children and create digital products (cartoons, games, educational videos, collages, map schemes, interior decoration elements);
- analyze the potential possibilities of digital services, tools, Internet resources for creative use in interaction with preschool children and predict its effectiveness.

The use of digital tools for studying the course «Art needlework» ensures the student's transition from general digital competences to the acquisition of skills in the use of digital tools in educational and professional activities.

The results of the conducted research allow us to note that in the distance learning format for mastering practical disciplines of a creative nature, synchronous online learning has significant advantages. This is explained by the fact that students have the opportunity, with the help of various DT, to interact with the teacher and with each other, which improves the quality of the educational process. Digital services diversify the forms of providing material to students, allow them to demonstrate and jointly practice the technique of making products, support dialogue between all participants. New opportunities for creating, systematizing and storing educational content in the form of technological maps, designs, projects, videos, sketches are important, which allows you to create a pedagogical portfolio with case studies for future professional activities. For applicants who enter a higher educational institution, the formation of ICT competence remains relevant not only in the use of DT for the organization of educational artistic and productive interaction with preschool children, but also in general digital culture and the ability to use DT for one's own learning. The use of DT by a teacher, encouraging students to use services to perform educational tasks, allows to form the professional competence of the teacher, and also improves the ICT competence of future teachers. This is relevant because the educational program of specialty 012 Preschool education lacks separate disciplines aimed at the formation of ICT competence.

References

1. About the approval of the professional standard «Preschool teacher» [Pro zatverdzhennia profesiinoho standartu «Vykhovatel zakladu doshkilnoi osvity»]. https://mon.gov.ua/ua/npa/pro-zatverdzhennya-profesijnogo-standartu-vihovatel-zakladu-doshkilnoyi-osviti. Accessed 25 Apr 2023
2. Akinci, M.S.: Pre-school teaching students' opinions on distance education: preferences and emotional states. Cypriot J. Educ. Sci. **16**(6), 2901–2915 (2021)
3. Bielienka, A., Polovina, O., Kondratets, I., Shynkar, T., Brovko, K.: The use of ICT for training future teachers: an example of the course on «art education of preschool children». In: ICTERI 2021 ICT in Education, Research and Industrial Applications. Integration, Harmonization and Knowledge Transfer, vol. XX, pp. 361–370 (2021)
4. Bykov, V., Spirin, O., Pinchuk, O.: Modern tasks of digital transformation of education. UNESCO Chair J. Lifelong Prof. Educ. XXI Century (1), 27–36 (2020). https://doi.org/10.35387/ucj.1(1).2020.27-36
5. Buynytska, O., Vasylenko, S.: Corporate standard of digital competence of a university teacher. Electronic scientific publication. Open Educ. E-Environ. Mod. Univ. (12), 1–20 (2022). https://doi.org/10.28925/2414-0325.2022.121. Accessed 25 Apr 2023
6. Velychko, V.Y., Fedorenko, E.G., Soloviev, V.N., Dolins'ka, L.V.: Creation of open educational resources during educational practice by means of cloud technologies. In: CTE Workshop Proceedings, vol. 9, pp. 278–289 (2022)
7. Davidaviciene, V., Zvirble, V., Daveiko, J.: Conducting creative workshops in the process of distance learning at higher education institutions. In: EDULEARN21 Proceedings, pp. 4477–4488 (2021)
8. Jalongo, M.R.: The effects of COVID-19 on early childhood education and care: Research and resources for children, families, teachers, and teacher educators. Early Childhood Educ. J. **49**(5), 763–774 (2021)
9. Kim, J.: Learning and teaching online during Covid-19: experiences of student teachers in an early childhood education practicum. Int. J. Early Childhood **52**(2), 145–158 (2020)
10. Kovalenko, O., Litichenko, O., Sopova, D.: ICT technologies in cooperation with parents of children in the first link of the education system: interaction for the sake of the child. In: Intellektuelles Kapital - die Grundlage für innovative Entwicklung: monograph. ScientificWorld-NetAkhatAV, Karlsruhe, Germany, pp. 24–37 (2021)
11. Kozlitin, D., Kochmar, D., Krystopchuk, T., Kozak, L.: The application of augmented reality in education and development of students cognitive activity. In: Proceedings of the 17th International Conference on ICT in Education, Research and Industrial Applications. Integration,

Harmonization and Knowledge Transfer. Volume I: Main Conference, PhD Symposium, and Posters, Kherson, Ukraine, 28 September-October 2021, pp. 345–352 (2021)

12. Kozlitin, D., Kochmar, D., Krystopchuk, T., Kozak, L.: Future educators' training for project activities using digital technologies. In: Proceedings of the PhD Symposium at ICT in Education, Research, and Industrial Applications co-located with 16th International Conference «ICT in Education, Research, and Industrial Applications 2020» (ICTERI 2020), pp. 31–41 (2020). http://www.scopus.com/inward/record.url?partnerID=HzOxMe3b&scp=850987 40352&origin=inward. Accessed 25 Apr 2023

13. Korobeinikova, T.I., et al.: Google cloud services as a way to enhance learning and teaching at university. In: CTE Workshop Proceedings, vol. 7, pp.106–118 (2020)

14. Litichenko, O., Harashchenko, L.: Preparation of bachelors of preschool education for artistic and creative activity in preschool educational institutions. Innov. Pedag. (28), 176–179 (2020)

15. Morze, N., Kommers, P., Smyrnova-Trybulska, E., Pavlova, T.: Using effective and adequate IT tools for developing teachers' skills. Int. J. Continuing Eng. Educ. Life-Long Learn. **27**(3), 219–245 (2017)

16. Morze, N.V., Vasylenko, S.V.: Report 1. Innovative learning and best practices: Ukrainian universities. The open educational e-environment of a modern university: special issue «Pedagogical higher education of Ukraine: analysis and research», pp. 1–68 (2020). https://ope nedu.kubg.edu.ua/journal/index.php/openedu/article/view/337. Accessed 25 Apr 2023

17. Morze, N.V., et al.: Description of the digital competence of the pedagogical worker. The open educational e-environment of a modern university: special issue, pp. 1–53 (2019). https://elibrary.kubg.edu.ua/id/eprint/27905/1/digital%20comp%20teacher%20M orze.pdf. Accessed 25 Apr 2023

18. Nezhyva, L.L., Palamar, S.P., Marienko, M.V.: Clouds of words as a didactic tool in literary education of primary school children. In: CTE Workshop Proceedings, vol. 9, pp. 381–393 (2022)

19. Ng, W.: New digital technology in education: Conceptualizing professional learning for educators (2015).https://doi.org/10.1007/978-3-319-05822-1, https://www.researchgate.net/ publication/283781901_New_digital_technology_in_education_Conceptualizing_professio nal_learning_for_educators. Accessed 25 Apr 2023

20. Pérez-Jorge, D., Rodríguez-Jiménez, M.D.C., Ariño-Mateo, E., Barragán-Medero, F.: The effect of covid-19 in university tutoring models. Sustainability **12**(20), 8631 (2020)

21. Proshkin, V.V., Drushlyak, M.G., Kharchenko, S.Y., Semenikhina, O.V.: Methodology of formation of modeling skills through GeoGebra cloud service based on a constructive approach. In: CEUR Workshop Proceedings (2879), pp. 458–472 (2020)

22. Palamar, S.P., Bielienka, G.V., Ponomarenko, T.O., Kozak, L.V., Nezhyva, L.L., Voznyak, A.: Formation of readiness of future teachers to use augmented reality in the educational process of preschool and primary education. In: AREdu 2021 Augmented Reality in Education 2021 Proceedings of the 4th International Workshop on Augmented Reality in Education (AREdu 2021) (2898), pp. 334–350 (2021)

23. Yaroshenko, O.G., Samborska, O.D., Kiv, A.E.: An integrated approach to digital training of prospective primary school teachers. In: CTE Workshop Proceedings, vol. 7, pp. 94–105 (2020)

A Bot-Based Self-report Diagnostic Tool to Assess Post-traumatic Stress Disorder

Vira Liubchenko[1,2][✉] (iD), Nataliia Komleva[1] (iD), and Svitlana Zinovatna[1] (iD)

[1] Odesa Polytechnic National University, 1 Shevchenko av., Odesa 65044, Ukraine
{lvv,komleva,zinovatnaya.svetlana}@op.edu.ua
[2] Fakultät Life Sciences, Hochschule für Angewandte Wissenschaften Hamburg, Ulmenliet 20, 21033 Hamburg, Germany

Abstract. This paper discussed using a bot-based self-report diagnostic tool to identify post-traumatic stress disorder. The authors present information technology for assessing the psychological state of individuals who have experienced traumatic events in a stressful environment. The technology uses the International Trauma Questionnaire (ITQ) and additional questions describing the current psychological environment. The study employed data analysis techniques to identify significant dependencies between respondents' answers to additional questions about the current environment and their ITQ scores. The bot interface provides a user-friendly platform for respondents to complete the questionnaire. The analytical system, which includes data collection, storage, and processing, allows for flexibility in modifying the questionnaire based on ongoing research.

Keywords: Post-Traumatic Stress Disorder · International Trauma Questionnaire · Bot · Data Analysis · Database Structure

1 Introduction

The full-scale invasion has caused irreparable damage to the people and economy of Ukraine. Observable damage includes dead and maimed people, destroyed infrastructure, mined areas, etc. But equally threatening is the unobservable damage, particularly the impact on people's mental health.

According to current estimates, 30% of Ukraine's population suffers from post-traumatic stress disorder (PTSD). This figure will increase over time. Stress affects health, and there is already talk of an explosive increase in heart attacks, strokes, and cancer.

PTSD is a mental disorder that can develop after exposure to exceptionally threatening or terrifying events. PTSD can occur after a single traumatic event or because of prolonged exposure to trauma. Individuals cannot recover from exposure to trauma [1].

In [2], it was explained how PTSD was first recognized and defined in DSM-III after pressure to acknowledge the psychological effects of war on Vietnam veterans and concentration camp survivors. Diagnostic categories are commonly used to study and treat mental illness, helping researchers identify those with the disorder based on specific

© The Author(s), under exclusive license to Springer Nature Switzerland AG 2023
G. Antoniou et al. (Eds.): ICTERI 2023, CCIS 1980, pp. 189–200, 2023.
https://doi.org/10.1007/978-3-031-48325-7_14

symptoms. PTSD includes four clusters of symptoms: intrusive memories, avoidance of trauma-related triggers, adverse changes in mood and thoughts, and changes in reactivity. The DSM-5 added new criteria for changes in mood and cognition seen in PTSD. Assessing trauma and correctly attributing symptoms to a specific event is essential, but multiple factors can influence response and recovery. Military personnel may experience trauma from combat duties or be in dangerous conditions, while civilians can be traumatized by witnessing others' suffering. However, no clear rules exist for determining when an event causes injury [3].

Unfortunately, people hide psychological problems and do not seek medical help. This is why we need a point-of-care diagnostic tool that is always at hand and against which there is no prejudice. The solution could be to use a bot.

As the ground for the bot, we propose to use the International Trauma Questionnaire (ITQ), which is a self-report diagnostic measure of PTSD and complex PTSD (CPTSD) [4]. However, the ITQ research was conducted under "stable" conditions. The current conditions of Ukrainians cannot be classified as stable. On the contrary, a challenging, stressful situation can lead to a disorder without a single traumatic event. Therefore, there is a need to investigate the properties of the ITQ for people on the home front living under stressful conditions but not sharply traumatized.

We formulated a working hypothesis: the ITQ retains the properties of PTSD recognition in the stressful environment of the home front of military conflict. The hypothesis can be tested using data analysis techniques.

The work aimed to propose an information technology approach for investigating the sensitivity of the ITQ in diagnosing PTSD and to conclude whether the questionnaire should be used to momentarily diagnose a person's mental condition in a stressful environment on the home front.

The paper is structured as follows. Section 2 provides an overview of the critical research on which our work is based. In Sect. 3, we described the information technology of ITQ properties studying. Section 4 describes the application of the proposed information technology. Finally, the general conclusions of the work are collected in Sect. 5.

2 Related Works

An inevitable outcome of information technology development became the development of medical software applications. The software for diseases diagnosing or predicting changes in patient's conditions has particular importance. Such software's results can directly impact a person's life.

The diagnostics are based on the study of the person's condition indicators. The primary task is to define a set of diagnosis indicators and decide whether to analyze the indicators statistically or dynamically.

In [5] overviewed such a category of software for diagnosis and prognosis of the patient as the various systems for EWS (Early Warning Scores), e.g., the Royal College of Physicians National Early Warning Scale (NEWS), Modified Early Warning System (MEWS), supplemented by other assessments, Pediatric Early Warning System. (PEWS).

Let us take the NEWS system, developed by researchers at the University of Portsmouth, as an example of software for predicting changes in a patient's condition—the development aimed to recognize a patient's deteriorating condition early to respond to it. The early warning assessment initiates a formal evaluation by the responsible clinician. NEWS is based on a fixed number of parameters and gives a cumulative score within certain limits. Patients are categorized into risk groups depending on the value of the cumulative score [5].

The system has evolved over a long period. The system processes' scheme and development are shown in Fig. 1, where *TS* describes the set of indicator values *PV* for patient *pt* at time *t*, and *pd* is the measurement period. The number and composition of the indicators remain unchanged. In the initial version [6], the integrated assessment was calculated without considering the disturbances in the values of the individual extreme indicators. In [7], the methodology was corrected to reduce additional work for the bedside nurse and treating physician, which was disproportionate to the benefit of more frequent detection of adverse outcomes.

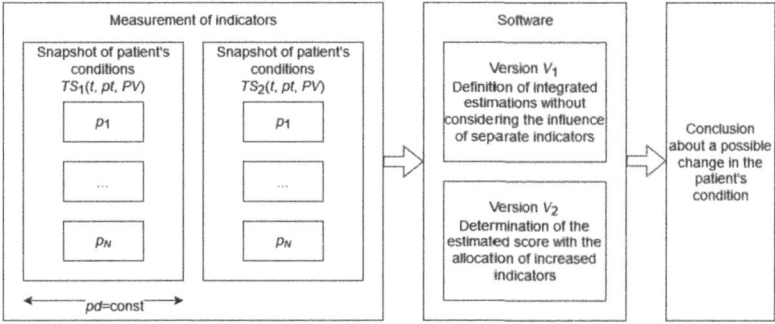

Fig. 1. Schema of NEWS system

A group of authors investigated the extent to which the results of commonly measured laboratory tests collected shortly after admission can be used to distinguish between in-hospital mortality [8].

The researchers also conducted a study applying this approach to surgical versus therapeutic patients. They showed that the developed EWS differentiated deterioration in surgical patients at least as much as in medicinal patients [9].

In [10], the application of the developed system was investigated in COVID-19 patients. It showed that adding new covariates or changing the weight of existing parameters is unnecessary when assessing patients with COVID-19.

The features of the considered systems are listed below:

- patients were hospitalized;
- measurements of parameters were made using appropriate tools and are objective;
- the frequency of measurements was guaranteed.

The issue may be caused by the short measurement period, which does not allow prediction before the patient's critical condition occurs.

Each mental illness has its characteristics. Diagnosis can be made using symptoms that are defined in different guidelines. In [11], the DSM-5 and ICD-11 (International Classification of Diseases) were compared. There was shown that different approaches to symptom management are used, "the DSM-5 encompassing a broad definition, and the ICD-11 instead proposing a narrow PTSD construct and introducing the new diagnosis complex PTSD (CPTSD), comprising PTSD in conjunction with ancillary symptoms." The DSM-5 diagnosis of PTSD consists of twenty symptoms, grouped into four groups: re-experiencing, avoidance, negative changes in cognition and mood, and changes in arousal and reactivity. The ICD-11 describes PTSD with six symptoms grouped into three groups: re-experiencing, avoidance, and heightened feelings of threat. Tools for assessing PTSD are highlighted in [12]: the PCL-5 post-traumatic stress checklist and the International Trauma Interview (ITI).

The PCL-5 is a 20-item self-assessment that assesses 20 symptoms of PTSD according to the DSM-5. The goals of the PCL-5 are to monitor symptom change during and after treatment, to screen people for PTSD, and to make a preliminary diagnosis of PTSD [13, 14].

In [2], there was estimated that 636,120 combinations of symptoms are possible for the DSM-5. As a result, it could be that DSM diagnoses do not include people for whom the "right" combination of symptoms is missing or those for whom the diagnosis does not fit. The conclusion is that "the desire for a diagnosis to encompass the broad spectrum of posttraumatic presentations led to a diagnosis in DSM-5 that was complex, even compared with other DSM disorders".

The International Trauma Interview (ITI) with Physician Assessment and the International Trauma Questionnaire (ITQ) were designed according to general ICD-11 principles; that is, a limited number of core characteristics (12 symptom indicators plus indicators of functional impairment) are used to identify the disorder [12, 15]. The ITQ is a self-report measure of diagnostic PTSD and complex PTSD (CPTSD) [4].

In [16], the need to simultaneously consider the type of trauma and severity of PTSD symptoms when studying emotion regulation in trauma survivors was investigated. The relationship between PTSD and difficulties in emotion regulation was confirmed.

Studies show that the language of implementation of surveys initially written in English and translated into 25 languages does not affect diagnostic results (e.g., [12] for Swedish, [17] for Chinese, [18] for Dari).

In [19], the application of machine learning (ML) and the evaluation of induction modeling approaches are reviewed. Hypotheses are considered as to whether ML can be used to see a statistical correlation between observed symptoms and PTSD exacerbation, whether the relevance of early symptoms used by medical professionals to predict PTSD, whether the period needed to predict PTSD exacerbation can be determined, and whether ML-induced models can be used to predict PTSD exacerbation. All this when considering the situation one month after injury: the authors showed that using ML methods unambiguously brings results for all hypotheses. Also, an essential finding of the study is that surveys using smartphones to self-report symptoms can be simplified. However, the features of the study are that a rigid period of one month after an injury is set and that people with an already confirmed diagnosis participated in the study.

The outcome of the published work analysis was the hypothesis that the ITQ translated into Ukrainian could be used to diagnose PTSD in the current context. However, it makes sense to examine the influence of external factors on the components of the integral assessment under ITQ.

3   Information Technology of Significant Dependencies Identification

The proposed information technology is based on the use of an ITQ. We must consider that the ITQ properties were investigated in a stable environment. How the stressful environment affects the findings of the questionnaire as a whole and the values of individual indicators, have not been studied before. Information technology, therefore, should analyze the properties of ITQ under continuous stress conditions.

3.1   Identification of Significant Dependencies

The ITQ is a brief, simply worded measure of the core features of PTSD, which is consistent with maximizing clinical utility and international applicability through a focus on a limited but central set of symptoms [4]. In our research, we used an 18-item version of ITQ. Two-factor Second-Order Model of ITQ, which is used in the study, includes two second-order latent symptoms of PTSD, explaining the covariation between reexperiencing (RE), avoidance (AV), and perceived threat (TH), and DSO (Disturbances in Self-Organization), explaining covariation between affective dysregulation (AD), negative self-concept (NSC) and relationship disturbances (DR).

The analysis of the Second-Order Model supports two types of results. The first one is binary and is based on the Boolean logic. For this purpose, each response Q_i is labeled as True or False based on the condition $Q_i \geq 2$. Disjunctions of respective question labels provide the labels for symptoms, and conjunctions of respective symptom labels provide the labels for PTSD and DSO diagnosis. The second one is numerical and is based on dimensional scoring. Scores are calculated for each symptom and summed in clusters to produce PTSD and DSO scores.

Published studies of the ITQ properties had been conducted on the condition that the inclusion criteria for the sample selection contained screened positive for at least one lifetime traumatic event. Usually, the list of traumatic events is provided by Life Events Checklist, which has only one item, "Combat or exposure to a war zone (in the military or as a civilian)", connected with war conditions.

In the current situation, describing the environment in detail is advisable. Practicing psychologists have been brought in to identify questions that might reflect a military conflict environment. This question set was added after the primary ITQ questions. The basic assumption for identifying significant dependencies was that a significant factor splits the responses set into classes with considerable differences.

The input data for the study are the answers to the questions in the questionnaire. The first block of responses corresponded to the items in the ITQ questionnaire; the second block of responses corresponded to questions about the factors whose influence is being studied. Figure 2 shows the scheme of the multi-block questionnaire using.

Let us describe the proposed technology for identifying significant dependencies.

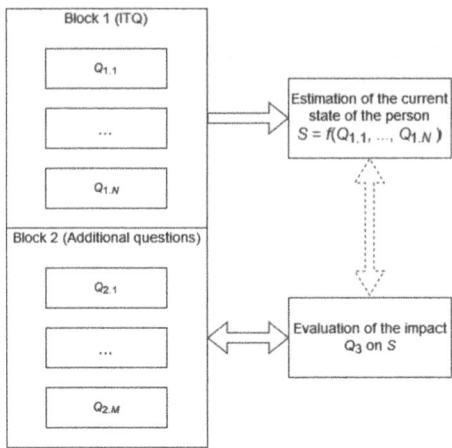

Fig. 2. Scheme of using the multi-block questionnaire

1. Move from answers to questions $Q_{1,1}$–$Q_{1,N}$ to integral estimates of indicators PTSD, RE, AV, TH, DSO, AD, NSC, and DR. In other words, object descriptions are moved from 18 features space to eight features space.
2. For each question from the set $Q_{2,1}$–$Q_{2,M}$
   2.1. Classify the objects represented by integral features by the variants of answers to the related question.
   2.2. Calculate the percentage distribution of the responses corresponding to high and low anxiety.
3. Analysis of the consistency of percentage distributions with ITQ results (integral estimations of PTSD and DSO).

Suppose the results of the analysis at step 3 agree with the results of the ITQ processing. In that case, it can be assumed that the ITQ is applicable for analyzing the condition of people in a stressful home front environment.

Figure 3 shows the activity diagram for the work process for information technology.

The work process begins with the experts, who supplement the ITQ questions with questions about the current environment to form the questionnaire. The questionnaire is then handed over to the data analysts, who prepare the software tool and data structures to efficiently handle the raw survey results. A link to the bot is sent out to potential respondents. After the respondent finishes answering the questionnaire, the bot displays an instant diagnostic result obtained with a standard algorithm for the ITQ. The data analysts prepare the collected data and aggregate the results for the questions in block 3. The results of the calculations are transmitted to the experts for review and semantic interpretation.

3.2 Bot Description

A survey bot is an automated tool that engages in interactive conversations with survey respondents to collect data and feedback.

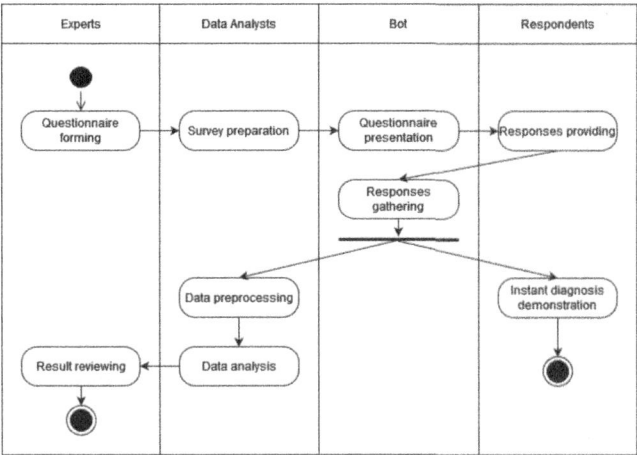

Fig. 3. Activity diagram for work process

To provide a user-friendly interface, the bot should have a clean and intuitive interface that guides respondents through the survey process. The bot follows a logical conversation flow, presenting questions to respondents with respect to the questionnaire's order. As the base of the questionnaire, we used the Ukrainian translation of the 18-item version of ITQ. Figure 4 shows the screenshots of the questionnaire.

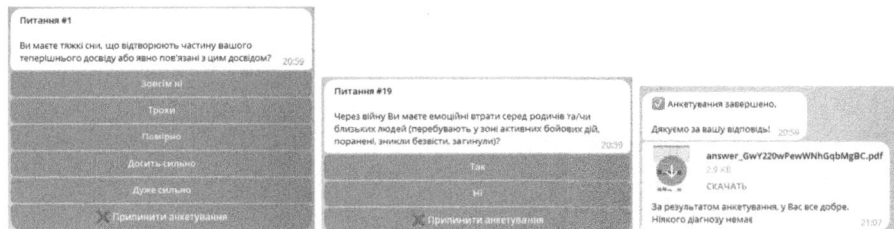

Fig. 4. Screenshots of the bot interface

The bot supports only the multiple-choice question type. Responses to the questions on the ITQ can be unambiguously translated into the Likert scale. The bot's algorithm was constructed so the participant could interrupt the interview at any time. This feature protects against the risk of discomfort for participants when asking about potentially traumatic events and psychological symptoms [11].

We focus attention on the sensitivity of gathered responses. So, the bot complies with relevant data protection regulations, ensures secure data transmission and storage, and provides clear information on how the collected data will be used and protected at the first run.

The developed bot is encapsulated in the analytical system.

3.3 Analytical System for Dependencies Identification

The development process of the analytical system can be represented as the set of activities shown in Fig. 5.

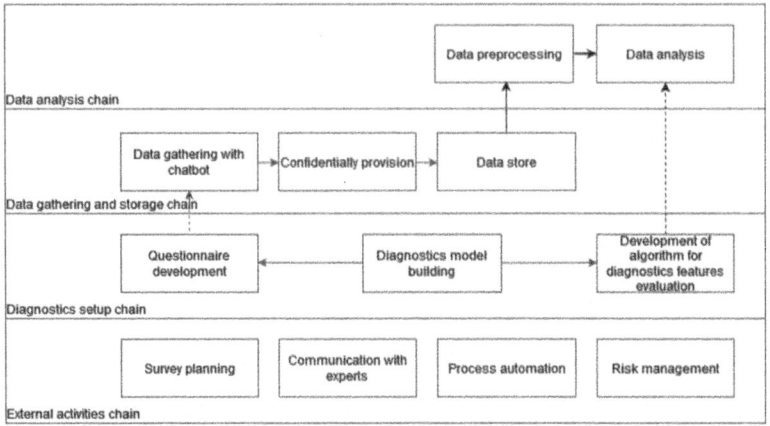

Fig. 5. Block diagram of the analytical system

The external activities chain includes survey planning, communication with experts, process automation, and risk management.

Survey planning consists of the following sub-activities:

- Defining the purpose of the survey: studying the psychological state of a person during the war, which is accompanied by intense stress and traumatic events for the population.
- Defining the target audience: people participating in the survey can be divided into separate categories (e.g., war veterans, those who were not directly affected by the military conflict but still experienced its impact, etc.). The civilians' category can be subdivided into subcategories: women, children, older adults, people with disabilities, etc.
- Setting up survey parameters: determining the time parameters, possible modes and deadlines for completing the survey, the number of questions, etc.

Communication with experts should include such sub-activities as:

- Finding professional psychologists who have a high level of knowledge, experience, and qualifications in the field of PTSD and can provide authoritative advice and recommendations on issues related to this field.
- Involvement of experts in the development of the questionnaire content.
- Involvement experts (if necessary) in consulting and decision-making based on the survey results.
- Ensuring effective communication implies clarity, mutual understanding, two-way interaction, trust, transparency, regular contacts, and reports to determine progress in resolving issues.

Automation of processes in PTSD diagnosis involves the use of various technologies and tools for automatic data collection and analysis, including the Telegram API, which reduces the time and cost of diagnosis, improves the accuracy and reliability of results, and increases the efficiency of resource use.

Risk management includes the identification, assessment, control, and minimization of possible risks associated with inadequate qualifications of experts, errors in the diagnostic methodology, errors in the structure of the questionnaire, algorithms for calculating diagnostic indicators, insufficient number of questions, and errors in the software.

Questionnaire development activity is closely connected with the development of database structure. Because we planned the questionnaire modification based on the result of the research, we should have provided flexibility in the data storage structure. We used the database structure, which made it possible to reconfigure the questionnaire "on the fly" and continue collecting data. For example, the fourth additional question in the first version of the questionnaire was phrased as "Do you hide during an air raid?" Then it was decided to change the question's wording and offer five possible answers.

The implemented analytical system was used to study the ITQ properties.

4 Case Study

The study was conducted on a non-clinical sample of 286 people between the ages of 16 and 60 who were in Ukraine at the start of the war. It was realized as an anonymous Internet survey using a Telegram bot. The link was distributed in a friend-to-friend way. We sent the link to the groups of students and refugees and colleagues and acquaintances. As well we asked our contacts to spread the link further. At the time of the survey, respondents were living both in the territory of Ukraine and abroad.

After removing the records with missing values, a final dataset consisted of $n = 212$ records. In the final set, 137 entries corresponded to respondents who manifested PTSD and disturbances in self-organization (DSO).

Initially, the number of questions in block 1 was $N = 18$, and in block 2 was $M = 9$. The data analysis showed that the personal feelings or circumstances revealed by the questions in block 2 correlated with the results determined by the standard algorithm for the ITQ, probably reinforcing the manifestations of PTSD.

We studied the relationship between the PTSD status identified using the standard algorithm for ITQ and the answers to questions in block 2. We can distinguish all responses in block 2 into two categories: K_1 is the category of responses for high anxiety, K_2 is the category of responses for low anxiety. We modeled the relevancy of the category responses as.

$K_j = \{Q_{2.1} <$"Answer1", "Answer2",$...>, ..., Q_{2.9} <$"Answer1", "Answer2",$...> \}$, where $Q_{2.i}$, $i = 1...9$ are the question numbers in block 2.

Figure 6 shows the percentage contributions of the two response categories.

As we can see, the answers to category K_1 are expressively dominant in all questions. As the survey results demonstrate the consistency of blocks 1 and 2, we can conclude that the ITQ is relevant when used in a stressful home front environment.

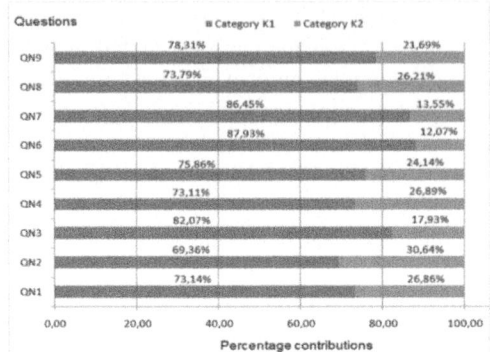

Fig. 6. Distribution of answers to questions in block 2 in the presence of PTSD

The list of questions in block two or options for answering individual questions could change as the Ukraine war situation changes. The proposed mechanism for obtaining raw data and subsequent calculations makes it possible to adjust the form of the questionnaire. The data obtained are helpful for psychologists to identify entry points for working with people who are potentially at risk of PTSD.

5 Discussion

Developing a bot as part of the proposed information technology for diagnosing PTSD holds significant potential. The bot provides a user-friendly interface that engages in interactive conversations with survey respondents, facilitating data collection and feedback. This section of the discussion focuses on the role and benefits of the bot in the diagnostic process.

One of the main advantages of using a bot is its accessibility and habituality. A bot is accessed through Telegram messenger, allowing respondents to conveniently engage with the diagnostic tool in a well-known environment. This is particularly important in diagnosing PTSD, as individuals may be hesitant or unable to seek traditional in-person medical help. The bot provides a non-judgmental and convenient avenue for individuals to assess their mental health and seek support.

Furthermore, the bot creates a comfortable space for respondents to share their experiences and symptoms. Many individuals may feel more at ease disclosing sensitive information to a bot than a healthcare professional in a face-to-face setting. The anonymity provided by the bot encourages respondents to be honest in their responses, leading to more accurate diagnostic results.

Additionally, the bot provides instant diagnostic results to respondents. This real-time feedback can be valuable for individuals seeking immediate insights into their mental health status. The instant results can also catalyze individuals to take proactive steps toward seeking appropriate treatment or support.

However, it is essential to note some limitations and considerations associated with using a bot for diagnosing PTSD. While the bot can provide valuable insights and initial diagnostic results, it should not replace a comprehensive assessment conducted by a qualified healthcare professional. The bot can serve as a screening tool to identify potential cases of PTSD, but a formal diagnosis and treatment plan should always be confirmed and provided by a trained professional.

6 Conclusion

The use of the bot in the messenger environment, familiar to many residents of Ukraine, is promising, as it allows to cover a large audience with surveys and supplement the standard questionnaire with questions that will enable considering the changes over time circumstances, tracking the dynamics of change in the mental mood of the person over time.

The ITQ is relevant in a stressful environment. It can be used to identify the psychological state of individuals who have been exposed to traumatic events. The proposed information technology makes it possible to adjust the questionnaire's list of questions and answers to reflect the current environment and obtain relevant data for diagnostics. The proposed method for assessing the psychological state of individuals can be seen as an effective way to help people suffering from PTSD and other mental disorders.

In the future, we can expand the bot with feedback to accept respondents' suggestions for changes to various additional questions. Also, over time, the data collected will allow us to predict the mental state of regular respondents.

Acknowledgements. This research was made possible through the UK-Ukraine R&I twinning grants scheme, funded by Research England with the support of Universities UK International and UK Research and Innovation.

References

1. Bisson, J., Cosgrove S., Lewis, C., Roberts, N.: Post-traumatic stress disorder. BMJ **351** (2015)
2. Galatzer-Levy, I.R., Bryant, R.A.: 636,120 ways to have posttraumatic stress disorder. Perspect. Psychol. Sci. **8**, 651–662 (2013)
3. Post Traumatic Stress Disorder (PTSD) and War-Related Stress, Veterans Affairs. https://www.veterans.gc.ca/eng/health-support/mental-health-and-wellness/understanding-mental-health/ptsd-warstress#Item3-1. Accessed 20 May 2023
4. Cloitre, M., et al.: The international trauma questionnaire: development of a self-report measure of ICD-11 PTSD and complex PTSD. Acta Psychiatr. Scand. **138**, 536–546 (2018)
5. Doyle, D.: Clinical early warning scores: new clinical tools in evolution. Open Anesthesia J. **12**, 26–33 (2018)
6. Prytherch, D.R., Smith, G.B., Schmidt, P.E., Featherstone, P.I.: ViEWS—towards a national early warning score for detecting adult inpatient deterioration. Clin. Pap. **81**, 932–937 (2010)
7. Jarvis, S., et al.: Aggregate national early warning score (NEWS) values are more important than high scores for a single vital signs parameter for discriminating the risk of adverse outcomes. Rapid Response Syst. **87**, 75–80 (2015)
8. Jarvis, S., et al.: Decision tree early warning scores based on common laboratory test results discriminate patients at risk of hospital mortality. https://pure.port.ac.uk/ws/portalfiles/portal/216265/LDT-EWS_poster_v5.pdf. Accessed 20 May 2023

9. Kovacs, C., et al.: Comparison of the national early warning score in non-elective medical and surgical patients. Br. J. Surg. **103**, 1385–1393 (2016)
10. Kostakis, I., Smith, G.B., Prytherch, D., Price, C., Chauhan, A.: The performance of the national early warning score and national early warning score 2 in hospitalised patients infected by the severe acute respiratory syndrome coronavirus 2 (SARS-CoV-2). Rapid Response Syst. **159**, 150–157 (2021)
11. Bondjers, K.: Post-traumatic stress disorder – assessment of current diagnostic definitions. Acta Universitatis Upsaliensis, Uppsala (2020)
12. Bondjers, K., Hyland, P., Roberts, N.P., Bisson, J.I., Willebrand, M., Arnberg, F.K.: Validation of a clinician-administered diagnostic measure of ICD-11 PTSD and complex PTSD: the international trauma interview in a Swedish sample. Eur. J. Psychotraumatol. **10**, 1665617 (2019)
13. Weathers, F.W., Litz, B.T., Keane, T.M., Palmieri, P.A., Marx, B.P., Schnurr, P.P.: The PTSD checklist for DSM-5 (PCL-5). https://www.ptsd.va.gov/professional/assessment/adult-sr/ptsd-checklist.asp. Accessed 20 May 2023
14. Blevins, C.A., Weathers, F.W., Davis, M.T., Witte, T.K., Domino, J.L.: The posttraumatic stress disorder checklist for DSM-5 (PCL-5): development and initial psychometric evaluation. J. Trauma. Stress **28**, 489–498 (2015)
15. Brewin, C.R., Cloitre, M., Hyland, P., et al.: A review of current evidence regarding the ICD-11 proposals for diagnosing PTSD and complex PTSD. Clin Psychol Rev **58**, 1–15 (2017)
16. Ehring, T., Quack, D.: Emotion regulation difficulties in trauma survivors: the role of trauma type and PTSD symptom severity. Behav. Ther. **41**(4), 587–598 (2010)
17. Ho, G.W.K., et al.: Translation and validation of the Chinese ICD-11 international trauma questionnaire (ITQ) for the assessment of posttraumatic stress disorder (PTSD) and complex PTSD (CPTSD). Eur. J. Psychotraumatol. **10**, 1–10 (2019)
18. Andisha, P., Shahab, M.J., Lueger-Schuster, B.: Translation and validation of the dari international trauma questionnaire (ITQ) in Afghan asylum seekers and refugees. Eur. J. Psychotraumatol. **14**, 1–12 (2023)
19. Wshah, S., Skalka, C., Price, M.: Predicting posttraumatic stress disorder risk: a machine learning approach. JMIR Ment. Health **6**(7), e13946 (2019)

Ph.D. Symposium

Stock Market Crashes as Phase Transitions

Andrii Bielinskyi[1,2,3(✉)] ⓘ, Vladimir Soloviev[1,3,4] ⓘ, Victoria Solovieva[2] ⓘ,
Andriy Matviychuk[3] ⓘ, Serhii Hushko[2] ⓘ, and Halyna Velykoivanenko[3] ⓘ

[1] Kryvyi Rih State Pedagogical University, Gagarin av. 54,
Kryvyi Rih 50086, Ukraine
[2] State University of Economics and Technology, Medychna St. 16,
Kryvyi Rih 50005, Ukraine
solovieva_vv@kneu.dp.ua, vice-rector@duet.edu.ua
[3] Kyiv National Economic University named after Vadym Hetman,
Peremogy av. 54/1, Kyiv 03057, Ukraine
bielinskyi@duet.edu.ua, editor@nfmte.com, ivanenko@kneu.edu.ua
[4] South Ukrainian National Pedagogical University named after K.D. Ushynsky,
Odesa, Ukraine
vnsoloviev@kdpu.edu.ua

Abstract. In this study, we apply the multifractal detrended fluctuation analysis concepts to financial time series that helps to determine the onset of a crash in the Dow Jones Industrial Average index. For the studied index we emphasize 4 the most influential stock market crashes: Wall Street Crash of 1929, Black Monday of 1987, Financial crisis of 2007–2008, and 2020 stock market crash. We present that economic crashes on a mapping with multifractality phenomena demonstrate a dynamic phase transition. Some of the presented multifractal measures appear to be analogues of free energy and specific heat, and can be used as indicators or indicators-precursors of the stock market crashes.

Keywords: stock crashes · phase transition · multifractal analysis · indicator-precursor · complex system

1 Introduction

The physics community has conducted extensive research on financial markets in general, and the underlying causes of market crashes in particular, in recent years. The motivation for studying market behavior arises from various factors, including the financial interests of investors and shareholders, as well as the severe societal and economic consequences of financial turbulence. The sudden and drastic collapse of complex systems is also of great interest to scientists, given its similarity to the collective rearrangement of physical systems at critical points. Consequently, econophysics has emerged as a new field of study [12,36, 39]. This interdisciplinary field employs a range of mathematical and physical tools, such as random matrix theory [31,40], clustering analysis [21], extreme and rare events [44], agent-based models [23], and network theory [35], among others, to address the complexities inherent in financial and economic systems.

© The Author(s), under exclusive license to Springer Nature Switzerland AG 2023
G. Antoniou et al. (Eds.): ICTERI 2023, CCIS 1980, pp. 203–214, 2023.
https://doi.org/10.1007/978-3-031-48325-7_15

Phase transitions refer to abrupt rearrangements in a system, triggered by a critical value of an external parameter such as temperature. At the critical point, the system exhibits scale invariance, resulting in a power-law behavior of the observables with critical exponents that can be analyzed using (multi-)fractal theory. To date, only two genuine sources of multifractality have been discovered [24]: (i) a wide (long or heavy-tailed) distribution in the system, and/or (ii) the existence of long-range dependence (such as nonlinear long-range correlations), resulting in a hierarchical arrangement of many scales. Additionally, it is widely believed that some stochastic or deterministic nonlinear combination of monofractals can generate multifractals [2, 34], with all of them capable of producing cascades that are at the heart of multifractality [15]. Multifractality is said to occur when fluctuations and/or dependencies emerge across a multitude of spatial and/or temporal scales according to various scaling laws, characterized by different scaling exponents, leading to a multiscaling phenomenon [27]. Additionally, those fluctuations in stock prices, returns, and foreign exchange rates, reveal substantial non-Gaussian deviations [42] associated with extreme occurrences and heavy-tailed statistical distributions.

The dynamics of financial markets remains a challenging area of investigation for both physicists and economists. Despite this ongoing research, the fundamental principles underlying stock market crashes and their worldwide dynamics are not yet fully comprehended by either discipline [43]. The renormalization group theory [16, 20, 22, 43] has suggested that such cooperative phenomena may stem from a critical phase transition. This proposition has motivated the exploration of constraints required to provide a fresh analytical perspective for understanding and describing such phase transitions.

Therefore, this article will be structured as follows: In the next section we will present a brief overview of the use of the theory of multifractals to identify phase transitions in both physical and economic systems. In Sect. 3, we will present a description of the procedure for the multifractal detrended fluctuation analysis and possible measures on its basis to identify crash phenomena as phase transitions. In Sect. 4, we will present the database and obtain empirical results based on the described methods. The conclusions will be presented in Sect. 5.

2   Literature Review

The identification of phase transitions using the multifractality paradigm is a rather relevant direction that has been taking place for more than a dozen years. One of the pioneering works belongs to Lee and Stanley [32]. According to their study, the multifractal spectrum of diffusion-limited aggregation exhibits evidence of a phase transition which was found with the "exact enumeration" approach. The results suggest the existence of a critical point, β_c, above which an infinite hierarchy of phases is observed, while below β_c, a single phase is found. The energy and specific heat exhibit singular behavior and fluctuations of all energy scales are observed at β_c. The maximum energy scales with system size

L as $E_{max}(L) \propto L^2/\ln L$. It was also found that for $\beta < \beta_c$, the partition function does not scale with L, which indicates the breakdown of the conventional moment expansion.

Then, according to the study conducted by Canessa [13], multifractal physics approach was applied to financial time series to identify the onset of crashes in the S&P500 stock index. The researcher found evidence of a dynamic phase transition in a simple economic model system based on a mapping with multifractality phenomena in random multiplicative processes. By applying former results obtained with a continuous probability theory for describing scaling measures, scientists was able to generate multifractal physics measures from a price equation derived from a nonlinear equilibrium model. The results indicated that the S&P500 price data displayed a shoulder to the right of the main peak as a function of time lags, resembling a classical phase transition at a critical point. The researcher explained this dynamic phase transition by a mapping with multifractality phenomena in random multiplicative processes. An analytical expression for the "analogous" specific heat $C(q)$ of the economic model system was derived.

Here [25], the authors thoroughly studied the statistical and thermodynamic properties of the anomalous multifractal structure of random interevent (or intertransaction) times, using the extended continuous-time random walk formalism. They applied this formalism to a financial market where heterogeneous agent activities can occur within a wide spectrum of time scales. They found the scaling or power-dependent form of the partition function, $Z(q')$, and its general exponent $\tau(q')$, which is the nonanalytic power of q' and is one of the pillars of higher-order phase transitions. Their most important finding is the third- and higher-order phase transitions, which can be interpreted as transitions between the phase where high frequency trading is most visible and the phase defined by low frequency trading. They used the pausing-time distribution as the central one, which takes the form of convolution and its integral kernel is given by the stretched exponential distribution. The specific order of the phase transition depends upon the shape exponent α defining the stretched exponential integral kernel. The authors also provide a practical hint for investors based on their findings.

In this study [27], the authors employed a modified multifractal detrended fluctuation analysis to investigate empirical time series of interevent or waiting times. They focused on the nonmonotonic behavior of the generalized Hurst exponent $h(q)$ and the consequent nonmonotonic behavior of the coarse Hölder exponent $\alpha(q)$ leading to multibranchedness of the spectrum of dimensions. They used the Legendre-Fenchel transform to reveal the thermodynamic consequences of the multibranched multifractality, which are expressed in the language of phase transitions between thermally stable, metastable, and unstable phases. These phase transitions are classified as of the first and second orders according to Mandelbrot's modified Ehrenfest classification. The authors considered the discovery of multibranchedness as a significant extension of multifractal analysis.

Xiao et al. [49] introduced a node-based fractal dimension and node-based multifractal analysis framework to uncover the generating rules and measure the scale-dependent topology and multifractal characteristics of dynamic complex networks. The proposed framework includes indicators for assessing the complexity, heterogeneity, and asymmetry of network structures, as well as the structural distance between networks. Through this approach, the authors gained new perspectives on the energy and phase transitions in networked systems and identified multiple generating mechanisms that govern network evolution.

Multifractal analysis has been used to investigate the presence of ramp-cliff patterns and their effect on the intermittency of scalar temperature fluctuations compared to longitudinal velocity fluctuations [18]. In a study of air temperature time series collected at a pine forest canopy top for different atmospheric stability regimes, the wavelet transform modulus maxima method was applied to show that the multifractal spectra exhibit a phase transition corresponding to the presence of strong singularities associated with sharp temperature drops or jumps. These singularities are found to be hierarchically distributed on a Cantor-like set under unstable or stable atmospheric conditions, and are suspected to enhance the internal intermittency of temperature fluctuations. However, no such phase transition is observed in the temperature multifractal spectra under near-neutral conditions, indicating that the statistical contribution of these singularities is not significant enough to account for the stronger intermittency of temperature fluctuations compared to longitudinal velocity fluctuations.

3   Multifractal Detrended Fluctuation Analysis

To investigate the scaling exponents of magnetization time series with different scaling behaviors, we utilize multifractal detrended fluctuation analysis (MF-DFA) [24], which is an extension of detrended fluctuation analysis (DFA) [38]. MF-DFA has been extensively employed in various fields including finance [45, 46], physiology [19], biology [17], traffic engineering [14], geophysics [47], and neuroscience [50] for characterizing the properties of non-stationary time series.

The MF-DFA method operates in the following way:

I Given a time series $\{x(i)\}, 1, ..., N$, we begin by integrating it to produce a profile $y(k) = \sum_{i=1}^{k} [x(i) - \langle x \rangle]$, where $\langle x \rangle$ is the average value of $\{x(i)\}$.

II Next, we divide the integrated series $y(k)$ into $l_s = \text{int}(l/s)$ non-overlapping segments of length s. For each of these l_s segments, we estimate the local trend by fitting a least-squares line and subtract it from $y(k)$ to remove the trend from the integrated series. Finally, we calculate the variance of each detrended segment, v:

$$F^2(v, s) = s^{-1} \sum_{i=1}^{s} \{y(v - 1 + i) - \hat{y}(v, i)\}^2 \tag{1}$$

in which $\hat{y}$ represents the fitted trend within segment $v = 1, ..., l_s$. Here, we utilize a first-order polynomial to fit the local trend.

III Proceeding from both the beginning and end of the time series, step (II) results in a total of $2l_s$ segments. We calculate the qth order fluctuation function by averaging over all segments:

$$F_q(s) = \left\{ (2l_s)^{-1} \sum_{v-1}^{2l_s} \left[F^2(v, s) \right]^{q/2} \right\}^{1/q}. \tag{2}$$

For a real number q, where q is positive, the fluctuation function $F_q(s)$ quantifies large fluctuations, whereas for negative q, it quantifies small fluctuations.

IV To obtain the fluctuation function $F_q(s)$ for different box sizes s, we perform the aforementioned calculation repeatedly. If $F_q(s)$ increases according to a power-law $F_q(s) \sim s^{h(q)}$, then the scaling exponents $h(q)$ (also known as generalized Hurst exponents) can be estimated by determining the slope of the linear regression of $\log F_q(s)$ versus $\log(s)$. The fluctuation parameters $h(q)$ depict the correlation structures of the time series at various magnitudes. The value of $h(0)$ cannot be determined using (2) due to the diverging exponent. The logarithmic averaging method should be utilized instead:

$$F_0(s) = \exp\left\{ (4l_s)^{-1} \sum_{v=1}^{2l_s} \ln \left[F^2(v, s) \right] \right\} \sim s^{h(0)}. \tag{3}$$

The multifractality of a time series can be measured by the generalized Hurst exponents $h(q)$ as a function of q. If the values of $h(q)$ are uniform for all q, the time series is monofractal. Conversely, if there is variation among $h(q)$ values, the time series is multifractal. The classical multifractal scaling exponents τ, defined using the standard partition function-based multifractal formalism, are directly linked to the generalized Hurst exponents $h(q)$ [1]:

$$\tau(q) = qh(q) - 1. \tag{4}$$

Utilizing the concepts of MF-DFA, we can gain fresh insight into viewing a time signal as a thermodynamic system. Within MF-DFA, the mass exponent $\tau(q)$, the Lipschitz-Hölder exponent α, the multifractal spectrum $f(\alpha)$, and the distortion exponent q can be seen as analogous to the free energy, energy, entropy, and temperature of a thermodynamic system, respectively. More specifically, the specific heat $C(q)$ can be defined as:

$$C(q) \equiv -\frac{\partial^2 \tau(q)}{\partial q^2} \approx \tau(q+1) - 2\tau(q) + \tau(q-1). \tag{5}$$

The specific heat, as a measure of the rate of energy variation, serves as an indicator of phase transition phenomena. In a thermodynamic system, a phase is characterized by uniform physical properties, and a phase transition refers to the discontinuous change of certain properties under a critical external condition. The study of phase transitions in the multifractal spectrum has been limited to simple systems, such as the Cantor set and logistic map. However, our analysis reveals the presence of phase transitions in complex networks and in the multifractal spectrum of financial systems. Specifically, the "energy" α exhibits significant fluctuations near q_c, which are reflected by the peak of the specific heat $C(q_c)$.

The overall degree of multifractality, total multifractal specific heat (C_{area}) can be expressed via the Eq. (5):

$$C_{area} = \int C(q)dq. \tag{6}$$

4   Empirical Results

In the presented paper, we consider the Dow Jones Industrial Average (DJIA) index. For the sake of presenting the validity of the described approaches for identifying edge events, for demonstration we identify 4 main crashes of the United States stock market, according to the list of stock market crashes and bear markets [48]: Wall Street Crash of 1929, Black Monday crash of 1987, Financial crisis of 2008, and COVID-19 crash of 2020.

We proceed to conduct multifractal analysis on financial crashes in the stock market, primarily employing the sliding window technique. The method involves selecting a sub-window of a pre-determined length w and performing MF-DFA analysis to obtain the relevant metrics, which are then appended to an array. Subsequently, the window is shifted by a fixed time step h, and the entire procedure is repeated until the entire time series has been analyzed. For our study, we use a window length of $w = 500$ and a time step of $h = 1$. Our findings and observations will be presented based on these parameters.

Based on previous research, it has been established that logarithmic values of returns that have been standardized exhibit multifractal properties. Consequently, we intend to compute additional measures pertaining to the standardized returns which are defined as follows:

$$G(t) = [x(t + \Delta t) - x(t)] / x(t) \quad \text{and} \quad g(t) \equiv [G(t) - \langle G \rangle] / \sigma, \tag{7}$$

where $x(t)$ is a price value at time t; Δt is a time shift ($\Delta t = 1$); $\langle G \rangle$ represents the average of returns G; σ is the standard deviation of G.

The mentioned multifractal measures were calculated for the following parameters:

1. sliding window $w = 500$ days for capturing more significant statistics for multifractal procedure;
2. $h = 1$ to take into account all the values of the studied signals;
3. in Eq. (1) we use first-order polynomial trend;
4. time scale s is defined in range from 20 to 500;
5. statistical q moments are defined in range from -10 to 10 with $\Delta q = 0.25$.
   These particular values of q provide a more comprehensive representation of

scales with varying degrees of fluctuation density, and offer a more refined assessment of phase transitions occurring at specific q moments. However, it is worth noting that investigations with wider or narrower ranges of q are also feasible.

The evolution pattern of the studied characteristics can serve as a prominent early-warning indicator of the critical transitions.

In Fig. 1 and Fig. 2 we present sliding window dynamics of C_{area} and $C(q)$, which we calculated for all the mentioned crashes.

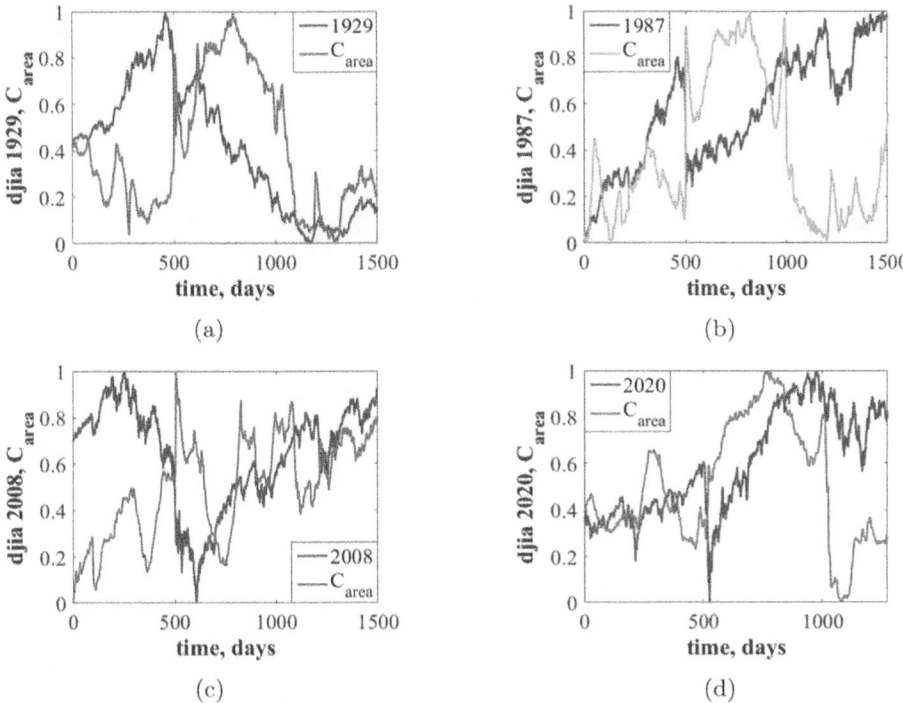

Fig. 1. Comparative dynamics of the total multifractal specific heat (C_{area}) and crashes of 1929 (a), 1987 (b), 2008 (c), and 2020 (d).

Figure 1 illustrates that during a crash event, the total multifractal specific heat begins to increase, which indicates an increase in the complexity of the system during this particular period, an increase in the nonlinearity and width of the multifractal spectrum. As already indicated, during crashes, the spectrum not only expands, but also acquires a left-sided asymmetry. It can also be observed that at the pre-crisis moment this indicator begins to fall, which indicates the streamlining and simplification of the system in the pre-crisis period. Thus, it can be used as an indicator-precursor.

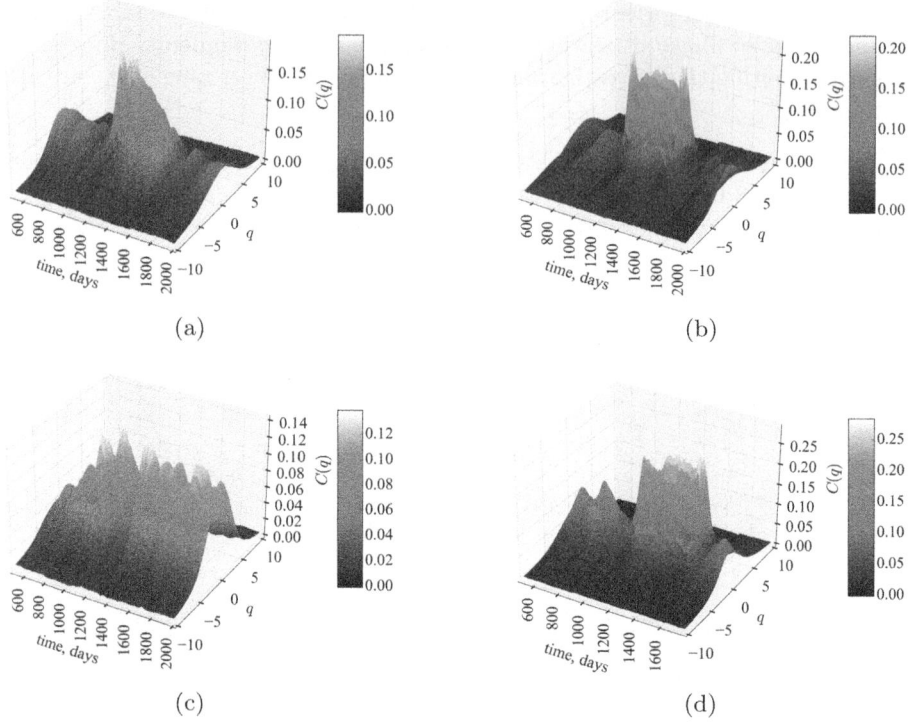

(a)   (b)

(c)   (d)

Fig. 2. Multifractal specific heat as a function of q and time calculated with the sliding window procedure for crashes of 1929 (a), 1987 (b), 2008 (c), and 2020 (d).

Figure 2 shows that the multifractal specific heat increases at the moment of the collapse event, which characterizes each of the crises as a quasi-phase transition. With respect to various statistical moments q, it can be seen that the specific heat looks symmetrical, but it is important to note that during the collapse it goes towards positive q values, which indicates the predomination of large fluctuations in the multifractal nature of each collapse.

5   Conclusions

In conclusion, we investigated the multifractality of the stock market using the example of the DJIA index and its 4 main crash events: the crash of 1929, the crash of 1987, the crash of 2008 and the coronavirus pandemic crisis. On the example of the mentioned crises, and taking into account the works considered in the review, we used the multifractal specific heat indicator as a harbinger indicator of crisis phenomena. It has been shown that this indicator, by analogy with its physical counterpart, makes it possible to identify "economic" quasi-phase transitions and identify changes in complexity in the dynamics of the system. Based on various statistical moments of q, this indicator allowed us to more clearly consider the influence of large and small fluctuations on the overall nonlinearity of the system and its resulting complexity. The indicator C_{area} obtained by us, which is an integral measure of the multifractal heat capacity, allows, by analogy with the width of the multifractal spectrum, to observe the variability of the weakly invariant characteristics of the system. The obtained indicators make it possible to more accurately identify events that are characterized by heavy tails and long-term dependencies, which makes them an excellent basis for building economic risk management models.

Since only the classical MF-DFA approach was used in this work, in the future we plan to use more advanced approaches of complex systems theory that consider approaches of network science, entropy methods, fuzzy techniques, irreversibility, chaos theory, etc. [3–11, 26, 28–30, 33, 37, 41, 49].

Acknowledgements. This work is part of the applied research "Monitoring, Forecasting, and Prevention of Crisis Phenomena in Complex Socio-Economic Systems", which is funded by the Ministry of Education and Science of Ukraine (projects No. 0122U001694).

References

1. Barabási, A.L., Vicsek, T.: Multifractality of self-affine fractals. Phys. Rev. A **44**, 2730–2733 (1991). https://doi.org/10.1103/PhysRevA.44.2730
2. Beck, C., Schögl, F.: Thermodynamics of Chaotic Systems: An Introduction. Cambridge Nonlinear Science Series, Cambridge University Press, Cambridge (1993). https://doi.org/10.1017/CBO9780511524585
3. Bielinskyi, A., Soloviev, V., Semerikov, S., Solovieva, V.: Identifying stock market crashes by fuzzy measures of complexity. Neuro-Fuzzy Model. Tech. Econ. **10**, 3–45 (2021). https://doi.org/10.33111/nfmte.2021.003
4. Bielinskyi, A., Semerikov, S., Serdyuk, O., Solovieva, V., Soloviev, V., Pichl, L.: Econophysics of sustainability indices, vol. 2713, pp. 372–392. CEUR-WS (2020). https://ceur-ws.org/Vol-2713/paper41.pdf
5. Bielinskyi, A., Soloviev, V., Semerikov, S., Solovieva, V.: Detecting stock crashes using levy distribution, vol. 2422, pp. 420–433. CEUR-WS (2019). https://ceur-ws.org/Vol-2422/paper34.pdf
6. Bielinskyi, A., Soloviev, V., Solovieva, V., Matviychuk, A., Semerikov, S.: The analysis of multifractal cross-correlation connectedness between bitcoin and the stock market. In: Faure, E., Danchenko, O., Bondarenko, M., Tryus, Y., Bazilo, C., Zaspa, G. (eds.) ITEST 2022. LNCS, vol. pp, pp. 323–345. Springer, Cham (2023). https://doi.org/10.1007/978-3-031-35467-0_21
7. Bielinskyi, A.O., Hushko, S.V., Matviychuk, A.V., Serdyuk, O.A., Semerikov, S.O., Soloviev, V.N.: Irreversibility of financial time series: a case of crisis, vol. 3048, pp. 134–150. CEUR-WS (2021). https://ceur-ws.org/Vol-3048/paper04.pdf
8. Bielinskyi, A.O., Matviychuk, A.V., Serdyuk, O.A., Semerikov, S.O., Solovieva, V.V., Soloviev, V.N.: Correlational and non-extensive nature of carbon dioxide pricing market. In: Ignatenko, O., et al. (eds.) ICTERI 2021. CCIS, vol. 1635, pp. 183–199. Springer, Cham (2022). https://doi.org/10.1007/978-3-031-14841-5_12

9. Bielinskyi, A.O., Serdyuk, O.A., Semerikov, S.O., Soloviev, V.N.: Econophysics of cryptocurrency crashes: a systematic review, vol. 3048, pp. 31–133. CEUR-WS (2021). https://ceur-ws.org/Vol-3048/paper03.pdf
10. Bielinskyi, A.O., Soloviev, V.N.: Complex network precursors of crashes and critical events in the cryptocurrency market, vol. 2292, pp. 37–45. CEUR-WS (2018). https://ceur-ws.org/Vol-2292/paper02.pdf
11. Bondarenko, M.: Modeling relation between at-the-money local volatility and realized volatility of stocks. Neuro-Fuzzy Model. Tech. Econ. **10**, 46–66 (2021). https://doi.org/10.33111/nfmte.2021.046
12. Bouchaud, J.P., Potters, M.: Theory of Financial Risk and Derivative Pricing: From Statistical Physics to Risk Management, 2nd edn. Cambridge University Press, Cambridge (2003). https://doi.org/10.1017/CBO9780511753893
13. Canessa, E.: Multifractality in time series. J. Phys. A: Math. Gen. **33**(19), 3637 (2000). https://doi.org/10.1088/0305-4470/33/19/302
14. Andjelković, M., Gupte, N., Tadić, B.: Hidden geometry of traffic jamming. Phys. Rev. E **91**, 052817 (2015). https://doi.org/10.1103/PhysRevE.91.052817
15. Cheng, Q.: Generalized binomial multiplicative cascade processes and asymmetrical multifractal distributions. Nonlinear Process. Geophys. **21**(2), 477–487 (2014). https://doi.org/10.5194/npg-21-477-2014
16. Sornette, D., Johansen, A., Bouchaud, J.-P.: Stock market crashes, precursors and replicas. J. Phys. I France **6**(1), 167–175 (1996). https://doi.org/10.1051/jp1:1996135
17. Duarte-Neto, P., Stošić, B., Stošić, T., Lessa, R., Milošević, M.V., Stanley, H.E.: Multifractal properties of a closed contour: a peek beyond the shape analysis. PLoS ONE **9**(12), 1–13 (2014). https://doi.org/10.1371/journal.pone.0115262
18. Dupont, S., Argoul, F., Gerasimova-Chechkina, E., Irvine, M.R., Arneodo, A.: Experimental evidence of a phase transition in the multifractal spectra of turbulent temperature fluctuations at a forest canopy top. J. Fluid Mech. **896**, A15 (2020). https://doi.org/10.1017/jfm.2020.348
19. Dutta, S.: Multifractal properties of ECG patterns of patients suffering from congestive heart failure. J. Stat. Mech: Theory Exp. **2010**(12), P12021 (2010). https://doi.org/10.1088/1742-5468/2010/12/P12021
20. Feigenbaum, J.A., Freund, P.G.: Discrete scale invariance in stock markets before crashes. Int. J. Mod. Phys. B **10**(27), 3737–3745 (1996). https://doi.org/10.1142/S021797929600204X
21. Fenn, D.J., Porter, M.A., Williams, S., McDonald, M., Johnson, N.F., Jones, N.S.: Temporal evolution of financial-market correlations. Phys. Rev. E **84**, 026109 (2011). https://doi.org/10.1103/PhysRevE.84.026109
22. Gluzman, S., Yukalov, V.I.: Renormalization group analysis of october market crashes. Modern Phys. Lett. **B 12**(02n03), 75–84 (1998). https://doi.org/10.1142/S0217984998000111
23. Gualdi, S., Tarzia, M., Zamponi, F., Bouchaud, J.P.: Tipping points in macroeconomic agent-based models. J. Econ. Dyn. Control **50**, 29–61 (2015). https://doi.org/10.1016/j.jedc.2014.08.003. crises and Complexity
24. Kantelhardt, J.W., Zschiegner, S.A., Koscielny-Bunde, E., Havlin, S., Bunde, A., Stanley, H.: Multifractal detrended fluctuation analysis of nonstationary time series. Phys. A **316**(1), 87–114 (2002). https://doi.org/10.1016/S0378-4371(02)01383-3
25. Kasprzak, A., Kutner, R., Perelló, J., Masoliver, J.: Higher-order phase transitions on financial markets. Eur. Phys. J. B **76**(4), 513–527 (2010). https://doi.org/10.1140/epjb/e2010-00064-y

26. Kiv, A., et al.: Irreversibility of plastic deformation processes in metals. In: Faure, E., Danchenko, O., Bondarenko, M., Tryus, Y., Bazilo, C., Zaspa, G. (eds.) Information Technology for Education, Science, and Technics. LNDECT, vol. 178, pp. 425–445. Springer, Cham (2023). https://doi.org/10.1007/978-3-031-35467-0_26

27. Klamut, J., Kutner, R., Gubiec, T., Struzik, Z.R.: Multibranch multifractality and the phase transitions in time series of mean interevent times. Phys. Rev. E **101**, 063303 (2020). https://doi.org/10.1103/PhysRevE.101.063303

28. Kmytiuk, T., Majore, G.: Time series forecasting of agricultural product prices using elman and jordan recurrent neural networks. Neuro-Fuzzy Model. Tech. Econ. **10**, 67–85 (2021). https://doi.org/10.33111/nfmte.2021.067

29. Kobets, V., Novak, O.: Eu countries clustering for the state of food security using machine learning techniques. Neuro-Fuzzy Model. Tech. Econ. **10**, 86–118 (2021). https://doi.org/10.33111/nfmte.2021.086

30. Kucherova, H., Honcharenko, Y., Ocheretin, D., Bilska, O.: Fuzzy logic model of usability of websites of higher education institutions in the context of digitalization of educational services. Neuro-Fuzzy Model. Tech. Econ. **10**, 119–135 (2021). https://doi.org/10.33111/nfmte.2021.119

31. Laloux, L., Cizeau, P., Potters, M., Bouchaud, J.P.: Random matrix theory and financial correlations. Int. J. Theoretical Appl. Finan. **03**(03), 391–397 (2000). https://doi.org/10.1142/S0219024900000255

32. Lee, J., Stanley, H.E.: Phase transition in the multifractal spectrum of diffusion-limited aggregation. Phys. Rev. Lett. **61**, 2945–2948 (1988). https://doi.org/10.1103/PhysRevLett.61.2945

33. Lukianenko, D., Strelchenko, I.: Neuromodeling of features of crisis contagion on financial markets between countries with different levels of economic development. Neuro-Fuzzy Model. Tech. Econ. **10**, 136–163 (2021). https://doi.org/10.33111/nfmte.2021.136

34. Mandelbrot, B.B.: Fractals and Scaling in Finance. Springer, New York (1997). https://doi.org/10.1007/978-1-4757-2763-0

35. Mantegna, R.: Hierarchical structure in financial markets. Eur. Phys. J. B **11**(1), 193–197 (1999). https://doi.org/10.1007/s100510050929

36. Mantegna, R.N., Stanley, H.E., Chriss, N.A.: An introduction to econophysics: correlations and complexity in finance. Phys. Today **53**(12), 70–70 (2000). https://doi.org/10.1063/1.1341926

37. Miroshnychenko, I., Kravchenko, T., Drobyna, Y.: Forecasting electricity generation from renewable sources in developing countries (on the example of Ukraine). Neuro-Fuzzy Model. Tech. Econ. **10**, 164–198 (2021). https://doi.org/10.33111/nfmte.2021.164

38. Peng, C.K., Buldyrev, S.V., Havlin, S., Simons, M., Stanley, H.E., Goldberger, A.L.: Mosaic organization of DNA nucleotides. Phys. Rev. E **49**, 1685–1689 (1994). https://doi.org/10.1103/PhysRevE.49.1685

39. Plerou, V., Gopikrishnan, P., Rosenow, B., Amaral, L.A., Stanley, H.: Econophysics: financial time series from a statistical physics point of view. Phys. A **279**(1), 443–456 (2000). https://doi.org/10.1016/S0378-4371(00)00010-8

40. Plerou, V., Gopikrishnan, P., Rosenow, B., Nunes Amaral, L.A., Stanley, H.E.: Universal and nonuniversal properties of cross correlations in financial time series. Phys. Rev. Lett. **83**, 1471–1474 (1999). https://doi.org/10.1103/PhysRevLett.83.1471

41. Soloviev, V.N., Bielinskyi, A.O., Kharadzjan, N.A.: Coverage of the coronavirus pandemic through entropy measures. vol. 2832, p. 24–42. CEUR-WS (2020). https://ceur-ws.org/Vol-2832/paper02.pdf

42. Sornette, D.: Critical Phenomena in Natural Sciences: Chaos, Fractals, Selforganization and Disorder: Concepts and Tools. Springer Series in Synergetics, Springer, Heidelberg (2003)
43. Sornette, D., Johansen, A.: Large financial crashes. Phys. A **245**(3), 411–422 (1997). https://doi.org/10.1016/S0378-4371(97)00318-X
44. Sornette, D., Ouillon, G.: Dragon-kings: mechanisms, statistical methods and empirical evidence. SSRN Electron. J. (2012). https://doi.org/10.2139/ssrn.2191590
45. Stošić, D., Stošić, D., Stošić, T., Stanley, H.E.: Multifractal analysis of managed and independent float exchange rates. Physica A: Stat. Mech. Appl. **428**, 13–18 (2015). https://doi.org/10.1016/j.physa.2015.02.055, https://www.sciencedirect.com/science/article/pii/S0378437115001612
46. Stošić, D., Stošić, D., Stošić, T., Eugene Stanley, H.: Multifractal properties of price change and volume change of stock market indices. Phys. A **428**, 46–51 (2015). https://doi.org/10.1016/j.physa.2015.02.046
47. Telesca, L., Lapenna, V.: Measuring multifractality in seismic sequences. Tectonophysics **423**(1), 115–123 (2006). https://doi.org/10.1016/j.tecto.2006.03.023. spatiotemporal Models of Seismicity and Earthquake Occurrence
48. Wikipedia: List of stock market crashes and bear markets (2023). https://en.wikipedia.org/wiki/List_of_stock_market_crashes_and_bear_markets
49. Xiao, X., Chen, H., Bogdan, P.: Deciphering the generating rules and functionalities of complex networks. Sci. Rep. **11**(1), 22964 (2021). https://doi.org/10.1038/s41598-021-02203-4
50. Zorick, T., Mandelkern, M.A.: Multifractal detrended fluctuation analysis of human EEG: preliminary investigation and comparison with the wavelet transform modulus maxima technique. PLoS ONE **8**(7), 1–7 (2013). https://doi.org/10.1371/journal.pone.0068360

Bibliometric Analysis of Adaptive Learning Literature from 2011-2019: Identifying Primary Concepts and Keyword Clusters

Liliia O. Fadieieva[✉]

Kryvyi Rih State Pedagogical University, 54 Gagarin Ave., Kryvyi Rih 50086, Ukraine
liliia.fadieieva@kdpu.edu.ua

Abstract. This paper presents a comprehensive bibliometric analysis of the adaptive learning literature from 2011 to 2019 in the social sciences domain. The study utilizes the Scopus database to identify relevant sources and employs cluster analysis based on keyword co-occurrence to categorize the primary concepts. The research focuses on understanding the key areas and trends within adaptive learning, shedding light on its development and impact over the specified period. The relevance of this analysis lies in the increasing importance of adaptive learning in modern education systems, especially in the context of integrating innovative technologies and addressing the challenges posed by the digital society. As the demand for quality education and skilled teaching staff grows, there is a need to explore and implement more student-centered approaches, such as adaptive learning, to enhance the learning experience and improve educational outcomes. The outputs of this research provide valuable insights into the main themes and areas of interest within the adaptive learning field during the selected timeframe. By identifying primary concepts and keyword clusters, the study offers a comprehensive overview of the key topics, theories, and technologies that have shaped the development of adaptive learning. This analysis can serve as a valuable resource for researchers, educators, and policymakers seeking to understand the current landscape of adaptive learning and explore potential avenues for future research and innovation.

Keywords: Adaptive Learning · Bibliometric Review · Adaptive Systems · Artificial Intelligence · VOSviewer

1 Introduction

Since Ukraine's policy is aimed at integration into the European Union (EU), we should also take into account the strategic directions for the digitalization of higher education in the EU represented in the Digital Education Action Plan for 2021–2027. It offers a strategy for European education, which includes improved

G. Antoniou et al. (Eds.): ICTERI 2023, CCIS 1980, pp. 215–226, 2023.
https://doi.org/10.1007/978-3-031-48325-7_16

quality and quantity of teaching concerning digital technologies, support for the digitalisation of teaching methods and pedagogies. The Action Plan emphasys to [2]:

- digitally competent and confident teachers and education and training staff [6];
- high-quality learning content, user-friendly tools and secure platforms which respect e-privacy rules and ethical standards [3];
- digital literacy, including tackling disinformation [7];
- good knowledge and understanding of data-intensive technologies, such as artificial intelligence (AI) [8].

Since 2020 (wide spread of the novel coronavirus), and especially since 2022 (Russia's invasion in Ukraine), the Ukrainian teaching staff challenges in performing and managing emergency distance education [5]. This rise a lot of issues both technical and organizing which made a drastic changes in the Ukrainian educators' digital competence. For the first time, the educational community has been self-organized to prevent a disruption of education on the all levels, from pre-school to tertiary. There is a growing interest in more flexible, innovative and sustained models of professional development, in particular where educators learn from their peers.

In addition, the emergence of new technologies such as AI, virtual or augmented reality and social robotics, challenge educators and requires them to take a more active role in the design and implementation of these tools to ensure their use is effective, desirable and inclusive [1].

Overall, there is a need to develop and test new pedagogies and techniques, also by investigating how emerging technologies can be smoothly integrated in existing teaching and learning practices. One of the prominent application of AI in education is a technology supported adaptive learning.

2 Method

To systematize available scientific knowledge, a bibliometric analysis was conducted using the VOSviewer [4]. In order to carry out the analysis, a selection of sources from the scientometric database Scopus was made upon request:

```
TITLE("adaptive learning") AND (LIMIT-TO(SUBJAREA, "SOCI"))
```

According to the request, the term "adaptive learning" appeared in the titles of articles, chapters, or books belonging to the subject area "social sciences".

3 Results and Discussion

As a result, 344 documents were received, and the distribution of works by year is presented in Fig. 1.

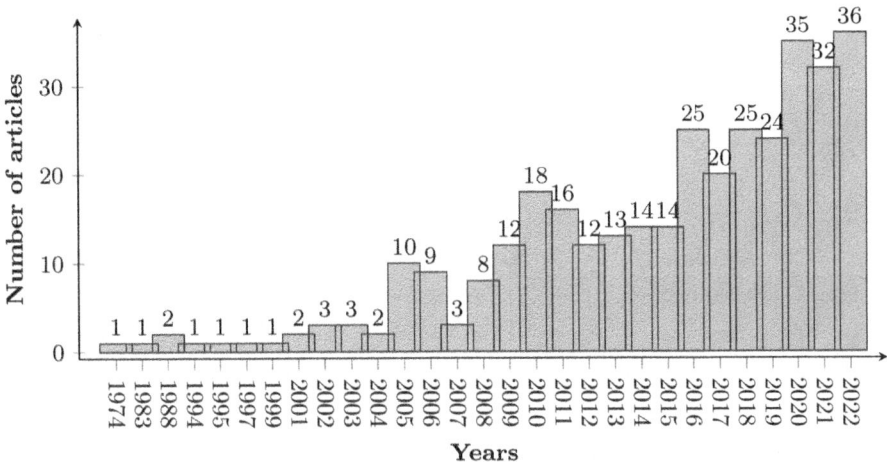

Fig. 1. Distribution by articles by years.

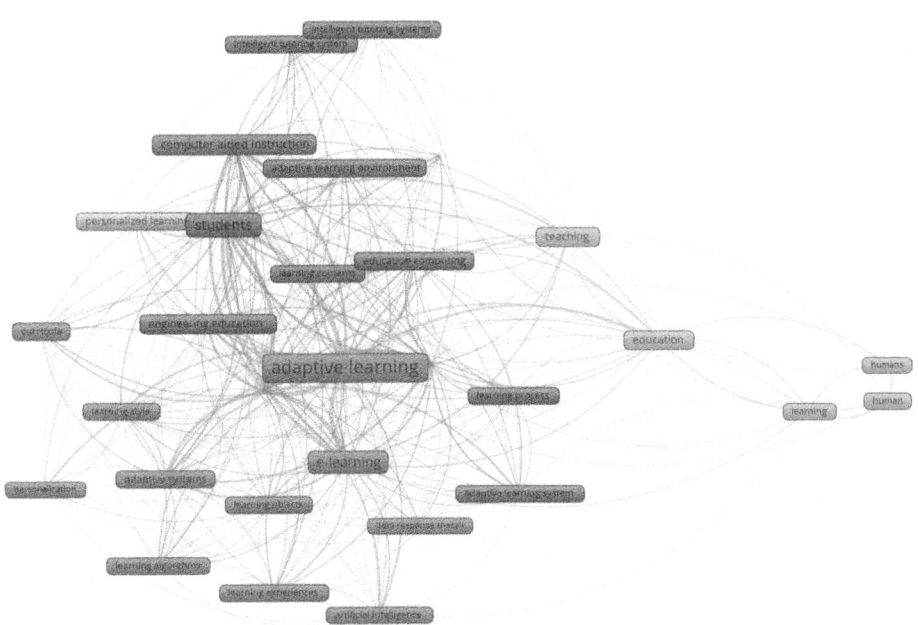

Fig. 2. Network visualization.

Table 1. Distribution of keywords by clusters.

Keyword	Cluster	Weight$_{Links}$	Weight$_{Total\ link\ strength}$	Weight$_{Occurrences}$	Score$_{Avg.\ pub.\ year}$	Score$_{Avg.\ citations}$	Score$_{Avg.\ norm.\ citations}$
adaptive learning	1	29	425	158	2016.0443	10.6076	0.966
learning systems	1	27	445	118	2014.6356	10.6949	0.8159
e-learning	1	27	280	70	2015.4571	6.6857	0.7592
adaptive systems	1	25	126	28	2013.6429	28.3929	1.1589
learning objects	1	22	46	12	2012	24.8333	1.6496
learning style	1	21	72	17	2013.4706	32.4118	1.5342
learning algorithms	1	20	53	13	2015.4615	34	1.6297
learning experiences	1	19	56	11	2015.4545	9.2727	1.1661
artificial intelligence	1	18	41	14	2016.6429	7.8571	0.8118
item response theory	1	18	33	10	2017.3	17.1	1.8503
curricula	1	17	78	21	2014.7143	7.9048	0.8894
personalization	1	14	27	10	2013.9	70.8	3.1478
students	2	26	338	75	2015.68	11.72	1.0332
computer aided instruction	2	25	209	40	2014.975	13.975	1.0588
adaptive learning environment	2	23	82	19	2014.4211	14.5263	1.4182
learning contents	2	20	49	11	2013.6364	11.6364	0.9208
learning performance	2	19	48	11	2014.1818	11.3636	0.8037
intelligent tutoring system	2	17	49	11	2014.5455	22.0909	1.4888
intelligent tutoring systems	2	15	34	12	2017.9167	15.6667	1.9702
adaptive learning systems	3	27	269	66	2015.1667	7.9697	0.8392
education computing	3	25	135	25	2015.96	4	0.5437
engineering education	3	23	121	29	2013.7931	6.8966	0.7189
learning process	3	20	48	12	2016.5833	4	0.5026
adaptive learning system	3	19	58	15	2016.4667	9.8667	0.8269
education	4	25	144	35	2011.8286	11.8286	0.7189
teaching	4	25	133	29	2015.7931	8.1379	0.6896
learning	4	12	38	17	2015.0588	5.9412	0.5421
human	4	5	29	11	2013	12.0909	0.8125
humans	4	5	27	10	2015.5	10.3	0.8126
personalized learning	5	20	79	21	2017.7619	14.5714	1.3967

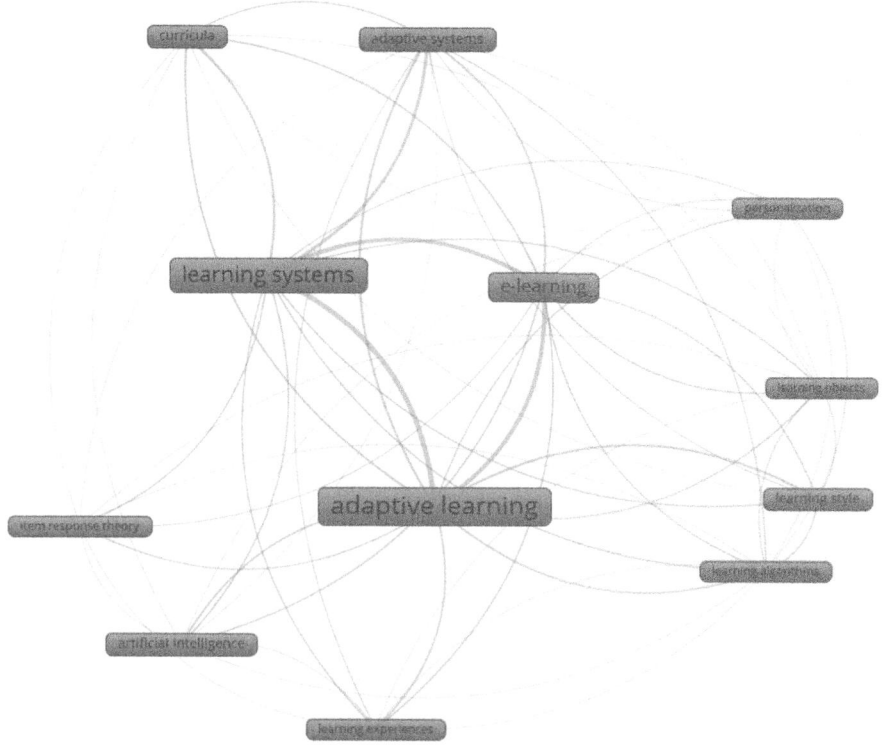

Fig. 3. A cluster of general concepts of adaptive learning in e-learning systems.

Cluster analysis by keyword co-occurrence was conducted: from 1836, keywords were selected, that appeared at least 10 times (Table 1).

The results of the cluster analysis are presented in Fig. 2.

According to Table 1 and Fig. 2, keywords were divided into five clusters. Let's analyze them in more detail.

The first cluster includes 12 keywords (Fig. 3), 5 of which are primarily related to the theory of adaptive learning: *adaptive learning, adaptive systems, curricula, learning style, learning experiences, learning algorithms, personalization*. Other concepts are related to adaptive testing (*item response theory*), which is implemented in *e-learning* systems – a type of *learning system* that operate with *learning objects* and can be automated by means of *artificial intelligence*.

The second cluster contains 7 keywords (Fig. 4) related to the practice of *computer aided instruction* of *students* (in particular, assessment of *learning performance*, and evaluation of *learning contents*) at *adaptive learning environment* (in particular, *intelligent tutoring systems*).

The third cluster contains 5 keywords (Fig. 5) that describe the implementation of *learning process* within *engineering education* by means of *education computing*, e.g. *adaptive learning systems*.

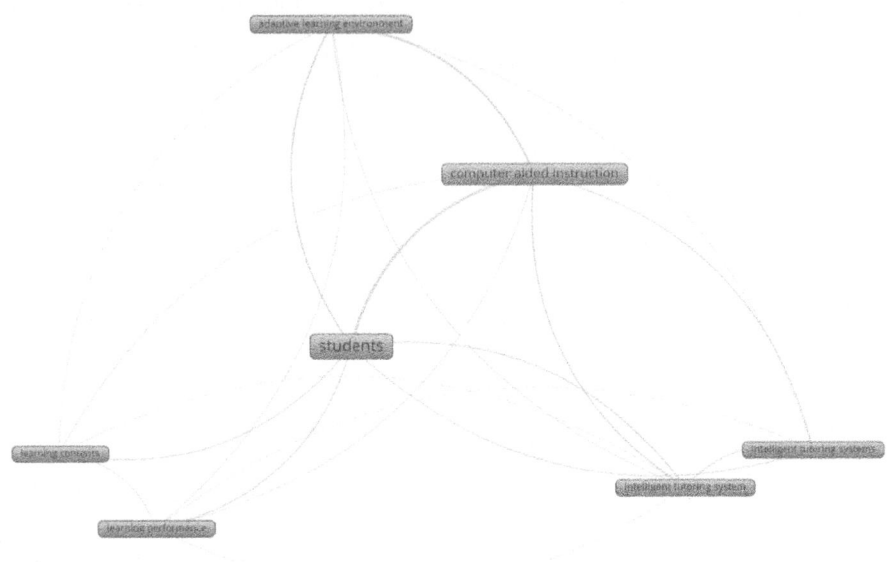

Fig. 4. A cluster of educational technology.

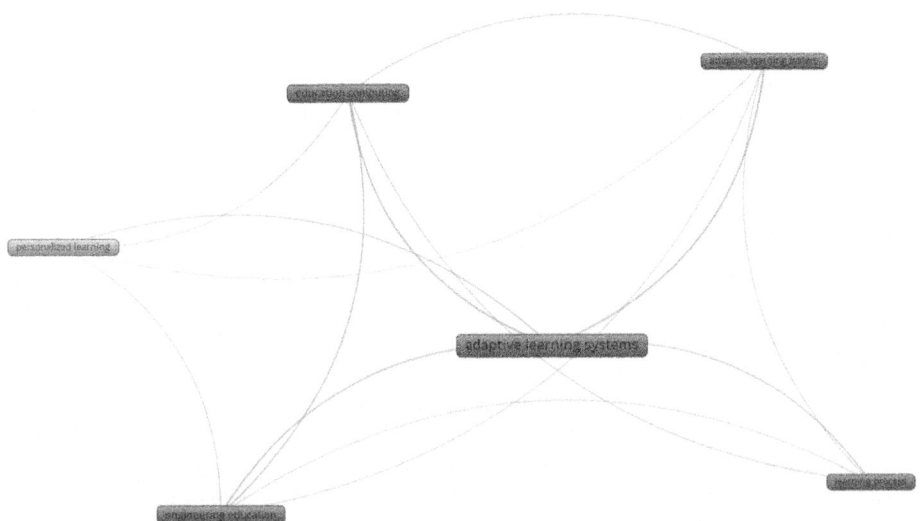

Fig. 5. A cluster of adaptive learning systems and education computing.

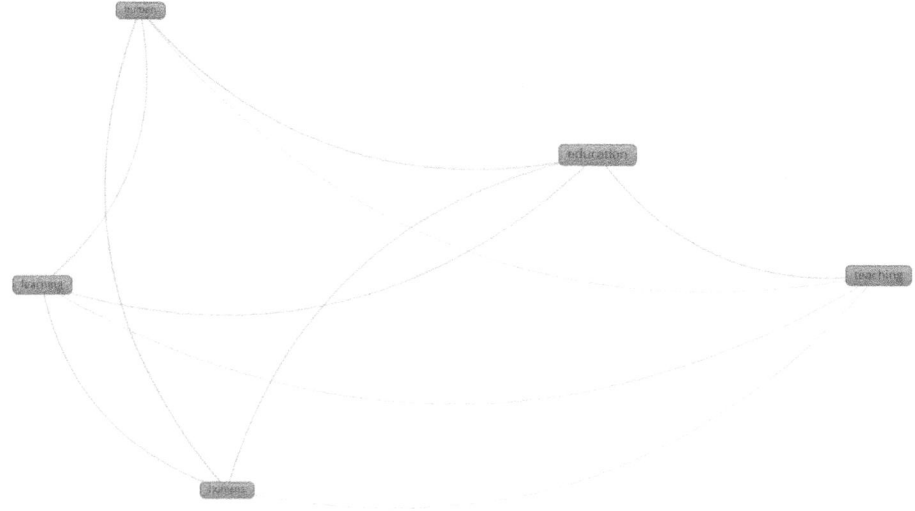

Fig. 6. A cluster of learning and education research.

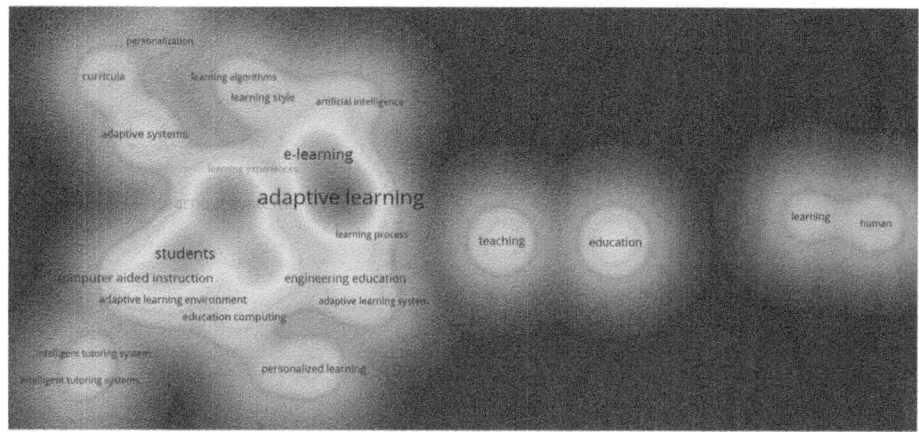

Fig. 7. Item density visualization.

The fourth cluster also includes 5 keywords (Fig. 6) that describe the didactic fundamentals: *human(s), education, teaching,* and *learning*.

The smallest cluster consists of only 1 keyword (Fig. 5) – *personalized learning*.

Another important criterion for source analysis is density. First, was analyzing the item's density (Fig. 7). From this visualization, the keywords "adaptive learning" (Weigth$_{\text{Total link strength}}$ = 425), "learning system" (Weigth$_{\text{Total link strength}}$ = 445), "students" (Weigth$_{\text{Total link strength}}$ = 338), and "e-learning" (Weigth$_{\text{Total link strength}}$ = 280) have the highest density. These items are the most interconnected (maximum value of total link strength).

In order to determine <u>primary concepts</u> (earliest keywords by time scale), let's show overlay data visualization by years. As is shown in Fig. 9, there are no fundamentally new concepts, their emergence, and spread have occurred at least since 2020. There are also no concepts that were widespread before 2000. This visualization gives us grounds for limiting the analysis years.

So, let's analyze these concepts from 2000 to 2020 (Fig. 10). Within these limits, we can see that such concepts as *"human(s)"*, *"education"*, and *"learning objects"* begin to stand out as those that were formed earlier. At the same time, such concepts as *"personalized learning"* and *"intelligent tutoring systems"* are distinguished as those highlighted later. And since most of the concepts were disseminated after 2000, to see their distribution more accurately, we will raise the lower limit from 2000 to 2010 (Fig. 11) (Fig. 8).

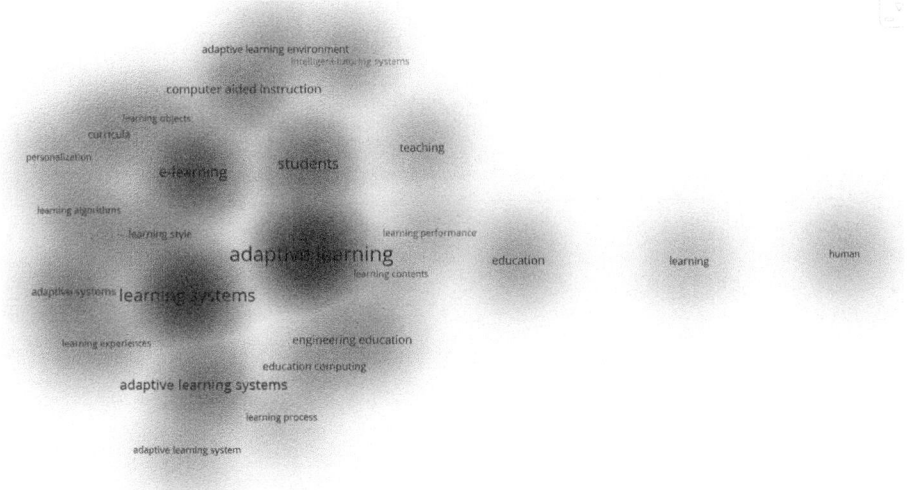

Fig. 8. Cluster density visualization.

When comparing Fig. 11 with Fig. 10, we observe that most of the concepts have changed color, but it is still unclear which concepts were discussed by researchers in the different years.

So, let's try to change both the upper and lower limits for 1 year, i.e. from 2011 to 2019 (Fig. 12). Now we can observe a more transparent distribution of concepts by time scale. From this figure, we can see that adaptive learning and artificial intelligence became disseminated later than those related to the use of ICT in education.

4   Limitations

The use of only the Scopus database and the social sciences section of this database instead of the entire range are the key restrictions on the research.

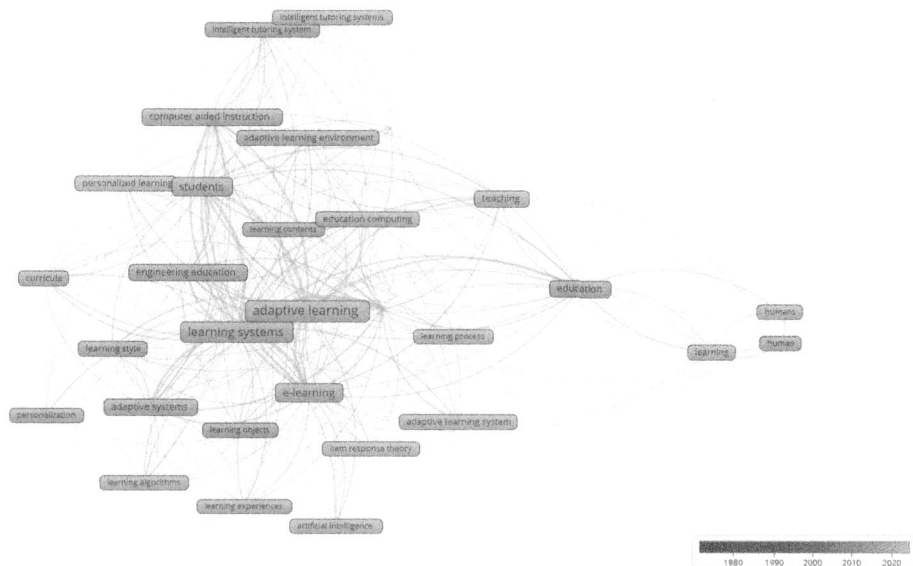

Fig. 9. Extension of terms from 1974 to 2022.

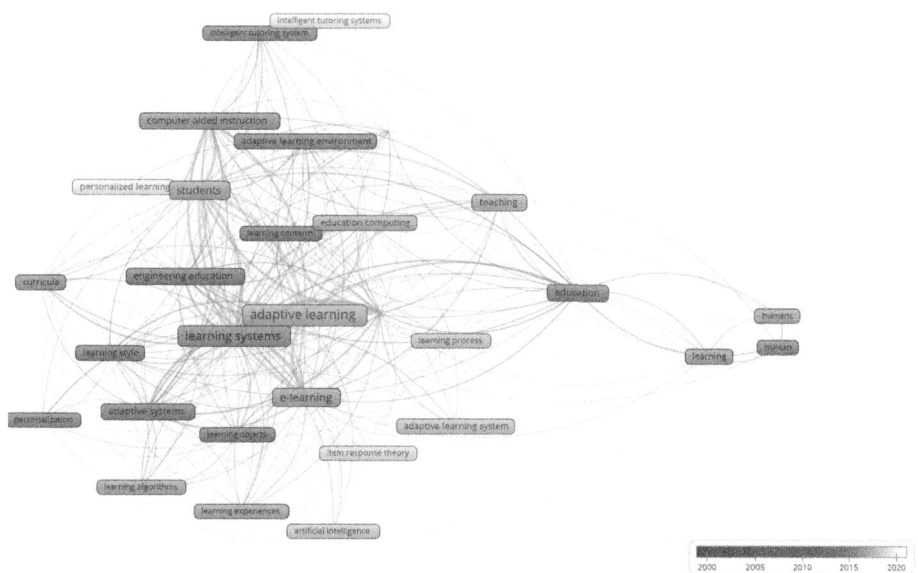

Fig. 10. Extension of terms from 2000 to 2020.

Additionally, sources like tech reports and Ph.D. theses that are not indexed by Scopus can be useful for this research. There are some restrictions with the VOSViewer tool: a clustering algorithm was applied with the default settings,

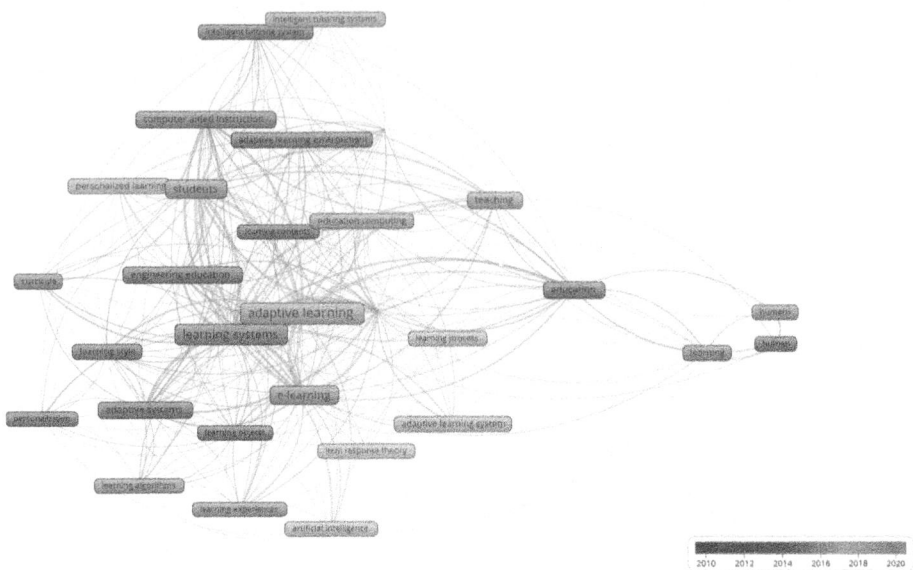

Fig. 11. Extension of terms from 2010 to 2020.

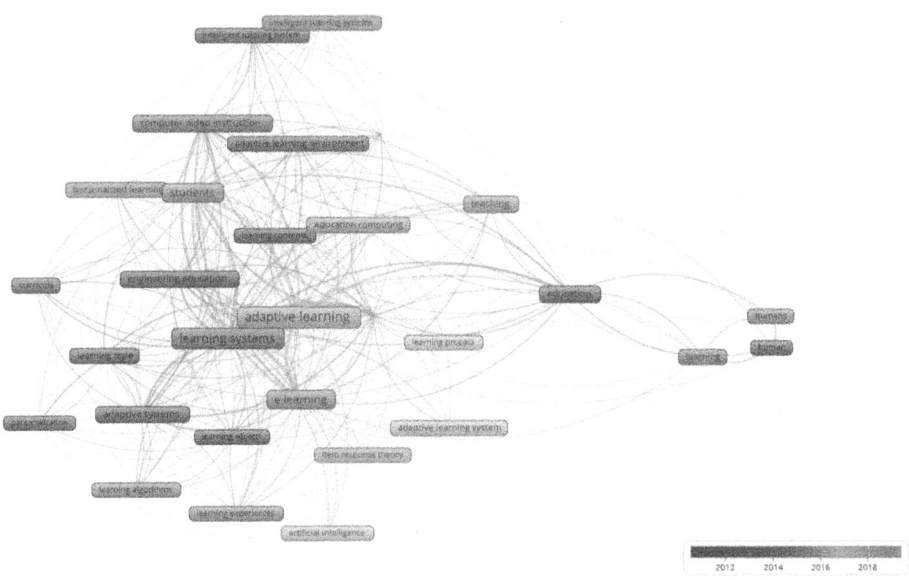

Fig. 12. Extension of terms from 2011 to 2019.

and the low limit for keyword occurrence was set at 10. The number of clusters can be decreased or increased depending on the clustering settings. Additionally, the third cluster can be combined with the fifth one because the fifth cluster only has one keyword (personalized learning).

5   Conclusions

Several important conclusions resulted from the bibliometric review. To begin, the research focused on publications from 2011 to 2019, a period that showed a large increase in interest in adaptive learning. Using cluster analysis, the study successfully identified the research problem domain.

Adaptive learning, learning systems, e-learning, adaptive systems, learning objects, learning style, learning algorithms, students, computer-aided instruction, adaptive learning environment, learning contents, adaptive learning systems, education computing, engineering education, learning process, education, teaching, and personalized learning were also established as key concepts central to the subject.

6   Future Work

The results of the bibliometric analysis have provided valuable insights into the current state and emerging trends in adaptive learning literature. Building on this analysis, several areas of future research can be explored to advance our understanding of adaptive learning and its applications in education:

1. *Systematic Review*: Conduct a comprehensive systematic review of the identified clusters to delve deeper into the current state of research in adaptive learning. This review can focus on specific themes such as the theory of adaptive learning, learner's models, the effectiveness of adaptive learning systems, and the integration of AI in adaptive learning platforms.
2. *AI-driven Adaptive Learning*: Investigate the role of artificial intelligence in the development and improvement of adaptive learning systems. Explore the use of advanced AI algorithms and approaches to enhance the adaptivity and personalization of learning experiences.
3. *Evaluation and Effectiveness*: Examine the implementation and effectiveness of adaptive learning systems in educational settings. Conduct empirical studies to assess the impact of adaptive learning on student performance, engagement, and satisfaction.
4. *Adaptive Learning Environments*: Explore the design and development of adaptive learning environments that cater to individual learners' needs and preferences. Investigate how adaptive systems can be integrated into existing educational platforms to create personalized learning experiences.
5. *Gamification and Adaptive Learning*: Investigate the potential of gamification techniques in adaptive learning systems. Explore how gamified elements can enhance student motivation, engagement, and learning outcomes.
6. *Learning Styles and Personalization*: Study the role of learning styles in adaptive e-learning hypermedia systems. Investigate how learning styles can be effectively integrated into adaptive learning platforms to cater to diverse learner preferences.

7. *Adaptive Learning in Mathematics Education*: Focus on the application of adaptive learning and intelligent tutoring systems in mathematics education. Explore how adaptive approaches can enhance student learning and problem-solving skills in this domain.

8. *Analytics and Adaptive Learning*: Explore the use of learning analytics in adaptive learning systems. Investigate how data-driven insights can be utilized to personalize learning pathways and support educators in making informed decisions.

9. *Application in Different Educational Settings*: Investigate the implementation of adaptive learning and analytics in various educational settings, including K-12 schools, higher education institutions, corporate training, and online learning platforms.

10. *Ethical and Privacy Considerations*: Examine the ethical and privacy implications of using adaptive learning systems, especially when leveraging AI and data-driven approaches. Address concerns related to data security, bias, and transparency.

References

1. Communication from the Commission to the European Parliament, the Council, the European Economic and Social Committee and the Committee of the Regions (2020). https://tinyurl.com/yc7v78a6

2. Digital Education Action Plan (2021–2027) (2020). https://education.ec.europa.eu/focus-topics/digital-education/digital-education-action-plan

3. Burov, O., Butnik-Siversky, O., Orliuk, O., Horska, K.: Cybersecurity and innovative digital educational environment. Inf. Technol. Learn. Tools **80**(6), 414–430 (2020). https://doi.org/10.33407/itlt.v80i6.4159 https://doi.org/10.33407/itlt.v80i6.4159 https://doi.org/10.33407/itlt.v80i6.4159

4. Centre for Science and Technology Studies, Leiden University, The Netherlands: VOSviewer - Visualizing scientific landscapes (2022). https://www.vosviewer.com/

5. Kovalchuk, V.I., Maslich, S.V., Movchan, L.H.: Digitalization of vocational education under crisis conditions. Educ. Technol. Q. **2023**(1), 1–17 (2023). https://doi.org/10.55056/etq.49 https://doi.org/10.55056/etq.49

6. Morze, N., et al.: System for digital professional development of university teachers. Educ. Technol. Q. **2022**(2), 152–168 (2022). https://doi.org/10.55056/etq.6

7. Vakaliuk, T., Pilkevych, I., Fedorchuk, D., Osadchyi, V., Tokar, A., Naumchak, O.: Methodology of monitoring negative psychological influences in online media. Educ. Technol. Q. **2022**(2), 143–151 (2022). https://doi.org/10.55056/etq.1

8. Yuskovych-Zhukovska, V., Poplavska, T., Diachenko, O., Mishenina, T., Topolnyk, Y., Gurevych, R.: Application of artificial intelligence in education Problems and opportunities for sustainable development. BRAIN Broad Res. Artif. Intell. Neurosci. **13**(1Sup1), 339–356 (2022). https://doi.org/10.18662/brain/13.1Sup1/322 https://doi.org/10.18662/brain/13.1Sup1/322

Data Analysis for Predicting Stock Prices Using Financial Indicators Based on Business Reports

Oleksii Ivanov and Vitaliy Kobets(✉)

Kherson State University, 27 Universitetska St., Kherson 73003, Ukraine
`oleksii.ivanov@university.kherson.ua, vkobets@kse.org.ua`

Abstract. Increased access to the stock market leads to a growing interest in investing and an increase in the number of new investors. However, investing in the stock market involves risks and requires good preparation and analysis. To protect and grow their investments, investors should carefully analyze the company they plan to invest in. Our paper aims to develop a model for predicting financial indicators from companies' business reports based on data analysis and machine learning using Python programming language.

Tasks of our research contain the collection of historical data from business reports or financial databases, data cleaning, and feature selection of relevant explanatory variables (such as net income, price-to-earnings ratio, total equity, operating margin, and gross margin) to predict the dependent variable (market price per share). Random Forest outperformed Multiple Linear Regression in predicting stock prices, displaying lower RMSE (9.53), Deviation Percentage (34.61%), and higher R-squared (97.22%).

The Deviation Percentage of 34.61% may seem relatively high, suggesting that there is still room for improvement in the model's precision in predicting stock prices accurately. It is essential to consider other factors that might affect stock prices beyond the selected financial indicators.

The conclusion emphasizes the importance of considering additional market factors, such as competition, economic conditions, and industry changes when making investment decisions. Although the random forest model is an effective tool for analyzing the dependence of stock prices on various variables, it does not account for all factors affecting stock prices. Therefore, investors should use market analysis to make the correct investment decisions.

Keywords: Financial indicators · Data analysis · Python · Business reports

1 Introduction

In today's world, the stock market is becoming increasingly popular. This is due to several factors, including the increased accessibility of the market for ordinary users. Nowadays, anyone in the world can easily purchase shares of any company. It is only necessary to open an account with a brokerage firm. Currently, some of the most popular brokerage firms for regular users are Interactive Brokers [1] and Robinhood [1]. Both applications provide broad access to the stock market and other financial instruments, including ETFs, bonds, and more. Based on statistics on Robinhood usage [2], we can see that over 15 million people use trading platform.

© The Author(s), under exclusive license to Springer Nature Switzerland AG 2023
G. Antoniou et al. (Eds.): ICTERI 2023, CCIS 1980, pp. 227–239, 2023.
https://doi.org/10.1007/978-3-031-48325-7_17

Increased access to the stock market leads to a growing interest in investing and an increase in the number of new investors. However, investing in the stock market is not without risks and requires good preparation and analysis. To protect and grow their investments, investors must carefully analyze the company they plan to invest in. Speaking of the US market, currently one of the largest markets in the global economy, all public companies in this market are regulated by the Securities and Exchange Commission (SEC), which requires companies to submit quarterly reports on their business; these reports are called Form 10-K [3]. These reports are public and available for investors and analysis, and based on the data provided in these reports, one can assess the state of affairs in the company, how well their business is going, and make decisions accordingly.

In this paper, we will discuss the possible use of Python and its tools for automating the extraction of key data from companies' quarterly reports (10K) and methods of predicting them based on historical indicators. We will also find relationships between stock prices and critical indicators. Thus, the goal of this paper is to find the relationship between stock prices and critical indicators based on business reports to predict stock prices for investment decisions.

The paper has the following structure: Sect. 2 considers literature reviews; Section 3 describes general financial ratios used for the analysis model; Section 4 demonstrates a data analysis methods and their implementation; the conclusion is the last section.

2 Related Works

In the field of financial data analysis and financial indicator forecasting using machine learning and data analysis methods, numerous studies have been conducted. We will consider some of them in this section.

Zanc et al. [4] investigated the application of deep neural networks for financial market forecasting, they used the stock index with exchange rate data and demonstrated that deep neural networks allow for high accuracy in forecasting. Wasserbacher et al. [5] explored the application of machine learning methods for analyzing and predicting financial data. Their research used company stock data and showed that machine learning methods could be effectively used for financial indicator forecasting. Doryab et al. [6] examined the application of regression analysis for predicting investment returns. In their research, they used company stock return data and demonstrated that regression analysis could be effectively used for investment return forecasting. Snihovyi et al. [7] predicted cryptocurrency prices using different ML algorithms. All these studies show that machine learning and data analysis methods can be effectively used for financial indicator forecasting. In contrast of previous researches, in this paper, we will apply these methods for analyzing financial data and predicting financial indicators based on business reports.

Mushtaq et al. [8] employ Natural Language Processing (NLP), a subdomain of AI, to predict the sentiments while analyzing 3729 annual 10-k financial reports of S&P 500 companies over the 2002–2019 years. They disclosed that there is no significant association between the firm's financial performance indicators and 10-ks positivity [8]. We believe that the more reports, the less correlation between 10-k reports and financial indicators, because stock markets react more strongly to negative results of reports than positive ones as described by Huang et al. [9].

A firm's annual reports help investors decide about the company's stocks. As a rule, investors analyze financial data to predict stock prices and future returns, volatility, and risks. At the same time, financial performance indicators may affect the massive text of the company's 10-k report.

10-k reports are a signal that can disclose the positive performance of the company, using complex and obfuscation narratives [10]. At the same time, financial indicators can reveal actual state of the company without manipulation from the sides of agents (executives) who try to save the company's positive image and their own positions in the company. Cohen et al. [11] revealed that 10-k reports are relevant for the firm's financial indicators, such as future earnings, profitability, and news announcements, and can predict firm-level bankruptcies.

A regression model with explanatory financial and control variables can measure their impact on dependent variables (e.g., positive or negative news as a qualitative variable, price of the stock as a quantitative variable, etc.) [8]. Among financial indicators, existing research contains return on assets (ROA), return on equity (ROE), Tobin's Q (TQ), and return on invested capital (ROIC). Control variables of the company consist of firm size, firm's assets tangibility, liquidity, financing needs or deficit, and financial leverage. Investors have different risk profiles that defines their propensity to different FI [12]. Clusters can combine investors with identical preferences who are interested in same business reports [13]. Neural networks are used to predict stock prices based on big data [14].

Our model includes quantitative variables, both dependent and explanatory ones. Our study, in contrast to existing ones, reveals only a statistically significant explanatory variable for the stock's price change 10-k report to reveal the direction of changes using 10-k reports and data analysis with API. After data analysis, stock price forecasts can be made based on only non-multicollinearity explanatory variables from the 10-k report, not all financial metrics of the report.

3   Financial Ratios of 10K Report

A 10-K Report is a report that companies registered in the United States and traded on American stock exchanges should submit to the United States Securities and Exchange Commission (SEC). The 10-K report contains detailed information about the company's financial condition, operations, management, risks, and strategy.

The 10-K report includes the following information:

1. Financial Statements: Financial indicators such as balance sheet, income statement, cash flow statement, and statement of changes in equity.Business overview: A description of the main aspects of the business, including products, services, industry, geographic markets, and major competitors.Risk factors: An analysis of potential risks that could negatively affect the company's performance, including competition, regulatory risks, and technological changes.
2. Management's Discussion and Analysis (MD&A): The company's analysis of its financial results, strategy, plans, and factors that could impact future results.

3. Corporate governance: Information on the company's board of directors and executive officers, as well as information on their compensation, stock holdings, and stock options.
4. Security ownership and additional expenses: A description of the shareholder capital structure, including various classes of shares and shareholder rights.
5. Legal proceedings: Information on any significant legal proceedings in which the company is involved.
6. Tax matters: A description of the company's tax obligations and any potential tax issues that may arise.
7. Significant agreements and contracts: A description of significant contracts and agreements that may be material to the company's business.

In the 10-K Report, vital financial indicators provide information about the company's financial condition and performance. These indicators are divided into several financial statements, such as the balance sheet, income statement, and cash flow statement.

The Income Statement, also known as the earnings report or the statement of performance, is a summary of a company's revenues and expenses for a specific period of time, usually a year or a quarter. It shows how the company converts revenue from the sale of goods and services into net profit, considering all expenses.

Here are the main sections and items of the Income Statement:

1. Revenue: Income from the sale of goods or services. Revenue can also be referred to as sales or turnover.
2. Cost of Goods Sold (COGS): The costs of producing or purchasing the goods or services that the company sells. COGS includes materials, labor, and overhead costs for production.
3. Gross Profit: The difference between revenue and cost of goods sold. Gross profit shows how much a company earns after paying its direct costs for producing goods or services.
4. Operating Expenses: Costs of managing the company that is not related to producing goods or services. Operating expenses include employee salaries, rent, advertising, depreciation, research and development, and other non-production costs.
5. Operating Income: The difference between gross profit and operating expenses. Operating income shows how much a company earns from its core business, excluding interest and taxes.
6. Interest and Other Financial Expenses: Costs of interest on debts and other financial expenses, such as fees for servicing debts.
7. Income Tax: The amount of taxes the company must pay on its income.
8. Net Income: Reflects the final profit after accounting for all revenues, expenses, interest, and taxes for a specific period of time, usually a year or a quarter. Net income is used to determine the success of a company and its ability to generate profits for shareholders.

The balance sheet is a snapshot of a company's financial position at a specific date. It includes assets (what the company owns), liabilities (what the company owes), and equity (the difference between assets and liabilities). The balance sheet consists of the following sections:

1. Assets: These are divided into current assets (e.g., cash, accounts receivable, inventories) and long-term assets (e.g., equipment, real estate, intellectual property).
2. Liabilities: These include current liabilities (e.g., accounts payable, short-term debt) and long-term liabilities (e.g., borrowed funds, pension obligations).
3. Equity: This is the sum of funds invested by shareholders and the company's accumulated earnings.

Cash Flow Statement: This report shows how a company generates and uses cash over a specific period of time, typically a year. It details changes in cash and cash equivalents, divided into three main categories:

1. Operating Cash Flow: reflects the net cash flow generated from the company's core activities, such as selling goods and services, paying suppliers, employee salaries, and taxes. Positive operating cash flow indicates that the company is successfully converting its profits into cash.
2. Investing Cash Flow: reflects cash flows related to investments in long-term assets, such as the purchase or sale of equipment, real estate, intellectual property, shares of other companies, and debt instruments. Negative investing cash flow may be associated with investments in the company's growth and development.
3. Financing Cash Flow: reflects cash flows related to the company's financing, including the issuance and repayment of shares and debts, payment of dividends to shareholders, and other financing-related operations. Negative financing cash flow may indicate the repayment of debts or dividends.
4. Based on data from this report, we can calculate the next important financial ratios (Table 1):

Table 1. Financial metrics of companies' 10-K reports

Financial metric and signals	Definition	Formula
Gross Margin Positive signal for stock price: the higher demand for the company's stock the more metric and share price	Presents how many percentage points of revenue remain after deducting costs related to the production or sale of goods and services	$GrossMargin = (Revenue - CostofGoodsSold)/Revenue$
Operating Margin Positive signal: the more profit per dollar of sales revenue the higher metric	Shows how many percentage points of profit a company receives from its operating activities, after deducting costs related to production, sales, administrative expenses, and taxes	$OperatingMargin = OperatingIncome/Revenue$

(continued)

Table 1. (*continued*)

Financial metric and signals	Definition	Formula
Net Profit Margin	shows how much profit a company earns in net terms after deducting all expenses, including taxes and interest on debt, as a percentage of its revenue	$NetProfitMargin = (NetProfit / Revenue) * 100\%$
Current Ratio Small positive or negative signal: the higher metric the more confidence of investors or more cash which is not used effectively	assesses a company's ability to meet its current liabilities based on its current assets	$CurrentRatio = CurrentAssets / CurrentLiabilities$
Debt-to-Equity Ratio Positive signal: the lower level of debt the lower metric	compares a company's total debt to its total equity	$DebttoEquityRatio = TotalDebt / TotalEquity$
Return on Assets (ROA) Positive signal: the higher profit the more metric	measures how efficiently a company uses its assets to generate profit	$ROA = NetIncome / TotalAssets$
Return on Equity (ROE)	measures the amount of net income a company generates as a percentage of the total amount of equity invested by shareholders	$ROE = Net\ Income / Shareholders'\ Equity$
Earnings per Share (EPS) Positive signal: the more metric the more earnings	measures the amount of profit a company generates on a per-share basis	$EPS = Net\ Income / Outstanding\ Shares$
Price-to-Earnings Ratio (P/E) Positive or negative signal: if metric increases then share price rises or stock will overvalued	compares a company's stock price to its earnings per share	$P/E\ Ratio = Market\ Price\ per\ Share / Earnings\ per\ Share$
Price-to-Book Ratio (P/B)	compares a company's market price per share to its book value per share	$P/B\ Ratio = Market\ Price\ per\ Share / Book\ Value\ per\ Share$

4 Data Analysis

We obtained historical data for the past 20 years (1997–2022) for Amazon Inc. [15] using a third-party service called Financial Modeling [16], which stores and provides annual and quarterly reports of companies through an API (Fig. 1). From this data, we extracted essential metrics such as Revenue, Cost of Revenue, Operating Income, Net Income, Current Assets, Current Liabilities, Total Debt, Total Equity, Shareholders Equity, Outstanding Shares, Market Price per Share, Earnings per Share, and Book Value per Share.

Using Python, we manipulated this data and created a quarterly dataset (Table 2). We analyze this dataset and identify relationships between stock price and financial indicators extracted from the quarter report using random forest and multiple linear regression algorithms.

We selected two algorithms to compare their performance, predicting stock prices based on financial indicators extracted from business reports. Random Forest uses multiple decision trees to make accurate predictions, preventing overfitting by averaging the results. Multiple linear regression analyzes the impact of various variables on stock prices. Comparing these algorithms will help us assess each variable's contribution to explaining stock price changes (Fig. 2).

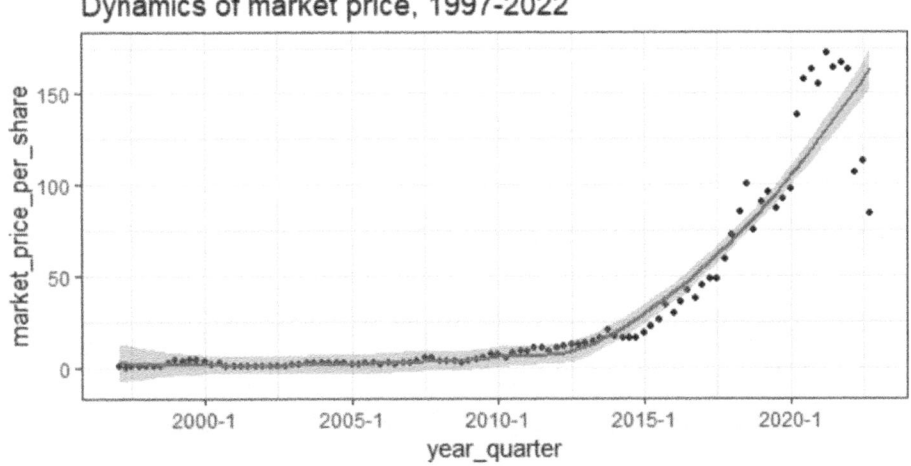

Fig. 1. Dynamics of market price per share for Amazon, Inc., 1997–2022

Therefore, before using multiple linear regression to analyze the relationship between stock prices and various variables, it is necessary to check for multicollinearity and ensure this phenomenon is absent in the data. If multicollinearity is detected, measures such as removing one of the explanatory variables or using Farrar–Glober algorithm may be taken to eliminate it. To remove correlated variables, we calculated the correlation coefficients between the independent variables, as shown in Fig. 3. Then, we set a correlation threshold of 0.8 and removed variables that were highly correlated.

Table 2. Summary statistics

Variable	Min	Q1	Median	Mean	Q3	Max
market_price_per_share,$	0.0771	1.98537	6.0945	30.5576	36.2091	172.01
Capitalization, bln $	0.2341	16.36	54.35	299.3	343.1	1737
revenue, bln $	0.016	1324	6917	26.09	33.46	149.2
cost_of_revenue, bln $	0.0118	0.9969	5.428	20.99	25.98	129.6
operating_income, bln $	−0.544	0.011337	0.173	0.99599	0.72225	8.865
net_income, bln $	−3.844	−0.00951	0.0975	0.7847	0.3109	14.32
current_assets, bln $	0.009	1.497	8.041	28.72	35.83	161.6
current_liabilities, bln $	0.009	0.9201	5.222	26.61	33.6	155.4
total_debt, bln $	0	1.258	2.145	17.84	8.252	140.1
total_equity, bln $	−1.478	0.03518	5.438	20.17	16.85	146.0
shareholders_equity,bln $	−1.478	0.03518	5.438	20.17	16.85	146.0
outstanding_shares bln $	0.9324	8.269	9.080	8.478	9.67	10.33
earning_per_share	−0.60	−0.00452	0.0101	0.0704	0.034942	1.4105
book_value_per_share	−0.199	0.0118	0.5988	2.0054	1.7423	14.333

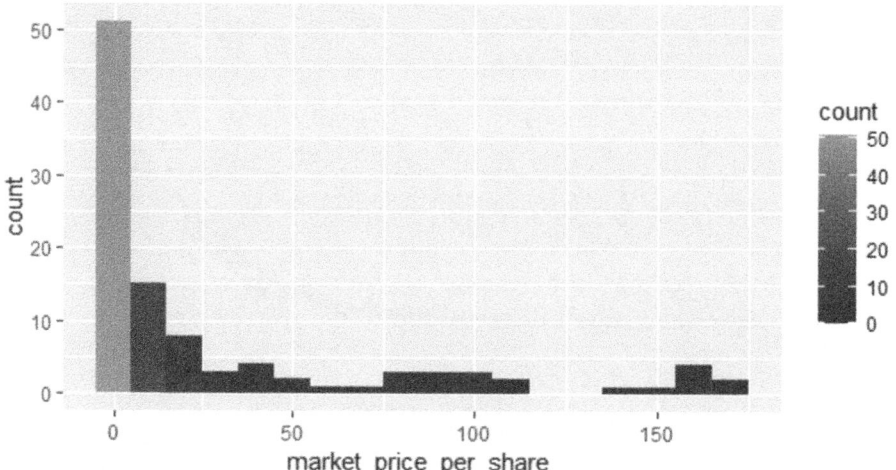

Fig. 2. Histogram of market price per share for Amazon, Inc., 1997–2022

After analyzing for correlation, we removed the correlated explanatory variables: earning_per_share, total_debt, total_assets, net_profit_margin, return_on_assets, price_to_book_ratio. We also removed three additional variables: debt_to_equity_ratio, current_ratio, and return_on_equity. We observed that after removing these variables, the deviation percentage for our model improved significantly. The improved deviation percentage indicated that the model's predictions were closer to the actual stock prices. And now we only have 5 independent variables: net_income, price_to_earning_ratio, total_equity, operating_margin, and gross_margin. To build multiple regression and random forest models, we used the sklearn library and pandas to create a dataset based on the previously processed data.

```
                              net_income   ...   price_to_book_ratio
net_income                      1.000000   ...              0.045761
total_equity                    0.600405   ...              0.069058
total_debt                      0.590331   ...              0.060904
total_assets                    0.608594   ...              0.071829
gross_margin                    0.122052   ...              0.056929
operating_margin                0.245492   ...              0.069011
net_profit_margin               0.294221   ...              0.075584
current_ratio                   0.172363   ...              0.091200
debt_to_equity_ratio            0.039714   ...              0.995438
return_on_assets                0.245485   ...              0.047480
return_on_equity                0.056283   ...              0.124220
earning_per_share               0.962601   ...              0.040639
price_to_earning_ratio          0.019554   ...              0.008598
price_to_book_ratio             0.045761   ...              1.000000
```

Fig. 3. Correlation analysis for explanatory variables

We split the original dataset into training and testing subsets to train the model and evaluate its performance. We followed a commonly used approach where the data was split in an 80:20 ratio, where 80% of the data was used for training the model, and the remaining 20% was used for evaluating its performance. It helped us estimate the accuracy and robustness of the model on new data and prevent overfitting. The dataset splitting into training and testing subsets was performed using the "train_test_split" function in the sci-kit-learn library. The code is shown in Fig. 4.

We used RMSE, Deviation Percentage, and R-squared to compare our models. RMSE measures accuracy, Deviation Percentage measures relative error, and R-squared measures how well the model fits the data. Together, these metrics provide a comprehensive assessment of the model's performance, because they are interpretable, evaluated thoroughly, and capture different aspects of prediction accuracy and model fit. Overall, they allow us to evaluate and communicate the effectiveness of our models in predicting stock prices. The results of our models are shown in Table 3.

```
df = pd.read_csv('data_AMZN/AMZN_v4.csv')
columns = ['price_to_earning_ratio', 'total_equity', 'gross_margin',
           'operating_margin', 'net_income']
x = df[columns]
y = df['market_price_per_share']
x_train, x_test, y_train, y_test = train_test_split(
    x, y, test_size=0.2, random_state=42
)
linear_model = LinearRegression().fit(x_train, y_train)
forest_model = RandomForestRegressor(random_state=42).fit(x_train, y_train)
linear_predictions = linear_model.predict(x_test)
forest_predictions = forest_model.predict(x_test)

def calculate_deviation_percentage(actual, predicted):
    deviation = predicted - actual
    return (deviation / actual) * 100

def calculate_metrics(actual, predicted):
    rmse = mean_squared_error(actual, predicted, squared=False)
    r_squared = r2_score(actual, predicted)
    return rmse, r_squared

results = pd.DataFrame(columns=['Algorithm', 'RMSE', 'Deviation Percentage', 'R-squared'])
algorithms = {
    'Linear Regression': linear_predictions,
    'Random Forest': forest_predictions,
}
results = []
for algorithm, predictions in algorithms.items():
    rmse, r_squared = calculate_metrics(y_test, predictions)
    deviation_percentage = calculate_deviation_percentage(y_test, predictions)
    average_deviation_percentage = sum(deviation_percentage) / len(deviation_percentage)
    results.append(
        [algorithm, rmse, f"{average_deviation_percentage:.2f}%", f"{r_squared *
100:)2f}%"]
print(tabulate(results,
               headers=["Algorithm", "RMSE", "Deviation Percentage", "R-squared"],
               tablefmt='psql', numalign='center'))
```

Fig. 4. Estimation of selected algorithms

Table 3. Coefficient comparison

Model	RMSE	Deviation Percentage	R-squared
Linear Regression	27.87	394.24%	76.22%
Random Forest	9.53	34.61%	97.22%

Based on the evaluation results, the Random Forest model predicts stock prices using financial indicators better than the Linear Regression model. The Random Forest model has a lower RMSE of 9.54 (compared to 27.87 for Linear Regression) and a Deviation Percentage of 34.61% (compared to 394.24% for Linear Regression), which means it has better predictive accuracy and fewer prediction errors. Additionally, the Random Forest model has a higher R-squared value of 97.22% (compared to 76.22% for Linear Regression), indicating that it has better overall explanatory power in capturing the variability in stock prices. However, caution should be exercised in overestimating the model, as additional factors may also affect the result. We can distinguish direct (Fig. 5) and polynomial (Figs. 6–7) dependences among market price per share and significant explanatory variables.

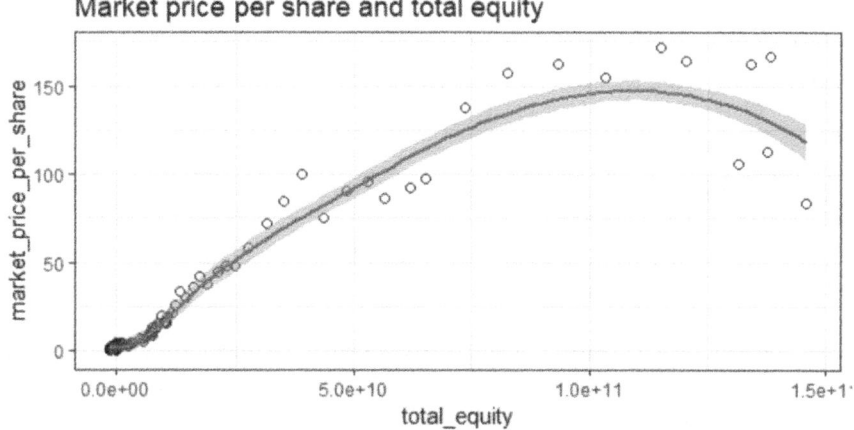

Fig. 5. Market price per share and total equity (graph)

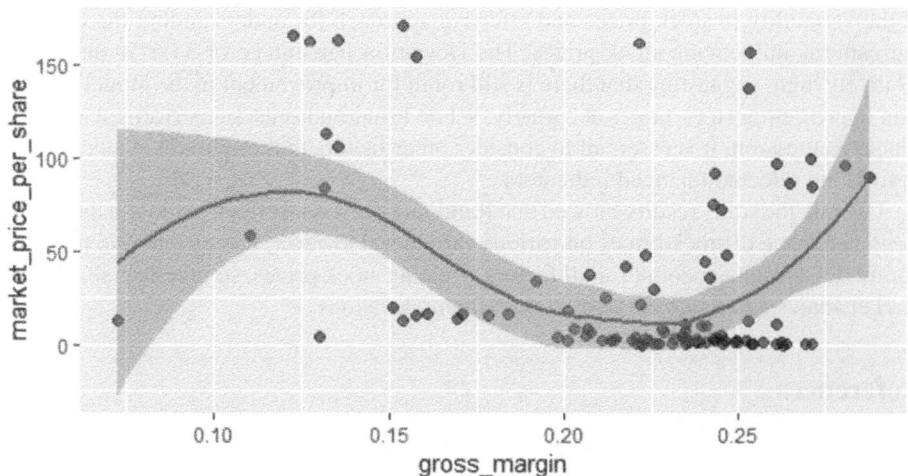

Fig. 6. Market price per share and gross margin (polynomial dependence)

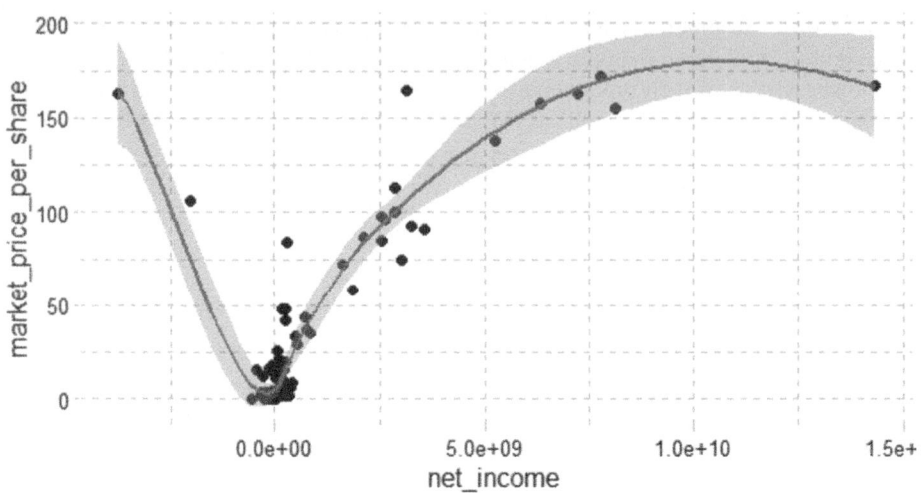

Fig. 7. Market price per share and net income (direct and polynomial dependences)

5   Conclusions

Based on the study, we can conclude that several factors, including total equity, gross margin, operating margin, price-to-earnings ratio, and net income, influence the market price per share. Taking into account the comparison of the two models, Random Forest exhibited superior performance in predicting stock prices compared to Multiple Linear Regression. It achieved significantly lower values for both RMSE (9.53) and Deviation Percentage (34.61%), indicating improved accuracy and reduced prediction errors.

Furthermore, the Random Forest model demonstrated a higher R-squared value of 97.22%, highlighting its exceptional explanatory power. This means that the financial indicators utilized in the model, making it a more reliable, can explain around 97.22% of the variability in stock prices and effective approach for understanding the underlying patterns influencing stock prices. The Deviation Percentage of 34.61% may seem relatively high, suggesting that there is still room for improvement in the model's precision in predicting stock prices accurately. While Random Forest outperformed Multiple Linear Regression, it is essential to consider other factors that might affect stock prices beyond the selected financial indicators.

Overall, the study results showed that Random Forest is an effective tool for analyzing the dependence of stock prices on various variables. However, it is essential to note that our model does not account for all factors affecting stock prices, so investors should use market analysis to make the correct investment decisions.

References

1. Interactive Brokers. https://www.interactivebrokers.com/en/home.php. Accessed 05 May 2023
2. Robinhood. https://robinhood.com. Accessed 05 May 2023

3. Rules and Use of Form 10-K. https://www.sec.gov/files/form10-k.pdf. Accessed 02 May 2023
4. Zanc, R., Cioara, T., Anghel, I.: Forecasting financial markets using deep learning. In: 2019 IEEE 15th International Conference on Intelligent Computer Communication and Processing (ICCP), Cluj-Napoca, Romania, pp. 459–466 (2019). https://doi.org/10.1109/ICCP48234.2019.8959715
5. Wasserbacher, H., Spindler, M.: Machine learning for financial forecasting, planning and analysis: recent developments and pitfalls. Digit Finan. **4**, 63–88 (2022). https://doi.org/10.1007/s42521-021-00046-2
6. Doryab, B., Salehi, M.: Modeling and forecasting abnormal stock returns using the nonlinear grey Bernoulli model. J. Econ. Finan. Adm. Sci. **23**(44), 95–112 (2018). https://doi.org/10.1108/JEFAS-06-2017-0075
7. Snihovyi, O., Ivanov, O., Kobets, V.: Cryptocurrencies prices forecasting with anaconda tool using machine learning techniques. In: Ermolayev, V. et al. (eds.) Proceedings of 14th International Conference ICTERI, pp. 453–456. CEUR-WS 2105, Aachen University (2018). https://ceur-ws.org/Vol-2105/10000453.pdf
8. Mushtaq, R., Gull, A.A., Shahab, Y., Derouiche, I.: Do financial performance indicators predict 10-K text sentiments? An application of artificial intelligence. Res. Int. Bus. Financ.Financ. **61**, 101679 (2022). https://doi.org/10.1016/j.ribaf.2022.101679
9. Huang, A.H., Zang, A.Y., Zheng, R., 2014. Evidence on the information content of text in analyst reports. Acc. Rev. **89**(6), 2151–2180 (2014). doi: https://doi.org/10.2308/accr-50833
10. De Souza, J.A.S., Rissatti, J.C., Rover, S., Borba, J.A.: The linguistic complexities of narrative accounting disclosure on financial statements: an analysis based on readability characteristics. Res. Int. Bus. Finance **48**, 59–74 (2019). https://doi.org/10.1016/j.ribaf.2018.12.008
11. Cohen, L., Malloy, C., Nguyen, Q.: Lazy prices. J. Finan. **75**(3), 1371–1415 (2020). https://doi.org/10.1111/jofi.12885
12. Kobets, V., Yatsenko, V., Mazur, A., Zubrii, M.: Data analysis of personalized investment decision making using robo-advisers Sci. Innov. **16**(2), 80–93 (2020). https://doi.org/10.15407/SCINE16.02.080
13. Kilinich, D., Kobets, V.: Support of investors' decision making in economic experiments using software tools. In: CEUR Workshop Proceedings, 2393, pp. 277–288 (2019). https://ceur-ws.org/Vol-2393/paper_273.pdf
14. Snihovyi, O., Ivanov, O., Kobets, V.: Implementation of Robo-advisors using neural networks for different risk attitude investment decisions. In: 9th International Conference on Intelligent Systems 2018: Theory, Research and Innovation in Applications, IS 2018 - Proceedings, art. no. 8710559, pp. 332–336 (2018). https://doi.org/10.1109/IS.2018.8710559
15. Amazon.com, Inc. (AMZN). https://finance.yahoo.com/quote/AMZN/. Accessed 05 May 2023
16. Financial Modeling. https://site.financialmodelingprep.com/. Accessed 02 May 2023

Artificial Intelligence Impact on Food Security of States in the World

Oleksandra Novak[1] (ID) and Vitaliy Kobets[2](✉) (ID)

[1] Taras Shevchenko National University of Kyiv, 36/1 Y. Illienka Street, Kyiv 04119, Ukraine
gs22.novak@clouds.iir.edu.ua
[2] Kherson State University, 27, Universitetska Street, Kherson 73003, Ukraine
vkobets@kse.org.ua

Abstract. Because food crisis is so wide-reaching, there is a strong a need in the transformation of world agriculture and food production sector. The innovative digital technologies, namely AI, are widely acknowledged as a solution for enhancing food crises management and agricultural productivity. The purpose of this paper is to research the linkage between food security and artificial intelligence against the backdrop of global digitalization processes by using cluster analysis (SOM algorithm). The level of impact of AI on food security is deployed in ascending order for clusters Absence, Starter, Adopter, Frontrunner. Countries with more developed digital infrastructure are better able to respond to current food security threats and build resilience for the future. Due to development of digital economy and AI solutions, the level of food security for clusters of Adopter, Frontrunner is largely higher than for countries with low level of digitalization and AI diffusion (clusters of Absence, Starter). Furthermore, the level of agriculture value added correlates with AI application and country's economic development. The more country's economy depends on agriculture, the lower is country's food security level and the slower is country's digitalization.

Keywords: Artificial intelligence · AI solutions · AgriTech · Agriculture 4.0 · Big data · Digital agriculture · Digitalization · Food security · Robotics · SOM algorithm

1 Introduction

The growing number of world population poses a series of challenges on the current agricultural model, namely the need to increase productivity, reduce costs, and preserve natural resources. The problem is exacerbated by climate change, extreme events are expected to jeopardize agricultural production. At the same time, the frequency and severity of shocks to food systems has increased due to increased number of socio-political (armed conflicts), climatic (extreme weather) and economic events. [1] Even before Russia's war against Ukraine disrupted crucial food supply chains, according to Food and Agriculture Organization (FAO) the level of global hunger had reached new records in 2021, with nearly 193 million people in acute food insecurity across 53 territories and in 2022 nearly 258 million people faced food insecurity in 58 countries

G. Antoniou et al. (Eds.): ICTERI 2023, CCIS 1980, pp. 240–251, 2023.
https://doi.org/10.1007/978-3-031-48325-7_18

[2]. The only alternative way to overcome all these challenges is to adopt emerging technologies in agriculture with a particular role of digital component and AI solutions.

According to FAO 'digitalization' of agriculture and the food value chain is ongoing [3] and is already improving access to information, inputs and markets, increasing production and productivity, streamlining supply chains and reducing operational costs. In other words, the world is witnessing the birth of next agricultural technology (agri-tech) revolution that promises to use resources efficiently and achieve food security at local level. According to 2023 WEF Markets of Tomorrow report, 29.7% percent of survey respondents from 126 countries confirmed that agriculture technologies rank first as the top technology of strategic importance globally [4].

The purpose of this paper is to research the linkage between food security and artificial intelligence against the backdrop of global digitalization processes.

We organise the remainder of our paper as follows: in Sect. 2 we consider related works and summarize the socio-economic impact of AI on food security. Section 3 is devoted to classifying the countries using self-organizing maps and machine-learning techniques into clusters in terms of their food security parameters, digitalization level and economic development. Finally, Section 4 concludes on the results achieved in the research paper.

2   Related Works

2.1   Agriculture 4.0 and AI Solutions Linkage

The technologies, acting in a synergistic and complementary way in agriculture, have the power of transformation that can be referred to as digital agriculture [5], also known as agriculture 4.0 [6], or the fourth agricultural revolution [7]. FAO explains digital agriculture as a process involving digital technologies that covers access, content and capabilities, which, if appropriately combined for the local context and needs within the existing food and agricultural practices, could deliver high agrifood value, and improve socioeconomic, and potentially environmental, impact [8]. Table 1 presents a conceptual comparison between current conventional farming and Agriculture 4.0, based on [5, 9, 10].

Table 1. Comparison between conventional agriculture and Agriculture 4.0

Conventional agriculture (Small-scale farm)	Agriculture 4.0 (Smart farm)
Analogical or mechanical Technology	Internet of Things (IoT)
No data or records	Big data
Manual labour	Robotics
Hand or animal power	Automated equipment
Farmer experience	Sensing technologies, satellite image and positioning

According to Silveira, F. D. (Fig. 1), there are 3 main levels under the "roof" of Agriculture 4.0 system. *First*, fundamental elements include basic pillars that guide the development of agriculture 4.0 (precision agriculture, smart farming, and digital farming) and without which it could not exist. *Second*, structuring elements cover key technologies that can revolutionize and impact the way commodities are produced, processed, traded, and consumed. *Third*, complementary elements encompass wider possibilities of action of agriculture 4.0 that address specific agricultural issues that require a certain degree of maturity with the structuring elements of agriculture 4.0.

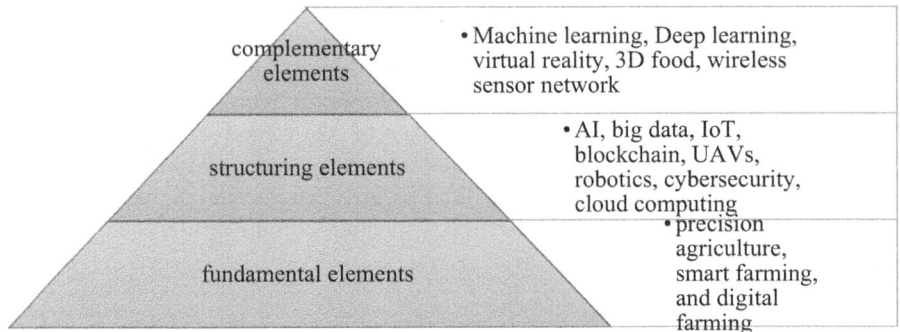

Fig. 1. The "House of Agriculture 4.0" [6]

In terms of digitalization of agriculture, IFAD experts define 6 categories of solutions: 1) advisory and information services; 2) market linkages; 3) supply chain management; 4) financial services; 5) macro-agricultural intelligence; and 6) encompassing integrated solutions [11]. In general, it is expected that technical improvements in new agricultural technologies should: optimize production efficiency (efficient control of machines, cost reduction); optimize quality (timely detection of diseases in crops); minimize environmental impact (efficient use of inputs and pesticides); minimize production-associated risks (more excellent knowledge of cultivated areas, blockchain technology adoption in value chains); build up resilience (ability of food systems to withstand shocks).

AI solutions (purely software or hardware-embedded) have become a mainstream in the global economy for the recent years. In general, AI allows computers and other machines (e.g. robots) to perform tasks previously thought to rely on human experience, creativity and ingenuity. It involves the ability of machines to function autonomously, and "learn" from large volumes of input data, without being explicitly programmed for the required task [5]. The market size of AI application in the global agriculture is expected to grow from USD 1.7 billion in 2023 to USD 4.7 billion in 2028 at CAGR of 23.1% during 2023–2028 period [12].

Moreover, there is an observable increase in investments in AI start-ups across all industries and in agrifood sector, in particular. According to AgTech report, global investment in foodtech and agtech (agrifoodtech) startups totaled $29.6 bn in 2022, a 44% decline on record-breaking 2021 levels ($51.7 billion) [13]. The reasons for such market crush are related to Russia's war against Ukraine, inflation, and continued (since COVID-19) supply chain disruptions. But the investment trend remains growing primarily due to the strong returns received by investors from AI capital and strong confidence in AI as a game changer in addressing food security challenges.

2.2 AI Role in Addressing Food Security Challenges

We consulted a number of studies investigating AI role in addressing food security challenges (Table 2).

Table 2. Research on AI solutions in addressing food security

Authors	Research focus
Bhagat P. and al	proved potential for the application of AI to attain sustainability, especially in predicting the yield, crop protection, climate control, crop genetic control, and produce supply-chain. [14]
Bobicev I., Koeleman E	importance of AI for dairy farming in developing countries to prove that farmers in Kenya who use local AI platform can increase milk production and significantly improve basic knowledge on insemination time and heat detection [5, p. 37]
von Braun J	broadly based policy agenda to include the poor and marginalized in opportunities of AI/R and to protect them from adverse effects. [15]
How M.L. and al	unified analysis of data from GFSI to illustrate how computational simulations can be used to produce forecasts of good and bad conditions in food security using multi-variant optimizations providing AI user-friendly approach. [16]
Deléglise H. and al	models that aim to predict two key indicators of food security: the food consumption score and the household dietary diversity score [17]
Hussain A. et al	policy recommendations for AI application in agri-food sector, including the need for exploitation and coordinated effort, proper regulation, multi-partner system of estimating AI effects and employment and schooling. [18]

Therefore, we decided to focus our research on investigating whether digitalization, as a whole, and AI solutions, in particular, give countries certain competitive advantages at the macro-level; and how the level of GDP dependence on agriculture correlates with AI application and country's economic development status.

2.3 The Socio-economic Impact of AI on Food Security

At the times of digital transformation era, the debate over socio-economic impact of applying AI in agriculture and food production (agri-food) sector is ongoing. The main discussion points are briefly summarized at Fig. 2.

Overall, AI solutions are aimed at increasing farming productivity and crop yield, in particular through predictive analytics-based techniques. Moreover, AI solutions are

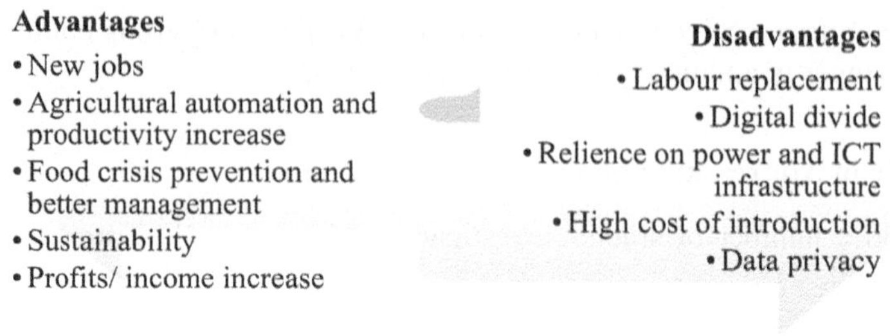

Advantages	Disadvantages
• New jobs	• Labour replacement
• Agricultural automation and productivity increase	• Digital divide
• Food crisis prevention and better management	• Relience on power and ICT infrastructure
• Sustainability	• High cost of introduction
• Profits/ income increase	• Data privacy

Fig. 2. Advantages and disadvantages of AI application in agri-food sector

helpful in soil monitoring, detection of pests and diseases, weather and temperature broadcasting which benefits the entire agri-food supply chain. Thus, these solutions are highly adopted for *first*, enhancing harvest quality in the agriculture industry, *second*, providing support services previously deemed too resource-intensive, expensive, or unavailable (e.g. due to lack of skills and expertise); *third*, driving down current operational costs by saving time and labour performed by agriculture workers. The most widely used AI solutions in agriculture include robotics, big data and sensing techniques (Table 3).

Table 3. Factors affecting the efficiency of most popular AI solutions in agriculture

Factor	Robotics (automation)	Big data (analytics)	Sensing techniques (drones, platforms)
Ownership and management of data	yes	yes	yes
Capacity of end users and data accuracy	yes	yes	yes
ICT infrastructure	yes	yes	yes
Purchase price	yes	yes	yes
Technical maintenance	yes	no	yes
Power asymmetry and dependency	no	yes	no

Elbehri, A. et al., Santos Valle et al. in their works define several factors negatively affecting the efficiency of most common AI solutions, namely, ownership and management of digital data (the absence/ presence of regulations), capacity of end users (technology adaption at the end user) and data accuracy, ICT infrastructure, purchase price, technical maintenance and servicing and power asymmetry and dependency (asymmetry of power between big data service providers and their clients). The first five are inherent to robotics, big data and sensing techniques, whereas power asymmetry and dependency is observed within big data solutions, and technical maintenance problems relate to robotics and sensing techniques.

We can observe that socio-economic impact of AI on food security has dual effect and the main issue is whether the positive effect outweigh the existing negative implications.

3   Main Results: Measuring the Impact of AI on Food Security of States

The main research question of our article is to define the impact of AI technologies on food security of states. *First,* we considered 4 food security parameters of Economist Impact Global Food Security Index (GFSI) 2022 data set. The index covers assessment of food security drivers for 111 countries ranked in GFSI rank 2022 under 4 food security pillars: Affordability, Availability, Quality and safety, Sustainability and adaptation. As of today, GFSI remains the major benchmarking model in terms of food security assessment, including 68 qualitative and quantitative food security drivers (Table 4).

Table 4. GFSI 2022 food security drivers

Affordability	Availability	Quality and Safety	Sustainability and adaptation
1.Change in average food costs FAO Consumer Production Index 2.Proportion of population under global poverty line 3.Inequality-adjusted income index 4.Agricultural trade 5.Food safety net programmes	1.Access to agricultural inputs 2.Agricultural research & development 3.Farm infrastructure 4.Volatility of agricultural production (FAO) 5. Food loss (FAO) 6. Supply chain infrastructure 7. Sufficiency of supply 8. Political and social barriers to access 9.Food security and access policy commitments	1.Dietary diversity 2.Nutritional standards 3.Micronutrient availability 4.Protein quality 5.Food safety	1.Exposure 2.Water 3.Land 4.Oceans, rivers and lakes 5.Political commitment to adaptation 6.Disaster risk management

Second, to account the impact of digitalization level (i.e. digital economy development, including AI solutions), we decide to choose the Global Connectivity Index (GCI) that evaluates the progress of 70 economies in deploying digital infrastructure and capabilities. GCI defines 3 categories of countries—Starter, Adopter, and Frontrunner and we will try to attribute this classification to the results of our analysis.

Third, in our research we included Agriculture value added (% of GDP) parameter that reflects the importance of agriculture sector development in country's GDP [19]. It also serves as a marker for country's level of economic development.

To sum up, to research the impact of AI on food security level we will build country clusters [20–22] to take into account 4 GFSI dimensions, GCI and Agriculture value added via unsupervised self-organizing maps with input layer of 6 neurons. All countries are self-organizing on the output layer neurons. The average distance to the nearest neurons after 100 iterations is decreased on almost third (Fig. 3).

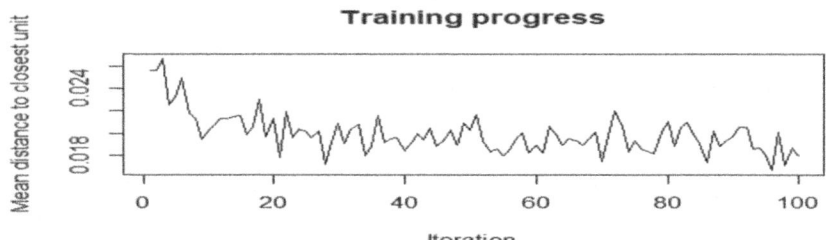

Fig. 3. Decrease in average distance to the nearest neurons after 100 learning iterations of the SOM network

The codes plot displays the value of 6 factors for each node, which corresponds to 111 countries. For the number of clusters k = 6, we have performed hierarchical clustering through SOM algorithm and have constructed the maps of the codes type. The results obtained are presented at Fig. 4.

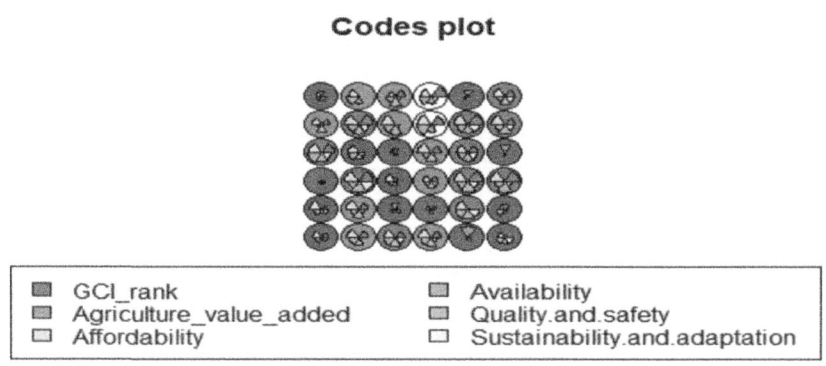

Fig. 4. Clustering of SOM map nodes

As a result of the analysis, we defined 6 country clusters as represented in Table 5 and classified them under 3 GCI categories (plus adding Absence category).

The sets of attributes of each country cluster are illustrated in Fig. 5.

As regards comparative advantages, food is most affordable and available in C3 and C6, the lowest affordable – in C4 and C5. The highest quality and sustainability is observed in C3, the very low quality food is in C4. The comparative advantages of each cluster are presented in Table 6.

Table 5. Clusters by countries

Clusters	Countries	GCI category
Cluster 1 (C1 - blue)	29 countries: Algeria, Azerbaijan, Bangladesh, Burkina Faso, Cambodia, Dominican Rep., Egypt, Ghana, Guatemala, Honduras, India, Indonesia, Jordan, Kenya, Laos, Myanmar, Nepal, Nicaragua, Pakistan, Panama, Philippines, Rwanda, Senegal, Sri Lanka, Tajikistan, Tanzania, Thailand, Tunisia, Uzbekistan	*Starter*
Cluster 2 (C2-orange)	28 countries: Argentina, Bahrain, Bolivia, Brazil, Bulgaria, Colombia, Ecuador, El Salvador, Greece, Hungary, Israel, Italy, Kuwait, Malaysia, Mexico, Morocco, Oman, Paraguay, Qatar, Romania, Saudi Arabia, Serbia, Slovakia, South Africa, Turkey, Ukraine, United Arab Emirates, Vietnam	*Adopter*
Cluster 3 (C3 - green)	26 countries: Australia, Austria, Belgium, Canada, Chile, Costa Rica, Czech Republic, Denmark, Finland, France, Germany, Ireland, Japan, Kazakhstan, Netherlands, New Zealand, Norway, Peru, Poland, Portugal, Spain, Sweden, Switzerland, United Kingdom, United States, Uruguay	*Frontrunner*
Cluster 4 (C4 - red)	20 countries: Benin, Burundi, Cameroon, Chad, Congo (Dem. Rep.), Côte d'Ivoire, Ethiopia, Guinea, Haiti, Madagascar, Malawi, Mali, Mozambique, Niger, Nigeria, Sierra Leone, Syria, Togo, Uganda, Yemen	*Absence*
Cluster 5 (C5 - purple)	5 countries: Angola, Botswana, Sudan, Venezuela, Zambia	*Absence*
Cluster 6 (C6 - white)	3 countries: China, Singapore, South Korea	*Frontrunner*

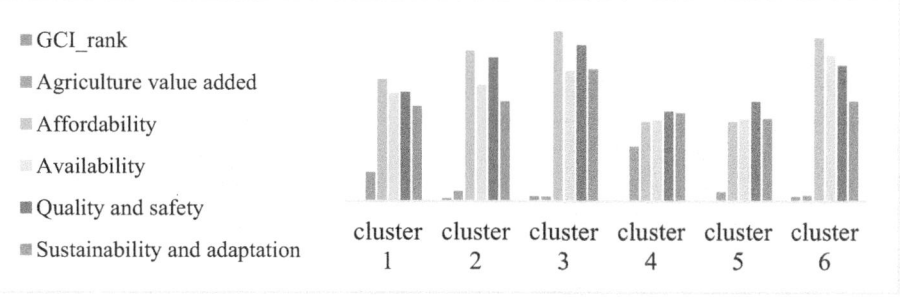

Fig. 5. Clusters by attributes

From the standpoint of our research the competitive advantages of digital technologies are the most interesting. Taking into account dependence between pillars of GFSI and GCI rank (level of digital development and AI) we can conclude that the more GFSI the more GCI rank (C2, C3, C6) and vice versa: the less GFSI the less GCI rank (C1, C4, C5). If we consider dependence between GFSI rank and Agriculture value added, we see that the more important is agriculture for country's economy, the less digitally developed it is and the more food insecure (C1, C4, C5) and vice versa: the more GFSI the less Agriculture value added (C2, C3, C6) and the more digitalised is the country.

Table 6. Clusters comparative advantages

Comparative advantages	Very high	Above average	Below average	Very low
Affordable food	C3, C6	C2	C1	C4, C5
Quality food	C3	C2, C6	C1, C5	C4
Digital development and AI	C3, C6	C2	x	C1, C4, C5
Available food	C6	C3	C4	C5
		C1, C2		
Sustainable food	C3		C1, C4, C5	x
		C2, C6		
Agriculture value added	C4	C1	C2, C5	C3, C6

The GCI categories were further used to perform the analysis of GFSI rank and agriculture value added for deferent level of AI development (Fig. 6) to prove AI comparative advantages. The countries with higher GCI rank (factor 3) have greater digital readiness and resilience, than countries with factor 1, thanks to strong digital infrastructure and as a result the potential of AI application. We can also observe that the greater the level of implementation of AI in a country, the higher the level of food security of the respective countries.

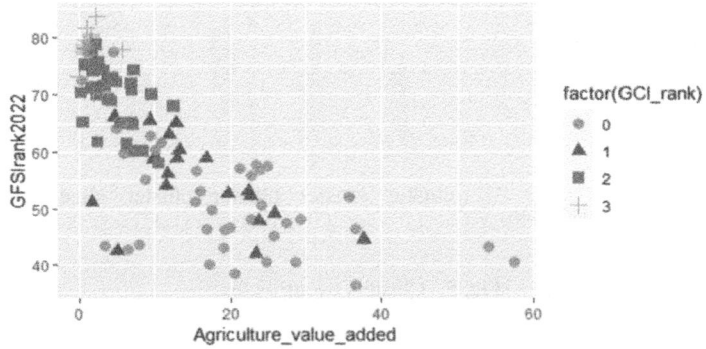

Fig. 6. GFSI rank and agriculture value added for deferent level of AI development

To check the validity of obtained results, we used the list prepared by Yahoo of 12 most advanced countries in agriculture technology (by number of agritech startups) [23]. And the results of our modelling confirm that the countries that have the biggest number of tech startups are situated in C3 and C6 with the lowest level of GDP dependency on agriculture and the highest food security level. These countries are (Australia (3), Canada (3), China (6), France (3), Germany (3), Israel (6), Japan (3), Netherlands (3), New Zealand (3), South Korea (6), UK (3), United States(3)). The majority of countries with developed agri-tech sector have two things in common – advanced economy status and high agricultural output. The latter has compelled these countries to invest in innovation in agri-technology to sustain and grow their outputs.

The regional scope of the obtained results is presented at Fig. 7. We start from defining C3, C6 as Industrial, Post-industrial economics with low level of agriculture value added in GDP, whereas other countries shall be regarded as Agrarian economies. The results obtained on countries in C1, C4 and C5 highly correlate with the 2023 FAO distribution of 45 countries in need of external assistance of food [24], therefore we shall call these clusters as Agrarian economies in Emergency.

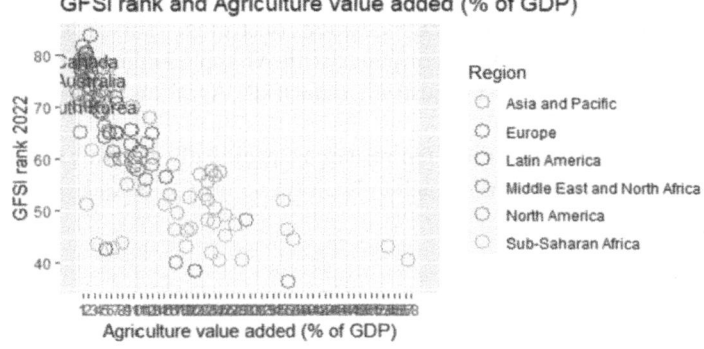

Fig. 7. Trade-offs between GFSI rank and agriculture value added by regions

Multiple regression between GFSI rank as dependent variable and explanatory variables (GCI rank and Agriculture value added) demonstrates that movement in clusters' countries from Absence to Starter, from Starter to Adopter, from Adopter to Frontrunner give rise to GFSI rank by an average of 5.6 positions. The more country's economy depends on agriculture, the lower the country's food security rating GFSI. If a country's agricultural value added increases by 1%, the country's GCI rating will decrease by 0.5 positions on average (Fig. 8).

Therefore, countries with more developed digital infrastructure are better able to respond to current food security threats, and build resilience for the future.

4   Conclusions

To sum up, we conclude that the transformative power of digital technologies, in general, and AI solutions, in particular, give countries certain competitive advantages to withstand current food security crisis. First, we found that the rise of GCI rank (level of digital

```
lm(formula = GFSIrank2022 ~ GCI_rank + Agriculture_value_added,
    data = data)

Residuals:
    Min      1Q   Median      3Q      Max
-21.2285  -3.8918  0.9052  3.7857  18.9536

Coefficients:
                          Estimate Std. Error t value Pr(>|t|)
(Intercept)                60.69808   1.67014  36.343  < 2e-16 ***
GCI_rank                    5.63230   0.72103   7.811 3.94e-12 ***
Agriculture_value_added    -0.50038   0.07135  -7.013 2.13e-10 ***
---
Signif. codes:  0 '***' 0.001 '**' 0.01 '*' 0.05 '.' 0.1 ' ' 1

Residual standard error: 6.652 on 108 degrees of freedom
Multiple R-squared:  0.7331,    Adjusted R-squared:  0.7281
F-statistic: 148.3 on 2 and 108 DF,  p-value: < 2.2e-16
```

Fig. 8. Multiple regression model of GFSI rank

development and AI) can increase food security index (GFSI rank) by an average of 5.6 positions. Therefore, due to development of digital economy and AI, the level of food security for clusters of Adopter, Frontrunner is largely higher than for countries with low level of digitalization and AI diffusion (clusters of Absence, Starter).

Second, we found that if a country's agricultural value added increases by 1%, the country's GCI rating will decrease by 0.5 positions on average. This proves that the level of GDP dependence on agriculture correlates with AI application and country's status of economic development (Post-industrial, Industrial, Agrarian economies; Agrarian economies in Emergency).

We are going to further continue our research, specifically, in terms of assessing the modern instruments (namely, AI) of achieving food security in already precarious state and constant threats.

References

1. Cottrell, R.S., et al.: Food production shocks across land and sea. Nat. Sustain. **2**, 130–137 (2019). https://doi.org/10.1038/s41893-018-0210-1
2. Food and Agriculture Organization of the United Nations (FAO). 2022 Global Report on Food Crises: Joint analysis for better decisions. https://www.fao.org/3/cb9997en/cb9997en.pdf. Accessed 31 July 2023
3. FAO, I., 2017. E-agriculture in action. FAO and ITU, 372
4. Markets of Tomorrow Report 2023: Turning Technologies into New Sources of Global Growth. https://www3.weforum.org/docs/WEF_Markets_of_Tomorrow_2023.pdf. Accessed 31 July 2023
5. Elbehri, A., Chestnov, R.: Digital agriculture in action – Artificial intelligence for agriculture. Bangkok, FAO and ITU (2021). https://doi.org/10.4060/cb7142en
6. da Silveira, F.D., Amaral, F.G.: Agriculture 4.0. Encyclopedia of smart agriculture technologies (2022). https://doi.org/10.1007/978-3-030-89123-7_207-1
7. Rose, D.C., Wheeler, R., Winter, M., Lobley, M., Chivers, C.-A.: Agriculture 4.0: making it work for people, production, and the planet. Land Use Policy **100**, 104933 (2021). https://doi.org/10.1016/j.landusepol.2020.104933
8. FAO, IFAD, United Nations, UNDP, UNICEF, WFP, WHO Regional Office for Europe and WMO. 2023. Regional Overview of Food Security and Nutrition in Europe and Central Asia (2022). Repurposing policies and incentives to make healthy diets more affordable and

agrifood systems more environmentally sustainable. Budapest. https://doi.org/10.4060/cc4 196en

9. Valle, S.S., Kienzle, J.: Agriculture 4.0 – Agricultural robotics and automated equipment for sustainable crop production. Integr. Crop Manage. **24**, 1–40 (2020). https://www.fao.org/3/cb2186en/CB2186EN.pdf

10. FAO, IFAD, UNICEF, WFP and WHO. The State of Food Security and Nutrition in the World 2022. Repurposing food and agricultural policies to make healthy diets more affordable. Rome, FAO (2022). https://doi.org/10.4060/cc0639en

11. Ceccarelli, T., Kannan, S., Cecchi, F., Janssen, S.: Contributions of information and communication technologies to food systems transformation. IFAD Res. Ser. **82** (2022)

12. Artificial Intelligence in Agriculture Market. https://www.marketsandmarkets.com/Market-Reports/ai-in-agriculture-market-159957009.html. Accessed 31 July 2023

13. AgFunder Global AgriFoodTech Investment Report 2023. https://agfunder.com/research/agf under-global-agrifoodtech-investment-report-2023/. Accessed 31 July 2023

14. Bhagat, P.R., Naz, F., Magda, R.: Artificial intelligence solutions enabling sustainable agriculture: a bibliometric analysis. PLoS ONE **17**(6), e0268989 (2022). https://doi.org/10.1371/journal.pone.0268989

15. von Braun, J.: AI and robotics implications for the poor. ZEF Working Paper Series **188**, 1–32 (2019)

16. How, M.-L., Chan, Y.J., Cheah, S.-M.: Predictive insights for improving the resilience of global food security using artificial intelligence. Sustainability **12**(15), 6272 (2016). https://doi.org/10.3390/su12156272

17. Deléglise, H., Interdonato, R., Bégué, A., d'Hôtel, E.M., Teisseire, M., Roche, M.: Food security prediction from heterogeneous data combining machine and deep learning methods. Expert Syst. Appl. **190**, 1–11 (2022). https://doi.org/10.1016/j.eswa.2021.116189

18. Hussain, A.A., Dawood, B.A., Altrjman, C., Alturjman, S., Al-Turjman, F.: Application of artificial intelligence and information and communication technology in the grid agricultural industry: business motivation, analytical tools, and challenges. In: Sustainable Networks in Smart Grid, pp. 179–205 (2022). https://doi.org/10.1016/B978-0-323-85626-3.00002-8

19. Kobets, V., Novak, O.: EU countries clustering for the state of food security using machine learning techniques. Neuro-Fuzzy Modeling Techn. Econ. **10**, 86–118 (2021). https://doi.org/10.33111/nfmte.2021.086

20. Kobets, V., Yatsenko, V., Voynarenko, M.: Cluster analysis of countries inequality due to it development through macros application. In: Ermolayev, V., Mallet, F., Yakovyna, V., Mayr, H.C., Spivakovsky, A. (eds.) ICTERI 2019. CCIS, vol. 1175, pp. 415–439. Springer, Cham (2020). https://doi.org/10.1007/978-3-030-39459-2_19

21. Kobets, V., Pilshchyk, E., Mykhaylova, V.: Spatial models of countries economic development under pandemic condition (2021). In: 2021 11th International Conference on Advanced Computer Information Technologies, ACIT 2021 - Proceedings, pp. 222–225. https://doi.org/10.1109/ACIT52158.2021.9548592

22. Kobets, V., Yatsenko, V., Voynarenko, M.: Cluster analysis of countries inequality due to IT development (2019). In: CEUR Workshop Proceedings, vol. 2393, pp. 406–421. https://ceur-ws.org/Vol-2393/paper_341.pdf

23. Twelve Most Advanced Countries in Agriculture Technology 2022. https://finance.yahoo.com/news/12-most-advanced-countries-agriculture-140128710.html. Accessed 31 July 2023

24. FAO. Crop Prospects and Food Situation – Quarterly Global Report No. 1 (2023). https://doi.org/10.4060/cc4665en

Increasing Investment Portfolio Profitability with Computer Analysis Trading Strategies

Serhii Savchenko and Vitaliy Kobets(⊠)

Kherson State University, 27, Universitetska st., Kherson 73003, Ukraine
savchenko.serhii@gmail.com, vkobets@kse.org.ua

Abstract. The paper is devoted to our research on the effectiveness of using different middle- and long-term trading strategies based on computer analysis (CA) indicators. This paper contains a brief overview of three technical analysis indicators that are usually used for getting buy or sell signals for some specific financial instruments (FIs). We have described the approach that allows using such signals not only for a single FI but for a whole investment portfolio. The initial investment portfolio is generated using Markowitz's Modern Portfolio Theory. The paper contains an overview of similar approaches presented by other researchers. During the experimental part of the research, we compared the effectiveness of using such CA indicators as moving average (MA), relative strength index (RSI), and support and resistance (S&R). The results prove that using certain CA strategies allows not only to increase the initial investment portfolio profitability on rising periods in the financial market but also may reduce loss during a global financial market recession.

Keywords: Investment Portfolio · Automated Financial Software · Robo-Advisor · Computer Analysis

1 Introduction

The inflation index (or consumer price index) is an indicator that characterizes changes in the general level of prices for goods and services that the population buys for personal consumption. In Ukraine, the inflation index was 5% and 10% in 2020 and 2021, respectively. At the end of 2022, the inflation rate exceeded 20%, and the Ukrainian government included an expected inflation index of 28% in the planning of the state budget for 2023 [1]. Traditional methods of protecting savings from inflation, such as bank deposits, bonds are unable to compensate for the devaluation of monetary savings of the population at such a high level of consumer price index growth.

Individuals who aim to at least preserve their existing savings are seeking alternative means to do so. One such solution is investing in securities, stocks, ETFs and other financial instruments (FIs). However, if a person lacks experience in the financial sector, there is a high risk of incorrectly composing an investment portfolio, such as failing to adhere to asset diversification rules. To assist in the composition of an investment plan

G. Antoniou et al. (Eds.): ICTERI 2023, CCIS 1980, pp. 252–264, 2023.
https://doi.org/10.1007/978-3-031-48325-7_19

for individuals without specialized skills and competences, special software tools called Robo-Advisors (RA) have been developed.

RA is a type of financial software that gives investment advice based on information (such as investment goals, risk preferences, budget, and desired investment assets) provided by a user. Unlike the classical financial consulting companies offering to compose a personalized investment portfolio for a quite high price, RA services charge a relatively small transaction fee and some of them only have fixed annual fees [2].

Technical analysis is a trading strategy that involves analyzing past market data, such as price changes and volume, to identify patterns and make predictions about future price movements. Technical analysis is widely used by traders and investors in various financial markets, including stocks, currencies, and other FIs. Technical analysis traders use various technical indicators, such as moving averages and oscillators, to identify potential buying and selling opportunities. Such indicators that calculate some numerical value based on past market data are also called computer analysis. Technical analysis as a rule is applied after fundamental analysis.

The purpose of this research is to develop an algorithm that allows using trading strategies based on computer analysis for an initial investment portfolio, to implement this algorithm in a software tool using Python programming language, and to analyze the effectiveness of this approach on real historical datasets.

Research tasks include examination on which indicators based trading strategies shows better performance versus simple buy-and-hold strategy (benchmarks strategy), providing the step-by-step algorithm for applying trading strategy not only for a single FI but also for a whole investment portfolio, and analysis of the obtained results during the experimental part of the research.

The paper is structured as follows: Sect. 2 contains a brief overview of an existed approaches of using technical analysis indicators, Sect. 3 describes the idea of using computer analysis trading strategies to enhance the profitability of an investment portfolio, Sect. 4 presents obtained practical results, the last part concludes.

2 Related Works

Algorithmic trading is a computer-based trading method that utilizes algorithms to execute trades in financial markets. Algorithmic trading uses predefined set of instructions to perform a trade [3]. This approach based on mathematical models and statistical analysis to identify trading opportunities, manage risk, and optimize trading strategies. It also allows reducing trading costs and increasing trading speed. With the rapid development of computer technology and the increasing availability of market data, algorithmic trading has become a very popular way of professional trading in recent years. The study by S. Baek et al. [4] describes the way of applying machine-learning techniques in algorithmic trading application for futures markets. Thus, algorithmic trading is a topic of interest and research in the academic and financial communities.

Technical analysis is a methodology that identifies certain patterns and trends by analyzing market data, such as price and volume, and suggests investment decisions. It based on the assumption that historical data can be used to predict future price values. Instruments for technical analysis can be divided into graphic tools (so-called charting) and indicators of technical analysis [5]. Traders which use technical analysis use charts and technical indicators to interpret market data and make trading decisions. In our research, we focused on three main indicators of technical analysis: Moving Average Convergence/Divergence (MACD), Support and Resistance (S&R), and Relative Strength Index (RSI) [6].

Ha et al. [7] proposed an optimal intraday trading algorithm to reduce overall transaction costs by absorbing price shocks when an online portfolio selection method rebalances its portfolio. The proposed trading algorithm optimizes the number of intraday trades and finds an optimal intraday trading path. Backtesting results from the historical data of NASDAQ-traded stocks show that the proposed trading algorithm significantly reduces the overall transaction costs when market liquidity is limited [7]. Intraday trading algorithm is applicable to portfolio rebalancing strategy.

Algorithmic trading is the computerized execution of FIs following pre-specified rules and guidelines. The proposed algorithm is much effective for large capital investment, it generates more benefit for more frequent rebalance, the lower transaction fees rate is, the more benefit [7].

Dai et al. [8] addresses the problem of the instability of forecasting stock price investment and the difficulty in determining investment proportion by proposing the trend peak price tracing which sets adjustable historical window width. It uses slope value to judge prediction direction to track price change, which uses Markowitz's mean-variance theory and Kelly's capital growth theory by the means of exponential MA and peak equal weight slope value. Online portfolio strategy using learning algorithm demonstrate advantages in balancing risk and return [8]. Asset allocation in a portfolio is the core issue that people are concerned about.

Bisht et al. [9] presents an integrated approach of portfolio construction based on sector analysis for investment in National Stock Exchange of India combining different evidences, such as Average directional index (ADX), Relative strength indicator (RSI), Simple moving average (SMA), Sector's relative strength (SRS) with different moving window. For optimizing of the constructed portfolio, an optimization function is constructed for minimum volatility, maximum Sharpe ratio, or maximum return under tolerable risk and simulated using a deep recurrent neural network, which is based on industry's performance [9]. Behavioral portfolio can demonstrate better results than Markowitz's model, when prevailing assumptions of overconfidence, gambler's fallacy, illusion of validity etc. [10].

Advancements in machine learning have revealed a wide range of new opportunities for using advanced computer algorithms, such as reinforcement learning in portfolio risk management. Ngo et al. [11] show superior performance of reinforcement learning models over traditional optimization models following the mean-variance framework in different financial market (e.g., ETFs) settings while optimizing the Sharpe ratio even for similar degree of investment portfolio diversification.

In our previous research [12] we have described the algorithm of generating next month's Close price prediction using Long Short-Term Memory neural network and proved that investment portfolio built using not only historical data but also a one-month prediction shows better results even during global stock market recession. We evaluated portfolio performance using simple buy-and-hold strategy. The aim of our current research is to expand that study and to propose an algorithm which applies trading strategy based on indicators of technical analysis for investment portfolio that includes more than one FI in contrast of existing researches.

3 Research Methodology

3.1 Model

The main objective of our research is to compare the profitability of investment portfolios that will change the distribution shares of FIs in accordance with buy or sell signals of some technical analysis indicator comparing to a simple buy-and-hold strategy. We used three technical analysis indicators: Moving Average Convergence/Divergence (MACD), Support and Resistance (S&R), and Relative Strength Index (RSI). All three indicators are based on price values for a certain time period in the past, but use different algorithms to calculate their numerical values.

Simple Moving Average (SMA) is a set of numeric values calculated by the arithmetic mean of a certain amount of previous price values (periods). We can use different numbers of periods that used to calculate current SMA value. When SMA values based on different periods diverge, it is called MA convergence or divergence. The periods for calculating SMA are divided into short-term and long-term. We used two SMA values to calculate MACD buy or sell signal. MACD strategy gives buy signal if 5-day SMA value is greater than 12-day SMA value and sell signal otherwise. The 5-day SMA reflects the short-term trend of price movement. Comparing the 5-day SMA to the 12-day SMA allows getting an early signal of a price trend change.

The Relative Strength Index (RSI) is used in financial analysis to measure the momentum of an asset's price movement developed by Wilder [13]. The RSI value is calculated by comparing the magnitude of an asset's recent price gains to its recent price losses for some previous timesteps number and oscillates between 0 and 100. The formal notation for calculating RSI signal is:

$$RSI(t) = 100 - \frac{100}{1 + RS(t)} \tag{1}$$

where t is previous timesteps number, and Relative Strength $(RS(t))$ is determined by dividing the average gain by the average loss for a given period. When the RSI of some FI reaches the value 70 and above there are a high probability that the price will come down soon. When the RSI value is 30 or less it is considered as a buy signal (Fig. 1).

Another technical analysis approach is Support and Resistance strategy (S&R), which consists of identifying key price levels where price action is likely to reverse in the opposite direction. These levels are referred to as support (when the price does not drop below some key level) and resistance (when the price does not rise above the key level) ones. The support and resistance levels can be identified through the use of various technical analysis tools such as trend lines, extremum values, moving averages, and Fibonacci retracements. In the research, we used S&R signals based on the determination of impulse price changes after a long-term consolidation.

The initial condition of the developed algorithm, which is described below, is a previously formed investment portfolio. In our previous work [12], we described a method

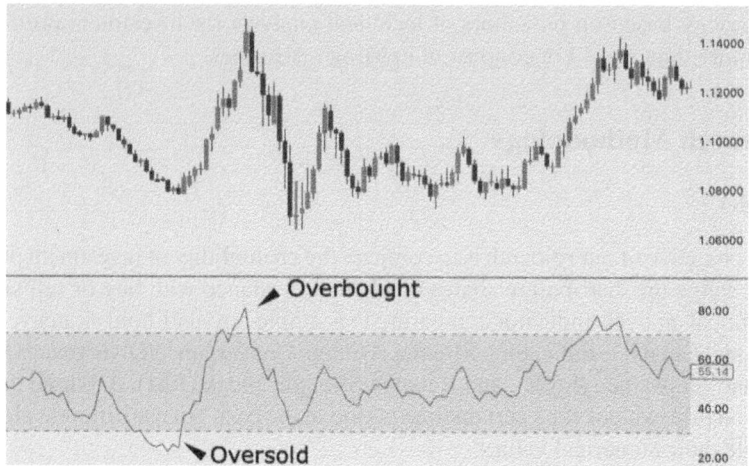

Fig. 1. RSI overbought and oversold signals on EUR/USD chart.

of building an initial investment portfolio that takes into account the investor's desire to minimize acceptable risk or maximize expected return using the Markowitz portfolio model. The basic investment portfolio was built for a risk-neutral investor, when the ratio of risk to expected return is minimized (2), and includes twelve stocks. Here N is a number of assets, R_p is an expected return of the portfolio, σ_p is a level of risk (standard deviation), and w_i is a percentage of asset i in portfolio p [12].

$$\begin{cases} \frac{\sigma_p}{R_p} \to min \\ w_1 + w_2 + \ldots + w_N = 1 \\ w_i \geq 0 \end{cases} \tag{2}$$

The algorithm for using technical analysis indicators for an investment portfolio consisting of several FIs is presented in Fig. 2.

Let us consider the algorithm presented above in more detail. The first step is to form an initial investment portfolio based on data on past price movements of FIs. To do this, we use the Markowitz method of investment portfolio formation. If the starting point of our experiment is, for example, January 2021, we will use data from the previous five years, i.e. from January 2016 to December 2020. After the initial portfolio is formed, we buy FIs according to their share in the portfolio at the opening price of 2021.

The experiment compares 4 portfolios. The first corresponds to the Buy-and-hold strategy. This means that the proportion of FIs in its allocation does not change. The other three portfolios can be rebalanced if buy or sell signals are received from the MACD, RSI, or S&R indicators. The task of the experiment is to study the dynamics of changes in the value of each portfolio over a period of one year.

The algorithm used to rebalance the portfolios is universal for all three indicators. The check and rebalancing is performed at the end of each week and consists of the following steps (Table 1):

Since we know the entire history of changes in the proportions of FIs in the portfolio and the history of price changes of each instrument, we can calculate the estimated value

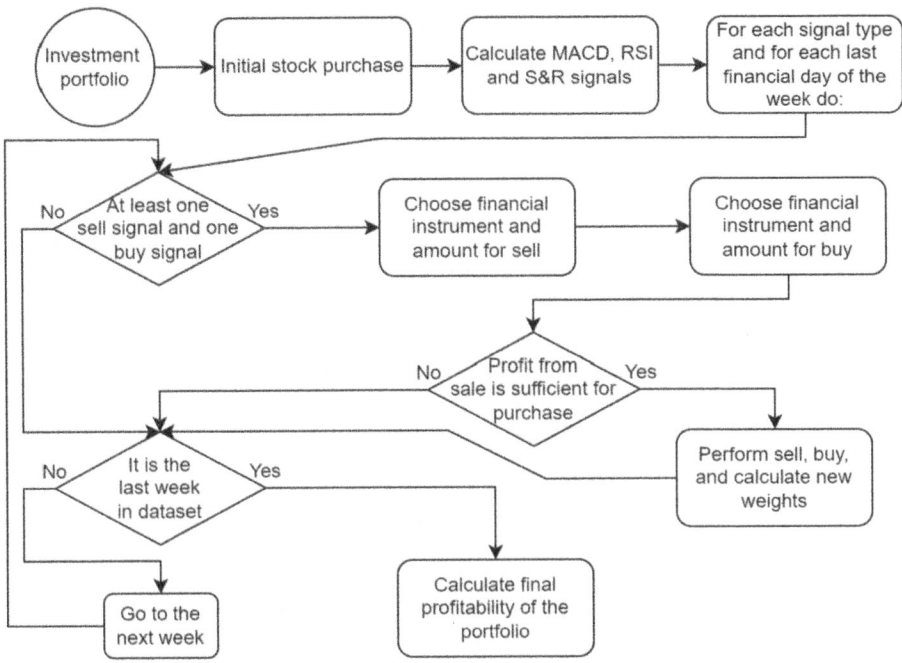

Fig. 2. The algorithm workflow.

of the portfolio at each point in time, for example, at the end of each month. We can also calculate the maximum and minimum values of the portfolio over time to further compare portfolios that have used different rebalancing strategies or that have used a buy-and-hold strategy.

3.2 Experimental Part

Listings 1–3 demonstrate some parts of the implementation of the algorithm described in Sect. 3.1. The full code of the experimental part can be found at https://github.com/serhii1savchenko/portfolio-strategies. To develop the experimental part, we used the Python, the Anaconda development platform, and the open-source libraries numpy, pandas, matplotlib, pypfopt, and yfinance.

For the practical experiments in this research, a selection of 12 stocks was made. This set includes stocks of companies from various sectors of the economy, such as high technology companies (Apple Inc., Alphabet Inc., Microsoft Corporation, Amazon.com, Inc.), microelectronics firms (Intel Corporation, Advanced Micro Devices, Inc., NVIDIA Corporation), engineering companies (Ford Motor Company, Tesla, Inc.), and financial institutions (JPMorgan Chase & Co, Morgan Stanley, Vanguard 500 Index Fund), which contributes to a well-diversified investment portfolio. The five-year dataset range allows us to take into account medium- and long-term trends in the price of FIs. Historical price data for all the listed companies can be sourced from the publicly available Yahoo Finance service [14]. The first step involves calculating the initial investment portfolio

Table 1. Rebalane portfolios algorithm.

Step	Description
1	Check whether there are FIs for which a sell signal and a buy signal have been received. Sell signals are taken into account only for those FIs that do not have a zero share in the current portfolio. If there are no such instruments, no rebalancing is performed
2	Among the FIs for which a sell signal has been received (candidates for sale), select one (ticker S): a. Build an investment portfolio based on data for the last five years. For each FI (among the candidates for sale), compare the share in the current portfolio with the share in the newly formed portfolio. If the share has not changed or has increased, the instrument is removed from the candidates for sale. If the share has decreased, you need to calculate how much of the FI should be sold. To do this, calculate by how many percent the share in the new portfolio is smaller than the share in the current portfolio b. If there are no candidates for sale after the above check, no rebalancing is performed. If there is at least one candidate for sale, the one with the highest profit from the sale of the relevant share is selected
3	Among the FIs for which a buy signal has been received (buy candidates), select one (ticker B): a. For each buy candidate, calculate the percentage change in value over the past five years b. Choose the FI whose value has increased the most or decreased the least
4	If the funds received from the sale of a certain share of FI S are sufficient to purchase at least one share of FI B, then we carry out the process of selling and purchasing the relevant FIs at current prices. Recalculate the shares of the distribution of FIs in the current portfolio
5	If it is not the last week of the test data set, proceed to step 1. Otherwise, calculate the final value of the portfolio using the last known prices of the FIs in the test data set

using historical daily closing price data. Once the initial portfolio is formed according to the classical Markowitz model [15] using the pypfopt library, we need to purchase shares according to the weights calculated in the portfolio. The initial balance in the experiment is $100,000.

Next, we calculate the values of the MACD, RSI, and S&R indicators for each FI. A value of -1 corresponds to a sell signal, 1 corresponds to a buy signal, and 0 means no signal. For the RSI oscillator, a value of 70 or higher is perceived as a sell signal, while a value of 30 or lower is perceived as a buy signal.

```
def check_tickers_for_sell(tickers_sell, now, current_weights):
    ticker_percent_sell = {}
    end_date = now.strftime('%Y-%m-%d')
    start_date = (now - relativedelta(years=5)).strftime('%Y-%m-%d')
    # Get data for all tickers for last 5 years
    data_for_tickers = yf.download(tickers, start_date, end_date, \
                                   progress=False)['Close'].dropna()
    # Generate portfolio on this data
    mu = mean_historical_return(data_for_tickers)
    S = CovarianceShrinkage(data_for_tickers).ledoit_wolf()
    ef = EfficientFrontier(mu, S, weight_bounds=(0,1))
    weights = ef.max_sharpe()
    new_weights = ef.clean_weights()
    # Filter tickers which have lower weight in new portfolio
    for ticker in tickers_sell:
        if new_weights[ticker] == 0:
            ticker_percent_sell[ticker] = 100
        elif new_weights[ticker] < current_weights[ticker]:
            delta = current_weights[ticker] - new_weights[ticker]
            sell_percent = (delta * 100) // current_weights[ticker]
            ticker_percent_sell[ticker] = sell_percent
    return ticker_percent_sell
```

Listing 1. Checking tickers for sale.

It is also necessary to define auxiliary functions that are used to select one FI for sale and one FI for purchase in case a sell or buy signal is received for several FIs at the same time. The check_tickers_for_sell function is used to filter the stocks for which a sell signal has been received. To do this, the investment portfolio is calculated based on the data for the last 5 years (the same period as used to build the initial portfolio). A FI is checked if the share of this FI in the new portfolio is less than its share in the current portfolio. The share to be sold is calculated as the percentage by which the share in the new portfolio is less than the share in the current portfolio (Listing 1). Among the instruments that have passed the check, the one that will give the highest profit from the sale of the corresponding share is selected (Listing 2).

Among the FIs for which a buy signal has been received, the one with the highest percentage of price increase or the lowest percentage of price decrease over the past 5 years is selected (Listing 3).

```
def select_ticker_for_sell(data, date, filtered_sell_tickers_with_percents, \
                           stocks_current):
    sell_ticker = ''
    sell_profit = 0
    stocks_to_sell = 0
    for ticker in filtered_sell_tickers_with_percents:
        no_of_stocks = stocks_current[ticker]
        cur_price = data[ticker][date]
        sell_percent = (filtered_sell_tickers_with_percents[ticker] / 100)
        potentail_profit = no_of_stocks * cur_price * sell_percent
        if potentail_profit > sell_profit:
            sell_profit = potentail_profit
            sell_ticker = ticker
            stocks_to_sell = no_of_stocks * sell_percent
    return [sell_ticker, sell_profit, stocks_to_sell]
```

Listing 2. Choosing tickers for sale.

```
def select_ticker_for_buy(tickers_buy, now):
    if len(tickers_buy) == 1:
        return tickers_buy[0]
    end_date = now.strftime('%Y-%m-%d')
    start_date = (now - relativedelta(years=5)).strftime('%Y-%m-%d')
    data_for_tickers_last_year = yf.download(tickers_buy, \
        start_date, end_date, progress=False)['Close'].dropna()
    ticker_percent_delta = {}
    for t in tickers_buy:
        first_price = data_for_tickers_last_year[t][0]
        last_price = data_for_tickers_last_year[t]\
                                [len(data_for_tickers_last_year[t]) - 1]
        delta = last_price - first_price
        delta_abs = abs(delta)
        percent = (delta_abs * 100) / first_price
        if delta > 0:
            ticker_percent_delta[t] = percent
        else:
            ticker_percent_delta[t] = -1 * percent
    sorted_by_delta = dict(sorted(ticker_percent_delta.items(), \
                                    key=lambda x:x[1], reverse=True))
    return list(sorted_by_delta.keys())[0]
```

Listing 3. Selecting ticker to buy.

After completing all portfolio rebalancing, the final version of the investment portfolio will be obtained, which takes into account all changes in the distribution of shares of FIs. A comparison of the results obtained by using the MACD, RSI, and S&R strategies relative to the basic buy-and-hold strategy is presented in Sect. 4.

4   Results

Table 2 shows a comparison of the dynamics of changes in the value of the four portfolios in 2021. In the first portfolio, the shares of FIs did not change throughout the entire period. The other portfolios used a weekly rebalancing strategy in accordance with the signals of the certain indicator (MACD, RSI or S&R).

Table 3 shows the minimum and maximum value of each portfolio and compares the maximum and final difference in the portfolio value relative to the portfolio in which the distribution of FIs did not change (buy-and-hold portfolio without rebalancing).

According to the results of the experiment, the strategy based on the S&R indicator signals showed the best performance (winner strategy). The final value of the portfolio that was rebalanced based on S&R signals was $10059.51 more than the value of the buy-and-hold portfolio. The final value of the portfolio that used RSI signals for rebalancing is almost the same as the value of the buy-and-hold portfolio. The worst result was obtained by the portfolio that used MACD signals for rebalancing, its final value is $9057.77 less than the buy-and-hold portfolio. A comparison of the dynamics of changes in the value of buy-and-hold, S&R, and other portfolios is shown in Fig. 3.

An experiment was also conducted using data for 2022. Results demonstrate a comparison of the dynamics of changes in the value of the four portfolios during 2022 (the first portfolio was not rebalanced, the other three used a certain indicator to initiate rebalancing). Table 4 shows the minimum and maximum portfolio values and a comparison

Table 2. Portfolios value changes (2021).

Date	Buy and hold	MACD	RSI	S&R
2021-01-29	100383.33	99617.89	100383.33	100383.33
2021-02-26	96859.97	95738.59	96859.97	96859.97
2021-03-31	94124.58	92538.74	94124.58	94251.82
2021-04-30	102026.81	99750.20	102026.81	102426.55
2021-05-28	100352.40	98422.86	100352.40	100698.61
2021-06-30	116128.21	111082.50	115941.38	117145.00
2021-07-30	119255.91	113388.36	119264.85	122469.00
2021-08-31	129365.25	121126.30	129096.94	131121.99
2021-09-30	123008.61	117461.52	123368.31	127097.62
2021-10-29	149742.30	149184.43	151142.82	158524.52
2021-11-30	179374.49	167608.01	171513.58	188390.81
2021-12-31	165761.09	156703.32	166938.95	175820.61

Table 3. Comparison of rebalancing strategies (2021).

Strategy	Final value	Max value	Min value	$|\Delta|_{max}$	Δ_{final}
MACD	156703.32	169608.5	84659.33	−7.6%	−5.46%
RSI	166938.95	173551.9	84853.55	−4.3%	+0.71%
S&R	175820.61	189987.09	84853.55	+7.4%	+6.06%
Buy and hold	165761.09	181497.87	84853.55	–	–

of the maximum and final difference in portfolio value relative to the portfolio in which the distribution of shares of FIs did not change.

According to the results of the experiment on the data on changes in the prices of FIs for 2022, the best final result was also shown by the portfolio that used the S&R indicator signals for weekly rebalancing. Its final value is $3226.44 higher than the buy-and-hold portfolio.

Unforeseen events, such as wars or natural disasters, have notable impact on macroeconomic indicators, often resulting in a collapse of prices across various FIs. However, it is crucial to recognize that investments in FIs typically target long-term growth, necessitating a broader examination of the overall dynamics of changes in their values over extended periods. For instance, the S&P 500 index graph over the past 15 years illustrates a consistent upward trend, outpacing the average inflation rate. Even during the challenging times of the COVID-19 pandemic in 2020, it took 27 weeks (from 10th February to 17th August) for the index's price to recover. Subsequently, the index demonstrated steady growth and reached an all-time high in late 2021. Presently, the following

Fig. 3. Comparison of portfolios value throughout 2021.

Table 4. Comparison of rebalancing strategies (2022).

Strategy	Final value	Max value	Min value	$|\Delta|_{max}$	Δ_{final}
MACD	54067.50	105072.29	51705.64	−5.8%	−5.08%
RSI	56390.98	105072.29	54296.01	−1.7%	−1%
S&R	60189.76	105072.29	57799.27	+7.9%	+5.66%
Buy and hold	56963.32	105072.29	54816.37	–	–

recession, which endured throughout 2022, displays signs of concluding, marking the beginning of a new period of growth.

The proposed algorithm for rebalancing the investment portfolio reduces the risk of a significant drawdown in the portfolio value when the value of certain FIs decreases. Experiments based on the price data for selected FIs for 2021 and 2022 show that the rebalancing algorithm performs better with the S&R and RSI indicators.

5 Conclusions

Thus, the paper provides a brief overview of technical analysis tools used by traders to maximize profits from buying and selling FIs. The advantage of such approaches lies in their high interpretability and adaptability to different types of FIs. The paper also presents a new algorithm that allows using technical analysis indicator signals to rebalance investment portfolios using different strategies. We also proposed enhanced criteria for selecting FIs for making buy and sell decisions. During the experimental part of the research, we developed a software application to test hypotheses on real historical data. The experimental part of the study confirmed the feasibility of using buy and sell signals from RSI and S&R indicators for periodic rebalancing of the investment portfolio.

In our further research, we are going to consider more technical analysis indicators, such as Commodity Channel Index, Weighted Moving Average, volume-based indicators and others and to investigate the dependence of the effectiveness of using a certain indicator depending on the type of FI (shares, precious metals and raw materials, cryptocurrency, etc.).

References

1. Consumer price indices for goods and services. State Statistics Service of Ukraine. https://www.ukrstat.gov.ua/operativ/operativ2010/ct/is_c/arh_isc/arh_iscm10_u.html. Accessed 05 May 2023
2. Savchenko, S., Kobets, V.: Development of software architecture and machine learning modules of robo-advisor system for personalized investment portfolio generation. In: Ermolayev, V., et al. (eds.) Information and Communication Technologies in Education, Research, and Industrial Applications. ICTERI 2021. CCIS, vol. 1698, pp. 153–179. Springer, Cham (2022). https://doi.org/10.1007/978-3-031-20834-8_8
3. Mathur, M., Mhadalekar, S., Mhatre, S., Mane, V.: Algorithmic trading bot. In: ITM Web of Conferences, vol. 40, p. 03041 (2021). https://doi.org/10.1051/itmconf/20214003041
4. Baek, S., Glambosky, M., Oh, S., Lee, J.: Machine learning and algorithmic pair trading in futures markets. Sustainability **12**, 6791 (2020). https://doi.org/10.3390/su12176791
5. Kolkova, A.: Comparison of trading systems based on technical analysis using real and random data. In: 21st International Scientific Conference Enterprise and Competitive Environment (2018)
6. Kobets, V., Petrov, O., Koval, S.: Sustainable robo-advisor bot and investment advice-taking behavior. In: Maślankowski, J., Marcinkowski, B., Rupino da Cunha, P. (eds.) Digital Transformation. PLAIS EuroSymposium 2022. LNBIP, vol. 465, pp. 15–35. Springer, Cham (2022). https://doi.org/10.1007/978-3-031-23012-7_2
7. Ha, Y., Zhang, H.: Algorithmic trading for online portfolio selection under limited market liquidity. Eur. J. Oper. Res. **286**, 1033–1051 (2020). https://doi.org/10.1016/j.ejor.2020.03.050
8. Dai, H.-L. Liang, C.-X. Dai, H-M. Huang, C.-Y., Adnan, R.M.: An online portfolio strategy based on trend promote price tracing ensemble learning algorithm. Knowl.-Based Syst. **239**, 107957 (2022). https://doi.org/10.1016/j.knosys.2021.107957
9. Bisht, K., Kumar, A.: A portfolio construction model based on sector analysis using Dempster-Shafer evidence theory and Granger causal network: an application to national stock exchange of India. Expert Syst. Appl. **215**, 119434 (2023). https://doi.org/10.1016/j.eswa.2022.119434
10. Majewski, S., Majewska, A.: Behavioral portfolio as a tool supporting investment decisions. Procedia Comput. Sci. **207**, 1713–1722 (2022). https://doi.org/10.1016/j.procs.2022.09.229
11. Ngo, V.M., Nguyen, H.H., Nguyen, P.V.: Does reinforcement learning outperform deep learning and traditional portfolio optimization models in frontier and developed financial markets? Res. Int. Bus. Financ. **65**, 101936 (2023). https://doi.org/10.1016/j.ribaf.2023.101936
12. Kobets, V., Savchenko, S.: Building an optimal investment portfolio with python machine learning tools. In: International Conference Information Technology and Interactions. CEUR Workshop Proceedings, vol. 3347, pp. 307–315 (2022). https://ceur-ws.org/Vol-3347/Short_1.pdf
13. Wilder, J. W.: New Concepts in Technical Trading Systems. Trend Research (1978)

14. Yahoo Finance. Stock market news, quotes, and information. Yahoo Finance. https://finance.yahoo.com/. Accessed 05 July 2023
15. Snihovyi, O., Ivanov, O., Kobets, V.: Implementation of robo-advisors using neural networks for different risk attitude investment decisions. In: 9th International Conference on Intelligent Systems 2018: Theory, Research and Innovation in Applications, IS 2018 - Proceedings, art. no. 8710559, pp. 332–336 (2018). https://doi.org/10.1109/IS.2018.8710559

Quality Assessment and Assurance of Machine Learning Systems: A Comprehensive Approach

Yurii Sholomii[✉] [iD] and Vitaliy Yakovyna [iD]

Lviv Polytechnic National University, Lviv, Ukraine
{yurii.y.sholomii,vitaliy.s.yakovyna}@lpnu.ua

Abstract. Machine learning (ML) is opening up new opportunities for the development of innovative systems across a wide range of industries. However, assessing and ensuring the quality of systems with ML components introduces unique challenges related to inherent characteristics of such components like data centricity and unpredictable behavior. Traditional software quality assessment and assurance methods may not be sufficient for ML systems: (1) they focus on software code, while ML systems' quality is influenced by the characteristics of the data and the algorithms used to create ML components; (2) they do not cover the emerging quality characteristics specific to ML systems, such as interpretability, explainability, fairness and trustworthiness. This PhD project aims to develop a comprehensive approach for assessing and assuring the quality of ML systems, with a focus on bias detection and prevention. The research will (1) explore the problem of bias in production ML systems; (2) analyze the gaps in existing software quality models and methods related to bias detection and prevention; and (3) propose an improved approach to quality assessment and assurance to address the challenges associated with bias in ML systems. The results of this PhD project are expected to contribute to the development of better models and methods for assessing and assuring the quality of ML systems, as well as have practical implications for industries that rely on ML systems to automate complex tasks, facilitate decision-making processes and gain insights from large amounts of data.

Keywords: Machine Learning Systems · Software Quality · Quality Assessment · Quality Assurance · Bias

1 Introduction

The rapid advancement of machine learning (ML) has opened new possibilities and created tremendous opportunities across industries like healthcare [1, 2], manufacturing [3], oil and gas [4], and supply chain [5]. As companies increasingly recognize the potential of ML, industry studies report that ML is seeing an increase in investments and adoption [6, 7]. As the demand for ML systems is expected to grow [8], it becomes increasingly important for individuals and organizations to possess the knowledge and tools necessary to assess and ensure the quality of these systems. Considering that many of these industries predominantly rely on supervised ML, quality concerns related to data-centric nature of ML systems are of paramount importance.

© The Author(s), under exclusive license to Springer Nature Switzerland AG 2023
G. Antoniou et al. (Eds.): ICTERI 2023, CCIS 1980, pp. 265–275, 2023.
https://doi.org/10.1007/978-3-031-48325-7_20

The purpose of this paper is to present and justify the PhD research project on quality assessment and assurance of ML systems. The project is based on a foundational hypothesis that existing software quality models and methods lack a comprehensive approach to bias in production ML systems. The project's goal is to research the problem of bias detection and prevention in production ML systems and develop a comprehensive approach that would bridge identified gaps in existing quality models and methods.

2 Terminology

For the purpose of this paper, the term "machine learning" or "ML" implies supervised ML unless specified otherwise. We define ML system as a software system which includes at least one component developed using ML. Our definition was informed by the following considerations: (1) ML is a subfield of AI, which is focused on automated model creation using algorithms as opposed to explicit manual programming [9–11]; (2) in industry applications, these (ML) models are deployed as components of larger systems (ML components), which also include non-ML components [12]; (3) such systems are commonly known as AI systems with ML components [13, 14]; (4) there's a consensus that AI systems, including those with ML components, are predominantly software-intensive systems [8, 13, 15–17]. Our definition of an ML system is also consistent with those found in related works [18].

In the context of ML systems, we define quality assessment as a process of evaluating the quality of ML system against a set of specified quality criteria. Our definition and understanding of this concept are derived from ISO [19] definitions for software quality evaluation process. In this paper we use terms quality assessment and quality evaluation interchangeably.

In the context of ML systems, we define quality assurance as a set of practices, which help to assure that ML system meets the specified quality requirements. As with previous definition, this definition and the understanding of the concept are based on ISO [20] definitions.

3 Background

3.1 Characteristics of ML Systems

Data Centricity. Data centricity [15], also known as data-dependent behavior [21, 22], is an inherent characteristic of ML systems, which reflects the central role that data plays in the development, performance, and functioning of these systems. Unlike traditional rule-based software, where systems are designed around code, i.e., explicit instructions and algorithms, ML systems are designed to learn from data and adapt their behavior based on patterns discovered within the data [10, 23].

Uncertainty. Uncertainty in an ML system refers to the level of unpredictability or lack of confidence in the model's predictions and is represented by the likelihood that a prediction may be incorrect or fall outside a specified range of accuracy [24]. In ML systems, uncertainty is the prevailing characteristic [8] which contributes significantly to the margin of error found in ML systems [25]. It is also one of the key challenging ML characteristics observed by software architecture practitioners [26].

There are two primary types of uncertainty: aleatoric and epistemic [27]. Aleatoric uncertainty, also known as data uncertainty, is the inherent and irreducible uncertainty in data that leads to prediction uncertainty [28]. On the other hand, epistemic uncertainty, or knowledge uncertainty, arises due to insufficient knowledge about the underlying system or process [28].

In ML systems, uncertainty arises from various sources such as inherent noise in data, variability in model parameters, suitability of model selection, and ambiguity due to extrapolation [29]. From model building and testing perspective, the following uncertainty sources can be identified [24]: (1) scope compliance, as ML models are usually built and tested within a specific context and can provide unreliable outputs if applied outside it; (2) data quality, since the uncertainty in an ML model's outcome depends on the data quality it is applied to; and (3) model fit, which occurs because ML techniques offer empirical models that approximate the actual relationship between input and output.

Complexity. Machine learning inherently exhibits complexity due to its data-driven nature and the intricacies of its underlying algorithms, especially in deep learning [30]. In addition to this, to build and run machine learning models, a substantial volume of supporting code needs to be developed [12, 31]. This complexity is further compounded by issues of model interpretability, the risk of overfitting, potential biases in data, and the significant computational demands of training advanced models.

Traditional Software Characteristics. Although having unique characteristics like data centricity or uncertainty, ML software remains a type of software [33]. ML is still a collection of algorithms, data structures, all supported by code that is designed and executed on computing machines to perform specific tasks. Just like traditional software, ML software is designed, developed, and deployed using programming languages, libraries, and tools.

Traditional software, AI software, and ML software can be seen as forming a hierarchy of characteristic inheritance (see Fig. 1). In this hierarchy, traditional software serves as the base entity, providing the foundational principles and techniques of software development, while AI and ML software inherit it and refine them.

3.2 Quality Assessment and Assurance Challenges in ML Systems

Trustworthiness. Trustworthiness is an encompassing challenge in ML systems, as it covers a wide range of factors that contribute to the credibility and reliability of these systems [34]. Key aspects of trustworthiness include: (1) data quality; (2) interpretability and explainability; (3) fairness and bias; (4) robustness and security. Trustworthiness is a comprehensive and multi-faceted challenge that requires addressing these and other aspects to ensure that ML systems are dependable and can be confidently used in real-world applications.

Data Quality. Machine learning systems are sensitive to the quality of their training data. Guaranteeing data quality, with respect to accuracy, consistency, and relevance,

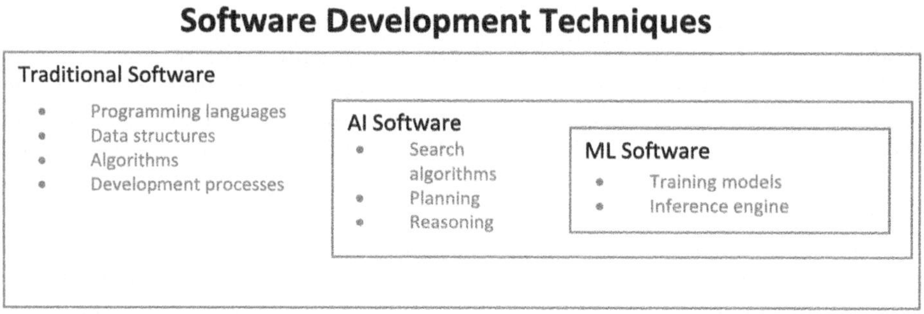

Software Development Techniques

```
Traditional Software

  • Programming languages
  • Data structures          AI Software
  • Algorithms                 • Search
  • Development processes          algorithms    ML Software
                               • Planning          • Training models
                               • Reasoning         • Inference engine
```

Fig. 1. The hierarchical representation of software development techniques for traditional, AI and ML software

is essential for maintaining the ML component's effectiveness and the overall system quality [35]. Data dependency is a significant contributor to the accumulation of shortcut cost, suboptimal decisions, and compromises, i.e., technical debt [12].

Assessing and assuring quality of ML systems should consider different dimensions of data quality. There are quality models specifically designed for data, which provide a structured approach to evaluate, manage, and improve the quality of data. A widely used data quality model [37] comprises several dimensions or attributes that describe various aspects of data quality, including accuracy, completeness, consistency, credibility, and curentness. In the context of supervised machine learning, accuracy refers to how closely the data points, both features and labels, match the real-world values. Completeness ensures that there aren't missing values in both the feature set and the target labels. Consistency refers to the alignment across all records, as inconsistent data can introduce bias into the learning process. Credibility defines the trustworthiness of the data source. Curentness, in the context of supervised ML, refers to how recent the data is and if it accurately represents the current state of what is being modeled or predicted.

The "garbage in, garbage out" (GIGO) principle is a concept in computing and data analysis that emphasizes the importance of high-quality input data. In the context of machine learning and data-driven systems, the GIGO principle highlights the crucial role that high quality data plays in building effective models [36]. If a model is trained on inaccurate or irrelevant data, it will likely produce inaccurate or nonsensical predictions and conclusions.

Another principle, "Changing Anything Changes Everything" (CACE), which is applicable to input signals, hyper-parameters, learning settings, sampling methods, convergence thresholds and data selection, underscores the interconnected nature of various elements within ML systems and the need for careful consideration when making changes [12].

Interpretability and Explainability. Both interpretability and explainability pose significant quality assessment and assurance challenges in ML systems. Although these concepts are closely related, the ideas behind them are not the same [38]. Interpretability refers to the degree to which a human can understand the inner workings of an ML model. Explainability, on the other hand, focuses on providing human-understandable explanations for individual predictions or decisions made by an ML model [39], even if the inner workings of the model itself are not easily interpretable.

Measuring interpretability ML predictions is challenging [40]. It's often difficult to quantify and evaluate how well people can grasp the complex relationships and patterns that an ML model has learned. Developing standardized metrics for interpretability remains a significant challenge in the field of machine learning. The tradeoff between explainability and performance is a major issue [41]. As models become more complex and accurate, they often sacrifice interpretability and explainability, making it difficult for users and stakeholders to understand their decision-making processes.

Fairness and Bias. Fairness and bias are additional concerns, related to explainability [41]. Although these quality concerns are often related to the equitable treatment of different groups of people by an ML model, in a more general context, bias refers to systematic errors that ML model exhibits for certain segments of data, while fairness concerns the model's consistency and accuracy across all data segments, e.g., the model's predictions or decisions do not disproportionately favor certain data categories based on their attributes. Bias can be introduced at various stages of the ML pipeline, including data collection, preprocessing, model training, and evaluation. As previously mentioned, training dataset preparation is a critical step in the context of supervised ML, which is also true from the bias perspective, which could be easily introduced to the dataset by unexperienced people. Considering the data-centric nature of ML systems, detecting, and addressing bias is a key step towards ensuring fairness and overall accuracy in ML model predictions and decisions.

Robustness and Security. Robustness and security challenges in ML systems arise due to their susceptibility to adversarial attacks [42], vulnerability to changing data distributions [43], and the need to protect sensitive data [44]. Machine learning models are vulnerable to adversarial attacks where minor alterations to input data can cause significant changes in the model's output or predictions [42, 45]. ML systems may experience undetected failures when encountering test examples that deviate from the training data distribution [43]. Preserving privacy is another significant challenge for ML systems, since ML models can be trained on sensitive data [44].

Software Architecture. Software architecture plays a crucial role in determining whether a system can attain its desired quality attributes. The software architecture of a computing system consists of structures that facilitate understanding the system, including software components, their relationships, and the properties of both [46]. ML systems components present unique architecting challenges, requiring researchers to reassess current software architecture practices and develop patterns for ML-specific quality attributes [47].

Architecture frameworks, architecting process, self-adaptive architecting, and architecture evolution are all considered important areas in architecting ML systems [48, 49]. Further exploration is needed on concepts like monitorability, co-architecting, and co-versioning to improve system maintenance and align development pipelines [50].

A study [26] has identified 18 software architecture challenges related to ML systems development, as seen from requirements, data, design, testing, and operations perspectives, including: (1) lack of functional requirements for ML components; (2) complexity of data (preparation) pipelines; (3) hard-to-test data quality; (4) uncertainty introduced by ML systems; (5) difficult validation of ML components.

Traditional Software Challenges. Traditional software architecture challenges, such as component coupling, observability, and maintainability, continue to be relevant in the context of ML systems [26]. As argued in [8], challenges like systems confidence, control and understanding of emergent behavior, formal specification, change isolation are also relevant for systems with ML components and are even intensified.

Another set of challenges was identified in [51], including limited transparency, troubleshooting, testing, dependency management, monitoring and logging, glue code and supporting systems.

4 Related Work

4.1 Quality Characteristics and Models

In [15] authors emphasize the importance of addressing key quality attributes such as security, privacy, data centricity, sustainability, and explainability in AI/ML systems. This is due to the uncertainty introduced by data elements and unique challenges posed by regulatory domains in the public sector. The paper provides examples from the healthcare domain and argues that these quality attributes pose new challenges to AI engineering and are crucial for the successful deployment of AI-enabled systems in the public sector.

The research at Booking.com [33] focuses on improving the quality of machine learning software by introducing a software quality model specifically for ML systems. This model provides a holistic view of the quality aspects in ML systems, allowing for more systematic and efficient improvement efforts. By understanding and implementing this quality model, the researchers aim to enhance the overall performance and reliability of ML systems.

The work [52] focuses on developing AI/ML quality models by defining and ordering characteristics of artificial intelligence systems. The objectives include creating principles for analyzing and constructing AI/ML quality models, offering quality models for evaluation, and demonstrating profiling for specific systems. The paper presents a general model of AI quality with a step-by-step construction procedure and a basic model with abbreviated sets of characteristics. The proposed quality model is open and can be adjusted to suit the specific purpose and scope of AI/ML systems.

The research [53] focuses on the need to adapt existing quality standards, like [54], to better suit the unique nature of ML systems. The authors propose a systematic process to construct quality models for ML systems, using an industrial use case. They present a meta-model for specifying quality models, reference elements for relevant views, entities, quality properties, and measures for ML systems. The study highlights the importance of following a systematic process to develop measurable quality properties that can be evaluated in practice. Future work aims to explore how quality differs between various ML systems and to develop reference quality models for evaluating ML system qualities.

4.2 Quality Assurance Models and Methods

The report [13] highlights the increasing use of ML in software systems and identifies two broad research areas: (1) software architecture for ML-based systems, and (2) ML for software architectures. Focusing on the former, the study aims to emphasize the various architecting practices for developing ML-based software systems. Based on the authors' experience, they identify four key areas of software architecture that require attention from both ML and software practitioners to establish standard practices for architecting ML-based software systems.

The research [55] focuses on the software engineering design patterns for ML techniques, aiming to address the software complexity and quality issues. Despite the popularity of ML techniques, there is no systematic collection, classification, and discussion of software engineering design patterns for ML systems. The study presents preliminary results of a systematic literature review on good/bad design patterns for ML, offering developers a comprehensive and organized classification of these patterns.

The work [56] addresses the challenge of maintaining stability in ML systems, where behavior is determined by both program code and input data. Common three-layer architectural patterns complicate troubleshooting due to tightly coupled functions. The study proposes a novel architectural pattern that separates components for business logic and ML, enabling easier troubleshooting. By breaking down failures into business logic and ML-specific parts, operators can rollback the inference engine independently if issues arise. The paper demonstrates the effectiveness of the proposed architecture through a practical case study.

The research [57] explores the challenges of deploying ML algorithms in real-life systems and the potential of Data-Oriented Architecture (DOA) to address these challenges. DOA, an emerging software engineering paradigm, promotes data-driven, loosely coupled, decentralized, and open systems. The study reviews the principles underpinning the DOA paradigm in the context of ML system challenges and examines the implementation of DOA principles in current ML-based real-world system architectures.

5 Limitations of Existing Software Quality Models and Methods

The foundational hypothesis for this PhD project is as follows: *existing software quality models and methods lack a comprehensive approach to bias in production ML systems.* It is decomposed into following hypotheses:

1. Bias in ML components contributes to more than 60% of incidents in production ML systems.
2. Existing quality assessment models and methods are not sufficient to cover bias detection needs in production ML systems.
3. Existing quality assurance models and methods are not sufficient to cover bias prevention needs in production ML systems.

These hypotheses are based on the authors' observations, experience in the field and analysis of existing literature, including the following: (1) traditional software quality models and methods focus on aspects like software code and explicit programming, while the quality of ML systems is heavily influenced by the data and the algorithms used in creating ML components; (2) ML systems have unique quality attributes that traditional methods do not cover, such as interpretability, explainability, fairness, and trustworthiness; and (3) bias is generally considered a major ML quality concern.

The PhD project seeks to explore the problem of bias detection and prevention in production ML systems in detail and to propose a set of improvements to existing quality assessment and assurance models and methods to address the gaps in bias detection and prevention in production ML systems.

6 Towards a Comprehensive Approach to Quality Assessment and Assurance of ML Systems

The goal of the project is to investigate the challenges of detecting and mitigating bias in operational ML systems, and to develop a comprehensive approach that addresses the shortcomings of current quality models and methods. By addressing the identified gaps in existing quality models and methods, the approach aims to enhance the overall quality of ML systems, ensuring they produce more reliable and unbiased results.

Research questions:

1. To what extent do biases in ML components lead to incidents in operational ML systems?
2. Where do the current quality assessment models and methods fall short in addressing bias detection in production ML systems?
3. What gaps or insufficiencies exist in current quality assurance models and methods when it comes to preventing bias in operational ML systems?

The first objective is to quantify the prevalence of bias-induced incidents in operational ML systems. Research agenda: (1) identify and categorize incidents in operational ML systems over a specified period; (2) determine which of these incidents have been influenced or caused by biases in ML components; (3) analyze the impact severity of each bias-induced incident, such as its effect on system operations, user experience, and organizational reputation.

The second objective is to critically analyze existing models and methods for quality assessment, and the third is to examine those for quality assurance, respectively. Research agenda for both is: (1) review the state-of-the-art quality models and methods; (2) evaluate the models and methods against controlled and real-world datasets to assess their efficiency; (3) document shortcomings of the current models and methods when exposed to different types and sources of biases; (4) collaborate with industry experts and stakeholders to understand practical challenges and needs in bias prevention in real-world operational settings.

7 Conclusion

The PhD research project presented aims to advance our understanding of quality assessment and assurance in ML systems by addressing the unique challenges they introduce. Throughout the course of this research, we expect to develop improved models and methods that can effectively detect and prevent bias in ML systems.

The anticipated findings of this research are expected to have practical implications for industries that rely on ML systems to automate complex tasks, facilitate decision-making processes, and gain insights from large amounts of data. By providing better tools for quality assessment and assurance, our research can help organizations build more reliable, efficient, and robust ML systems.

Overall, this PhD research project has the potential to make a lasting impact on the field of ML system quality assessment and assurance, benefiting both the academic community and industry practitioners. By enhancing our ability to evaluate and ensure the quality of ML systems, we can facilitate the development of more trustworthy, efficient, and robust ML systems, ultimately leading to better outcomes for businesses and users alike.

References

1. Parashar, G., Chaudhary, A., Rana, A.: Systematic mapping study of AI/machine learning in healthcare and future directions. SN Comput. Sci. **2**, 461 (2021). https://doi.org/10.1007/s42979-021-00848-6
2. Egger, J., Gsaxner, C., Pepe, A., et al.: Medical deep learning – A systematic meta-review. Comput. Methods Programs Biomed. **221**, 106874 (2022). https://doi.org/10.1016/j.cmpb.2022.106874
3. Kim, S.W., Kong, J.H., Lee, S.W., et al.: Recent advances of artificial intelligence in manufacturing industrial sectors: a review. Int. J. Precis. Eng. Manuf. **23**, 111–129 (2022). https://doi.org/10.1007/s12541-021-00600-3
4. Sircar, A., Yadav, K., Rayavarapu, K., et al.: Application of machine learning and artificial intelligence in oil and gas industry. Pet. Res. **6**(4), 379–391. https://doi.org/10.1016/j.ptlrs.2021.05.009
5. Younis, H., Sundarakani, B., Alsharairi, M.: Applications of artificial intelligence and machine learning within supply chains: systematic review and future research directions. J. Model. Manag. **17**(3), 916–940 (2022). https://doi.org/10.1108/JM2-12-2020-0322
6. The state of AI in 2021. https://www.mckinsey.com/capabilities/quantumblack/our-insights/global-survey-the-state-of-ai-in-2021. Accessed 01 Apr 2023
7. The state of AI in 2022—and a half decade in review. https://www.mckinsey.com/capabilities/quantumblack/our-insights/the-state-of-ai-in-2022-and-a-half-decade-in-review. Accessed 01 Apr 2023
8. Carleton, A., Klein, M., Robert, J., et al.: Architecting the Future of Software Engineering: A National Agenda for Software Engineering Research & Development. Carnegie Mellon University, Software Engineering Institute (2021)
9. Sarker, I.H.: Machine learning: algorithms, real-world applications and research directions. SN COMPUT. SCI. **2**, 1–21 (2021). https://doi.org/10.1007/s42979-021-00592-x
10. Janiesch, C., Zschech, P., Heinrich, K.: Machine learning and deep learning. Electron Markets **31**, 685–695 (2021). https://doi.org/10.1007/s12525-021-00475-2
11. Kühl, N., Schemmer, M., Goutier, M., et al.: Artificial intelligence and machine learning. Electron Markets **32**, 2235–2244 (2022). https://doi.org/10.1007/s12525-022-00598-0
12. Sculley, D., Holt, G., Golovin, D. et al.: Hidden technical debt in machine learning systems. In: Cortes, C., Lawrence, N., Lee, D., Sugiyama, M., R. Garnett, R. (eds.) Proceedings of the 28th International Conference on Neural Information Processing Systems, pp. 2503–2511 (2015)
13. Horneman, A., Mellinger, A., Ozkaya, I.: AI Engineering: 11 Foundational Practices. Carnegie Mellon University, Software Engineering Institute (2019)

14. ISO/IEC 23053:2022. https://www.iso.org/standard/74438.html. Accessed 01 May 2023
15. Pons, L., Ozkaya, I.: Priority quality attributes for engineering AI-enabled systems (2019). https://arxiv.org/abs/1911.02912
16. Bosch, J., Crnkovic, I., Holmström Olsson, H.: Engineering AI systems: a research agenda (2020). https://arxiv.org/abs/2001.07522
17. The AI Act. https://artificialintelligenceact.eu/the-act/. Accessed 01 May 2023
18. Siebert, J., Joeckel, L., Heidrich, J. et al.: Towards guidelines for assessing qualities of machine learning systems (2020). https://arxiv.org/abs/2008.11007
19. ISO/IEC 25040:2011, https://www.iso.org/standard/35765.html, last accessed 2023/05/01
20. ISO 9000:2015, https://www.iso.org/standard/45481.html, last accessed 2023/05/01
21. What is Really Different in Engineering AI-Enabled Systems?, https://apps.dtic.mil/sti/tre cms/pdf/AD1155001.pdf. Accessed 01 May 2023
22. Xu, X., Wang, C., Wang, Z. et al.: Dependency tracking for risk mitigation in machine learning (ML) systems. In: 2022 IEEE/ACM 44th International Conference on SE: Software Engineering in Practice (ICSE-SEIP), pp. 145–146, Pittsburgh, PA, USA (2022)
23. Goodfellow, I., Bengio, Y., Courville, A.: Deep Learning. MIT Press, Cambridge (2016)
24. Kläs, M.: Towards identifying and managing sources of uncertainty in AI and machine learning models - an overview (2018). https://arxiv.org/abs/1811.11669
25. Ozkaya, I.: What is really different in engineering AI-enabled systems? IEEE Softw. **37**(4), 3–6 (2020). https://doi.org/10.1109/MS.2020.2993662
26. Serban, A., Visser, J. An empirical study of software architecture for machine learning (2021). https://arxiv.org/abs/2105.12422
27. Hüllermeier, E., Waegeman, W.: Aleatoric and epistemic uncertainty in machine learning: an introduction to concepts and methods. Mach. Learn. **110**, 457–506 (2021). https://doi.org/10. 1007/s10994-021-05946-3
28. Abdar, M., Pourpanah, F., Hussain, S., et al.: A review of uncertainty quantification in deep learning: techniques, applications and challenges. Inf. Fusion **76**, 243–297 (2021)
29. Jalaian, B., Lee, M., Russell, S.: Uncertain context: uncertainty quantification in machine learning. AI Mag. **40**(4), 40–49 (2019)
30. Hu, X., Chu, L., Pei, J., et al.: model complexity of deep learning: a survey (2021). https:// arxiv.org/abs/2103.05127
31. Scaling Big Data Mining Infrastructure: The Twitter Experience, https://www.kdd.org/explor ation_files/V14-02-02-Lin.pdf. Accessed 01 May 2023
32. Architectural Components in ML-Enabled Systems. https://ckaestne.medium.com/architect ural-components-in-ml-enabled-systems-78cf76b29a92. Accessed 01 May 2023
33. A Quality Model for Machine Learning Systems. https://booking.ai/a-quality-model-for-mac hine-learning-systems-892118be9e19. Accessed 01 May 2023
34. Thuraisingham, B.: Trustworthy machine learning. IEEE Intell. Syst. **37**(1), 21–24 (2022). https://doi.org/10.1109/MIS.2022.3152946
35. Ghahramani, Z.: Probabilistic machine learning and artificial intelligence. Nature **521**, 452–459 (2015). https://doi.org/10.1038/nature14541. PMID: 26017444
36. Geiger, R.S., Cope, D., Ip, J., et al.: Garbage in, garbage out revisited: what do machine learning application papers report about human-labeled training data? Quant. Sci. Stud. **2**(3), 795–827 (2021). https://doi.org/10.1162/qss_a_00144
37. ISO/IEC 25012:2008. https://www.iso.org/standard/35736.html. Accessed 01 May 2023
38. Interpretability versus explainability. https://docs.aws.amazon.com/whitepapers/latest/ model-explainability-aws-ai-ml/interpretability-versus-explainability.html. Accessed 01 May 2023
39. Explainable AI (XAI). https://www.ibm.com/watson/explainable-ai. Accessed 01 May 2023
40. Schmidt, P., Felix Biessmann, F.: Quantifying interpretability and trust in machine learning systems (2019).https://arxiv.org/abs/1901.08558

41. Rawal, A., McCoy, J., Rawat, D. et al.: Recent advances in trustworthy explainable artificial intelligence: status, challenges and perspectives. https://www.techrxiv.org/articles/preprint/17054396. https://doi.org/10.36227/techrxiv.17054396.v1

42. Goldblum, M., Schwarzschild, A., Patel, A. et al.: Adversarial attacks on machine learning systems for high-frequency trading (2021). https://arxiv.org/abs/2002.09565

43. Piratla, V.: Robustness, evaluation and adaptation of machine learning models in the wild (2023). https://arxiv.org/abs/2303.02781

44. Chew, Y.J., Wong, K.-S., Ooi, S.Y.: Privacy protection in machine learning: the state-of-the-art for a private decision tree. (2017)

45. Kurakin, A., Goodfellow, I., Bengio, S.: Adversarial examples in the physical world (2016). https://arxiv.org/abs/1607.02533

46. Bass, L., Clements, P., Kazman, R.: Software Architecture in Practice, 4th edn. Addison-Wesley Professional, Boston (2021)

47. Lewis, G.A., Ozkaya, I., Xu, X.: Software architecture challenges for ML systems. In: IEEE International Conference on Software Maintenance and Evolution (ICSME), pp. 634–638, Luxembourg (2021). https://doi.org/10.1109/ICSME52107.2021.00071

48. Malavolta, I., Muccini, H., Ozkaya, I.: Software architecture and artificial intelligence. J. Syst. Softw. 193, 111436 (2022). https://doi.org/10.1016/j.jss.2022.111436

49. Muccini, H., Vaidhyanathan, K.: Leveraging machine learning techniques for architecting self-adaptive IoT systems (2020)https://doi.org/10.1109/SMARTCOMP50058.2020.00029

50. Overcoming Software Architecture Challenges for ML-Enabled Systems. https://apps.dtic.mil/sti/pdfs/AD1150241.pdf. Accessed 01 May 2023

51. Arpteg, A., Brinne, B., Crnkovic-Friis, L. et al.: Software engineering challenges of deep learning. https://arxiv.org/abs/1810.12034

52. Kharchenko, V., Fesenko, H., Illiashenko, O.: Basic model of non-functional characteristics for assessment of artificial intelligence quality. Radioelectron. Comput. Syst. 2, 131–144 (2022). https://doi.org/10.32620/reks.2022.2.11

53. Siebert, J., Joeckel, L., Heidrich, J., et al.: Construction of a quality model for machine learning systems. Software Qual. J. 30, 307–335 (2022). https://doi.org/10.1007/s11219-021-09557-y

54. ISO/IEC 25010:2011. https://www.iso.org/ru/standard/35733.html. Accessed 01 May 2023

55. Washizaki, H., Uchida, H., Khomh, F. et al.: Studying software engineering patterns for designing machine learning systems (2019). https://arxiv.org/abs/1910.04736

56. Yokoyama, H.: Machine learning system architectural pattern for improving operational stability. In: 2019 IEEE International Conference on Software Architecture Companion (ICSA-C), pp. 267–274, Hamburg, Germany (2019). https://doi.org/10.1109/ICSA-C.2019.00055

57. Cabrera, C., Paleyes, A., Thodoroff, P. et al.: Real-world machine learning systems: a survey from a data-oriented architecture perspective (2023). https://arxiv.org/abs/2302.04810

Vulnerability Detection of Smart Contracts Based on Bidirectional GRU and Attention Mechanism

Oleksandr Tereshchenko$^{(\boxtimes)}$ ⓘ and Nataliia Komleva ⓘ

Odessa National Polytechnic University, 1, Shevchenko Ave, Odesa 65044, Ukraine
alexandr.tereschenko2014@gmail.com, komleva@op.edu.ua

Abstract. The paper is devoted to methods of detecting vulnerabilities of smart contracts using machine learning. The purpose of the study is to improve the accuracy of detecting the reentrancy vulnerability of smart contracts by implementing new machine learning models. A thorough analysis of the current literature was performed and the shortcomings of the existing tools for detecting vulnerabilities of smart contracts were identified. In particular, insufficient accuracy and low adaptability of existing models to new vulnerabilities were noted. To solve these problems, a new model based on a kind of recurrent neural networks with a gating mechanism, namely a bidirectional GRU with an attention mechanism, was proposed to detect the reentrancy vulnerability at the Solidity code level. Using the Word2vec model, the source code of smart contracts was transformed into an array of vectors and used as input to the neural network. Precision, recall and F-beta score were used to evaluate the developed model. 500 smart contract source codes from the Ethereum blockchain were used to train the model, 250 of which had the reentrancy vulnerability. The constructed model was compared with Simple RNN, LSTM, BLSTM, BGRU and BLSTM-ATT models. The obtained results showed that the developed model is ahead of the listed models. The closest values of the metrics were obtained by the BLSTM-ATT model, while the developed BGRU-ATT model uses significantly fewer parameters that need to be optimized, which reduces the training time of the model to detect new vulnerabilities.

Keywords: smart contract · blockchain · deep learning · vulnerability detection

1 Introduction

The concept of blockchain was first proposed by Satoshi Nakamoto in 2008 and gave rise to the development of decentralized cryptocurrencies [1]. A cryptocurrency is a digital currency that is not controlled by a central bank but is decentralized using blockchain technology. The security of transactions on the blockchain network is guaranteed by consensus mechanisms that allow maintaining the integrity of data on different nodes.

Smart contracts have become one of the most successful applications of blockchain. Smart contracts are programs that run on blockchains and can execute transactions without third parties [2]. Today, the most popular platform supporting the deployment of smart contracts is Ethereum.

© The Author(s), under exclusive license to Springer Nature Switzerland AG 2023
G. Antoniou et al. (Eds.): ICTERI 2023, CCIS 1980, pp. 276–287, 2023.
https://doi.org/10.1007/978-3-031-48325-7_21

The range of applications of smart contracts is quite wide, including international payments, loans and mortgages, financial data registration, supply chain management, insurance, etc. One of the key features of smart contracts, compared to conventional programs, is that they cannot be changed after deployment in the blockchain, due to the immutable nature of the blockchain [3]. On the one hand, it guarantees users that the terms of execution of the smart contract will not change. On the other hand, this makes them more vulnerable, as errors in the program logic cannot be fixed. Nowadays, more and more transactions on the Ethereum platform are executed automatically using smart contracts.

Due to the fact that smart contracts manage valuable resources, they have become attractive to attackers. The most popular high-level programming language for creating smart contracts is Solidity. Solidity is a fairly new programming language, which is why many programmers make smart contract code full of vulnerabilities. Attacks on vulnerable smart contracts can lead to significant economic losses [4]. Therefore, more and more researchers are immersed in solving the problem of identifying vulnerabilities in smart contracts.

This paper proposes method based on deep learning, namely a bidirectional recurrent neural network method with Gated Recurrent Units (GRU) gating mechanism and an attention mechanism, to detect the most common vulnerability of smart contracts written in Solidity language. Gating mechanisms in recurrent neural networks allow to store long sequences in memory, which is especially relevant when working with smart contract code fragments, which can consist of thousands of lines. In turn, the attention mechanism helps to focus on those areas of code fragments that are more prone to vulnerabilities.

2 Literature Review and the Problem Statement

Neural networks are just beginning to be actively used to solve the problem of detecting vulnerabilities in smart contracts. The article [5] proposes a neural network architecture for the task of text translation based on recurrent neural networks with a Gated Recurrent Unit (GRU) gating mechanism. Gating mechanisms are widely used in neural network models, where they allow gradients to easily backpropagate in time. Bidirectional Gated Recurrent Unit (BiGRU) is a neural network consisting of the output state connection layer of forward GRU and reverse GRU.

Recently, deep neural networks have been actively used to detect vulnerabilities in program code. In the paper [6], research was started on the use of a method based on deep learning to detect vulnerabilities in C/C++ code that uses Long Short-Term Memory (LSTM).

The paper [7] explores the potential of deep learning for program analysis by embedding the nodes of the abstract syntax tree representations of source code and training a tree-based convolutional neural network (TBCNN) for simple supervised classification problems.

The article [8] proposes an approach of using CNN, which combines a model with an attention mechanism, to realize the detection of smart contract vulnerabilities.

In the article [9], a deep learning-based approach, namely Bidirectional Long Short-Term Memory with Attention Mechanism (BLSTM-ATT) is used to accurately detect reentrancy vulnerability. In addition, the authors propose a method for extracting code fragments for smart contracts, which helps to capture important semantic information.

One of the main reasons why classical methods, in particular fuzzing and symbolic execution, were used to solve the problem of vulnerability detection for a long time, is that there was a lack of smart contracts for creating datasets for training neural networks. The GRU and LSTM architectures, as modifications of recurrent neural networks, have become popular when working with sequential data, such as smart contract code in the Solidity language. One of the advantages of the GRU architecture over LSTM is that it uses fewer parameters for training due to having only two gates as opposed to LSTM which has three gates. Thus, when training the network, parameters need to be optimized in 6 weight matrices for each cell in GRU, while in LSTM – 8. It follows that GRU-based models use less memory and learn faster. That is why the use of GRU is more appropriate than LSTM when working with small training datasets [10]. In our task, the dataset for training the neural network is also small enough. For example, out of 20,000 unique source smart contracts collected, only 250 of them were found to contain the reentrancy vulnerability.

3 Reentrancy Vulnerability

This work is aimed at detecting the most common vulnerability of smart contracts, namely the reentrancy vulnerability. According to the Solidified platform, 17% of all detected vulnerable smart contracts are smart contracts with the reentrancy vulnerability [11].

The execution of smart contracts is not atomic and consistent, due to which there are certain gaps in their security. Attackers can re-enter the called function during the current execution of the program [12]. Like most programming languages, smart contracts use cross-functional or cross-contract calls to handle business logic. But the difference is that such calls aim to transfer some valuable assets. Attackers can manipulate the balance of smart contracts that have the reentrancy vulnerability. The balance of a smart contract refers to the amount of Ether stored at the contract address in the blockchain.

Calling the *transfer* function in the sender's contract will inevitably trigger the fallback function in the recipient's contract. When a smart contract performs a cross-contract money transfer operation, attackers can intercept such an external call and perform some malicious operations [13]. An example of such an operation is when an attacker injects malicious code into his fallback function, which implements a recursive entry into the victim's contract to re-call the *transfer* function to steal Ether (see Fig. 1). The reentrancy vulnerability led to the largest security incident in the history of smart contracts (the attack on "The DAO"), which not only resulted in a loss of almost $60 million but also caused the Ethereum hardfork.

One approach to preventing the reentrancy vulnerability is to change the user's balance before performing any interactions with other smart contracts (see Fig. 2).

Since the user's balance in the secure function is set to zero in advance, when the *withdraw* method is recursively called, the cryptoassets will not be re-sent to the attacker's address [14].

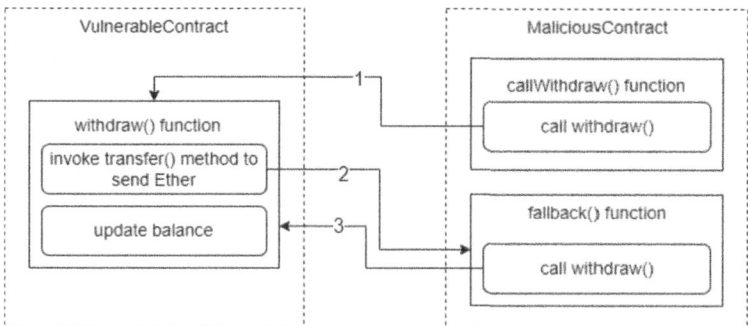

Fig. 1. Reentrancy attack.

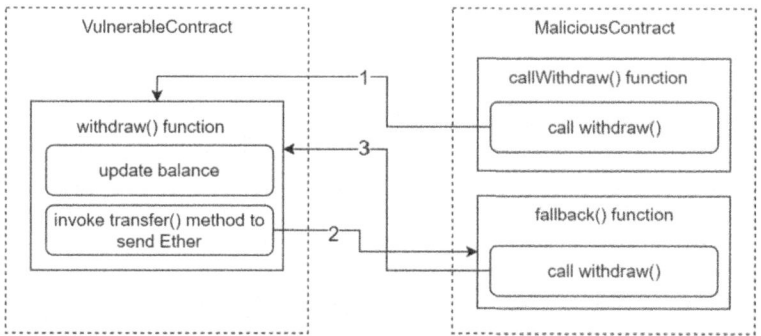

Fig. 2. Reentrancy attack prevention.

4 Methodology of Building a Model for Detecting Vulnerabilities of Smart Contracts

4.1 Vector Representation of Smart Contract Code

Smart contracts on the Ethereum blockchain are written in the high-level Solidity language, which is a set of lines of code. Since neural networks usually take vectors as input, it is necessary to convert the source code of smart contracts into an array of numerical vectors. Natural language processing methods were used for this purpose. Natural Language Processing (NLP) is a branch of information technology, artificial intelligence and linguistics, the purpose of which is to study the problems of computer analysis and synthesis of natural language. A significant part of NLP technologies works thanks to deep learning, in which algorithms try to automatically extract the best features from raw input data. Manually created features may be too specialized, incomplete, and time-consuming to create and approve.

The transformation of text into numerical vectors for use in machine learning algorithms can be implemented in various ways, for example, using the Bag-of-Words model [15], the N-Gram language model [16], and the Embedding layer.

The Word2vec model was used to convert the source code of smart contracts into vectors. Word2vec accepts a large text corpus of smart contract source codes, in which each word in a fixed dictionary is represented as a vector. The algorithm passes through each position *t* in the text, which represents the central word *c* and the context word *o*. Next, the similarity of the word vectors for *c* and *o* is used to calculate the probability of *o* given *c* (or vice versa), and word vectors are adjusted to maximize that probability. For example, the words denoting the data types *uint32* and *uint128* will be located close to each other but far from the words denoting the *onlyMinter* and *onlyOwner* function modifiers (see Fig. 3).

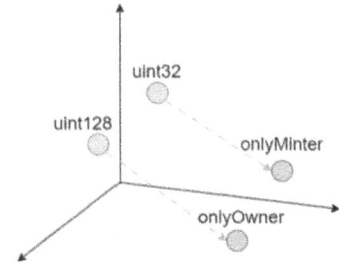

Fig. 3. In Word2vec, words of similar meaning are located closer in space, which indicates their semantic similarity.

To achieve a better performance of Word2vec, unnecessary stop words are removed from the dataset, which helps to increase the accuracy of the model and reduce the training time. Stop words are words that do not significantly affect the semantics. In our case, these will be the names of variables and functions.

4.2 Recurrent Neural Networks

A Recurrent Neural Networks (RNN) is a type of deep neural network adapted to sequential time-ordered data. For example, such data can be sentences consisting of words, video as a set of frames, etc. The main feature of recurrent neural networks is that they can "remember" the elements of the sequence that have already been considered [17]. These models can be trained and achieve good performance on complex sequential learning tasks, such as speech recognition, machine translation, and natural language processing (see Fig. 4). Thus, recurrent neural networks are suitable for solving the problem of detecting vulnerabilities of smart contracts at the Solidity code level.

Standard RNNs are difficult to train properly in practice. The main reason the model is so uncontrollable is that it suffers from both exploding and vanishing gradients. Both problems are due to the periodic nature of RNNs.

Such neural networks can have three configurations, namely "Sequence to sequence", "Sequence to one" and "One to sequence". The "Sequence to sequence" configuration is used in the task such as text translation when it is necessary to convert a sequence of one data into a sequence of other data using the network. The "Sequence to one" configuration is used to solve classification tasks, for example, text or video classification, when a sequence of words (or images) is taken as an input of the network, and a vector of probabilities for classes is returned as an output. The "One to sequence" configuration

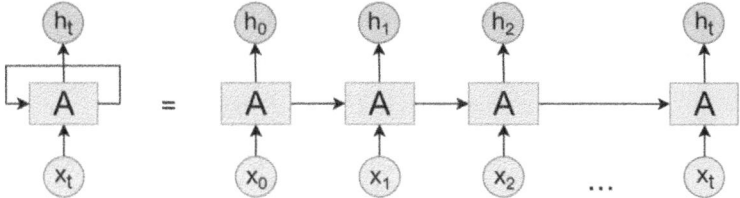

Fig. 4. A recurrent neural network and its unfolding in time.

works the other way around: one element is taken as an input and a sequence is obtained as the output. In this way, it is possible, for example, to generate descriptions for a photo transmitted to the network input.

For our task, the "Sequence to one" configuration is used, since the task of detecting vulnerable smart contracts is reduced to the task of classification.

Gated Recurrent Unit. The problem with standard RNN cells is that they cannot keep in memory too long sequences. This is because when we pass gradients through a long enough sequence, we run into one of two problems: either the gradients decrease so much that errors at the end of the sequence no longer affect the beginning of the sequence, or the gradients increase and the process diverges. Therefore, varieties of RNN, including LSTM and GRU, have emerged to combat this problem by adding another internal cell state.

The GRU architecture is similar to LSTM but has fewer gates (update gate z_t and reset gate r_t). Together, these gates control how new information updates the state. The update gate is responsible for determining the amount of previous information that needs to be transferred to the next state. A reset gate is used in the model to decide how much of the past information should be discarded (see Fig. 5). If r_t is equal to zero, the cell forgets its previous state [18].

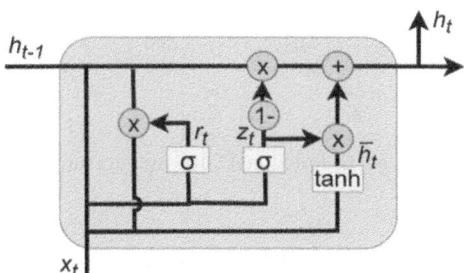

Fig. 5. The structure of GRU cell.

All connections in GRU can be described using the following formulas:

$$z_t = \sigma(W_z x_t + U_z h_{t-1} + b_z) \tag{1}$$

where z_t – update gate, σ – sigmoid function, W_z and U_z – weight matrices, x_t – input data, h_{t-1} – information from the previous t-1 unit, b_z – bias.

$$r_t = \sigma(W_r x_t + U_r h_{t-1} + b_r) \tag{2}$$

where r_t – reset gate, σ – sigmoid function, W_r and U_r – weight matrices, x_t – input data, h_{t-1} – information from the previous t-1 unit, b_r – bias.

$$\overline{h}_t = tanh(W_h x_t + r_t \odot (U_h h_{t-1}) + b_h) \tag{3}$$

where $\overline{h}_t$ – current memory content, $tanh$ – tanh function, W_h and U_h – weight matrices, r_t – reset gate, x_t – input data, h_{t-1} – information from the previous t-1 unit, b_h – bias.

$$h_t = (1 - z_t) \odot h_{t-1} + z_t \odot \overline{h}_t \tag{4}$$

where h_t – final memory at current time step, z_t – update gate, h_{t-1} – information from the previous t-1 unit, $\overline{h}_t$ – current memory content.

It is advisable to use the GRU architecture to solve the problem of vulnerability detection, as models built using it show better accuracy than models based on standard RNNs.

To increase the accuracy of the model, a backward layer consisting of GRU cells was added to the neural network architecture. Unlike a standard GRU, a bidirectional GRU receives input in both directions and is able to use information from both sides (see Fig. 6). It also provides opportunities for modeling sequential dependencies between words and phrases in both directions of the sequence.

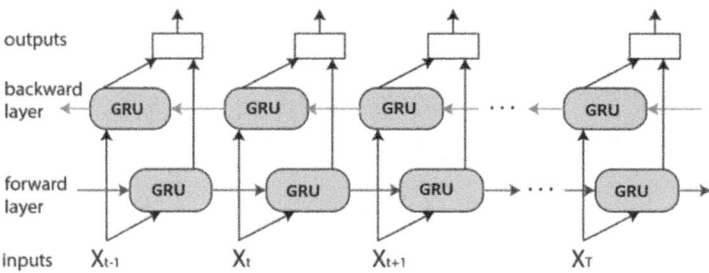

Fig. 6. Bidirectional GRU architecture.

Attention Mechanism. The standard GRU architecture cannot reflect the different importance of the source information at each moment. To solve this problem, an attention mechanism was introduced, which allows searching for relationships between different parts of input and output data [19].

Not all token vectors have the same effect on the classification of vulnerable smart contracts. Therefore, more attention should be paid to more useful vectors [20]. The attention mechanism layer is used to learn the weights for each hidden layer obtained at time step t by the bidirectional GRU layer.

The formula for calculating α_t is as follows:

$$\alpha_t = \frac{exp(u_t^T u_w)}{\sum_t exp(u_t^T u_w)} \tag{5}$$

where u_w- weight matrix.

The formula for calculating u_t is as follows:

$$u_t = tanh(W_w h_t + b_w) \tag{6}$$

where W_w – weight matrix; b_w – bias; $tanh$ – nonlinear activation function.

Having obtained the value of the probability distribution of attention at each moment, the feature vector v containing text information is calculated, and the calculation formula is as follows:

$$v = \sum_t \alpha_t h_t \tag{7}$$

4.3   Dataset for Training

SmartBugs Wild Dataset [21] was used to prepare the dataset for neural network training. Since the developed tool focuses on detecting the reentrancy vulnerability, 250 smart contracts with reentrancy vulnerability were selected. 250 smart contracts without vulnerabilities were also selected. Thus, the total volume of the dataset for neural network training was 500 smart contracts.

In order to transform the contract fragments into the common input form of our sequential models, they are vectorized using the word2vec tool, which is widely used in vector representation of words.

4.4   Architecture of the Model for Detecting Smart Contract Vulnerabilities

The built model receives the source code of the smart contract as input, selects code fragments that could potentially have vulnerabilities, and converts them into an array of vectors using vector embeddings. The number of tokens in the vector representation of the contract fragments is set to 100 (which was determined to be the best among the values [50, 100, 150]), while the dimensionality of the vectors is set to 300 (which was determined to be the best among the values [200, 300, 400]). In the next step, these vector representations are passed to the input of the Bidirectional GRU layer. This layer consists of 600 GRU blocks. After that, a layer with an attention mechanism is applied, the output of which is transferred to the linear layer. This layer has two outputs and uses the Softmax activation function (see Fig. 7).

When training the models, binary cross-entropy was used as a loss function. Adam's gradient descent algorithm was used as an optimizer. The value of the learning rate was chosen to be 0.0005 (it was determined to be the best among the values [0.0005, 0.0002, 0.001]), and as the dropout rate – 0.2 (it was determined to be the best among the values [0.2, 0.3, 0.4]). The batch size is set to 64. This model architecture based on neural networks was determined to be optimal after a series of empirical research.

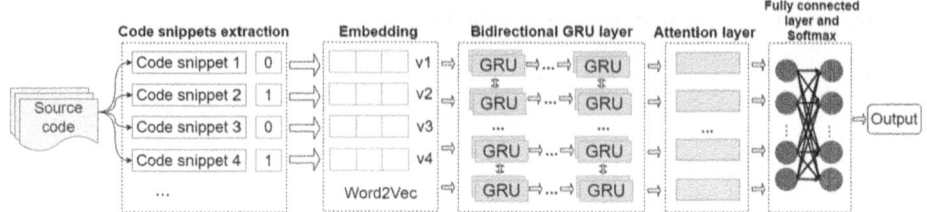

Fig. 7. Architecture of Bidirectional GRU with attention mechanism.

5 Evaluation of the Model

The performance of each model is evaluated using the F1-score, precision and recall metrics. Each metric is described in detail below.

5.1 Basic Values

The results of true positives (TP), true negatives (TN), false positives (FP) and false negatives (FN) are the baseline values for calculating other metrics.

5.2 Precision and Recall Metrics

The precision metric describes the ratio of true positive values to all positive predictions. This metric makes it possible to assess the reliability of the classifier's positive forecast.

The recall (or sensitivity) metric shows the proportion of actual positive results that are correctly classified. The formulas for calculating these two metrics are as follows:

$$Precision = \frac{TP}{TP + FP} \tag{8}$$

$$Recall = \frac{TP}{TP + FN} \tag{9}$$

5.3 F-Beta Score

The F-beta score provides a more balanced characterization of the model than the precision and recall metrics. The F-beta score is calculated as the weighted harmonic mean of precision and recall:

$$F_\beta = (1 + \beta^2) \frac{Precision * Recall}{(\beta^2 * Precision) + Recall} \tag{10}$$

The best and worst F-beta score values are 1 and 0, respectively. The parameter β determines the ratio of recall importance to accuracy importance.

5.4 Hardware and Software

The models were tested on a Lenovo Legion 5-17ACH6 server with the Windows 10 operating system. The server is equipped with an AMD Ryzen 7 5800H processor and has 8GB of RAM and 1TB of disk space. The sequential model proposed in the paper was implemented using the Keras high-level neural network API with a TensorFlow backend running on Python.

5.5 Evaluation Results

Table 1 shows the results of evaluating various models using the main metrics. For this, the following models were chosen: based on standard RNN (Simple RNN), LSTM, bidirectional LSTM (BLSTM), bidirectional GRU (BGRU) and bidirectional LSTM with attention mechanism (BLSTM-ATT).

Table 1. Evaluation results.

Model	Precision, %	Recall, %	F-score ($\beta = 2$), %
Simple RNN	66.34	64.85	65.14
LSTM	73.28	76.33	75.70
BLSTM	86.12	87.97	87.59
BGRU	86.05	88.10	87.68
BLSTM-ATT	89.87	90.66	90.50
BGRU-ATT	90.03	91.76	91.41

As it can be seen in Table 1, the model based on Simple RNN, which has an F-score of 65.14%, showed the worst results. Next is the model based on LSTM, which has slightly better performance than the model based on Simple RNN. Models based on bidirectional LSTM and GRU are significantly ahead of previous models due to the fact that the input data comes in both directions. Thus, for the BGRU-based model, the F-score is 87.68%. Modifications of these models by adding an attention mechanism also significantly improved the characteristics. The obtained metrics for the BLSTM-ATT and BGRU-ATT models turned out to be close in value, although the BGRU-ATT model has a slightly better recall metric, due to which the F-score increased to 91.41%.

6 Conclusion

A new method for detecting vulnerabilities of smart contracts at the Solidity code level was presented in this paper. The proposed method uses one of the modifications of recurrent neural networks, namely Bidirectional GRU with an attention mechanism. We have demonstrated that the BGRU-ATT model has high performance metrics compared to existing models. The implementation of the attention mechanism in the network architecture made it possible to focus the attention of the neural network on areas of the code that are more prone to vulnerabilities. The BGRU-ATT model has a higher F-score than the BLSTM-ATT model, while using less parameters to be trained. This allows to train a neural network to detect new types of vulnerabilities faster, which is an especially important characteristic of the method in conditions when new vulnerabilities of smart contracts that cannot be detected by most existing tools are constantly appearing.

References

1. Nakamoto, S.: Bitcoin: a peer-to-peer electronic cash system, Bitcoin (2009). https://bitcoin.org/bitcoin.pdf
2. Zheng, Z.: An overview of blockchain technology: architecture, consensus, and future trends. In: 2017 IEEE International Congress on Big Data (BigData Congress), pp. 557–564. Boston (2017)
3. Chen, T., Li, X., Luo, X., Zhang, X.: Under-optimized smart contracts devour your money. In: 2017 IEEE 24th International Conference on Software Analysis, Evolution and Reengineering (SANER), pp. 442–446. IEEE, Austria (2017)
4. Chen, W., et al.: Detecting Ponzi schemes on Ethereum: towards healthier blockchain technology. In: Proceedings of the 2018 World Wide Web Conference, pp. 1409–1418. (2018)
5. Cho, K., Merrienboer, V., Gulcehre, C.: Learning phrase representations using RNN encoder-decoder for statistical machine translation. In: Proceedings of the 2014 Conference on Empirical Methods in Natural Language, pp. 1724–1734, Doha (2014). https://doi.org/10.48550/arXiv.1406.1078
6. Li, Z., et al.: VulDeePecker: a deep learning-based system for vulnerability detection. In: 25th Annual Network and Distributed System Security Symposium, San Diego (2018). https://doi.org/10.14722/ndss.2018.23158
7. Mou, L., Li, G., Jin, Z.: TBCNN: a tree-based convolutional neural network for programming language processing. In: AAAI 2016: Proceedings of the Thirtieth AAAI Conference on Artificial Intelligence, pp. 1287–1293 (2016)
8. Yuhang, S., Lize, G.: Attention-based machine learning model for smart contract vulnerability detection. In: Journal of Physics: Conference Series (2021)
9. Peng, Q., Zhenguang, L., Qinming, H., Zimmermann, R., Wang, Z.: Towards automated reentrancy detection for smart contracts based on sequential models. IEEE Access **8**, 19685–19695 (2020). https://doi.org/10.1109/ACCESS.2020.2969429
10. Cahuantzi, R., Chen, X., Güttel, S.: A comparison of LSTM and GRU networks for learning symbolic sequences (2021)
11. Manoj, P.: Most common smart contract bugs of 2020. Solidified (2020). https://medium.com/solidified/most-common-smart-contract-bugs-of-2020-c1edfe9340ac
12. Tikhomirov, S., et al.: SmartCheck: static analysis of Ethereum smart contracts. In: IEEE 1st International Workshop on Computer Society, pp. 9–16. IEEE, Gothenburg (2018)
13. Komleva, N.O., Tereshchenko, O.I.: Requirements for the development of smart contracts and an overview of smart contract vulnerabilities at the solidity code level on the Ethereum platform. Herald Adv. Inf. Technol. **6**(1), 54–68 (2023). https://doi.org/10.15276/hait.06.2023.4
14. Hajdu, A., Jovanovic, D.: Solc-verify: a modular verifier for solidity smart contracts. In: Chakraborty, S., Navas, J.A. (eds.) VSTTE 2019. LNCS, vol. 12031, pp. 161–179. Springer, Cham (2020). https://doi.org/10.1007/978-3-030-41600-3_11
15. Wisam, A., Musa, M., Bilal, I.: An overview of bag of words; importance, implementation, applications, and challenges. In: 2019 International Engineering Conference (IEC), pp. 200–204. Erbil (2019). https://doi.org/10.1109/IEC47844.2019.8950616

16. Cavnar, W., Trenkle, J.: N-Gram-based text categorization. In: Proceedings of the Third Annual Symposium on Document Analysis and Information Retrieval (2001)
17. Sherstinsky, A.: Fundamentals of recurrent neural network (RNN) and long short-term memory (LSTM) network. In: Physica D: Nonlinear Phenomena, vol. 404 (2020). https://doi.org/10.1016/j.physd.2019.132306
18. Junyoung, C., Caglar, G., KyungHyun, C., Yoshua, B.: Empirical evaluation of gated recurrent neural networks on sequence modeling. In: NIPS 2014 Deep Learning and Representation Learning Workshop (2014). https://doi.org/10.48550/arXiv.1412.3555
19. Soydaner, D.: Attention mechanism in neural networks: where it comes and where it goes. Neural Comput. Appl. **34**, 13371–13385 (2022). https://doi.org/10.1007/s00521-022-07366-3
20. Kalra, S., Goel, S., Dhawan, M., Sharma, S.: ZEUS: analyzing safety of smart contracts. In: The Network and Distributed System Security Symposium, California (2018). https://doi.org/10.14722/ndss.2018.23082
21. Ferreira, J., Cruz, P., Durieux, T., Abreu, R.: SmartBugs: a framework to analyze solidity smart contracts. In: 35th IEEE/ACM International Conference on Automated Software Engineering (ASE 2020). Melbourne (2020). https://doi.org/10.48550/arXiv.2007.04771

Modeling the Resource Planning System for Grocery Retail Using Machine Learning

Bohdan Yakymchuk$^{(\boxtimes)}$ and Olena Liashenko

Taras Shevchenko National University of Kyiv, Kyiv, Ukraine
bogdan.yakymchuk3@gmail.com, olenalyashenko@knu.ua

Abstract. The reach of online grocery services has expanded to encompass new customer segments in recent years. During the early stages of the COVID-19 outbreak, when delivery slots were limited and customer demand was high, click-and-collect models became increasingly popular. In order to keep pace with evolving customer behavior, it is crucial for retailers to maintain a high degree of operational process efficiency within their business model. This research paper proposes a resource planning system for grocery retail delivery services that utilizes machine learning techniques. The system aims to optimize the allocation of resources, such as delivery drivers, and reduce transport costs, improving the overall efficiency and profitability of the delivery operations. The system is designed to capture and analyze data from various sources, including delivery orders, traffic patterns, weather conditions, and driver schedules. The proposed research demonstrates the potential of machine learning techniques to transform resource planning in grocery retail delivery services and highlights the importance of Data-Driven decision-making in today's highly competitive retail landscape.

Keywords: food delivery · grocery retail · resource model · machine learning · mathematical optimization

1 Introduction

The COVID-19 pandemic and subsequent Russian full-scale invasion of Ukraine have dramatically altered consumer behavior and created new trends in the grocery retail market, making market dynamics less predictable and adding a new level of complexity to operational and strategic planning tasks. As a result, it is crucial not only to forecast the impact of internal initiatives and strategies but also to consider the influence of stochastic exogenous factors, including periods of blackouts and air alarms, to keep operational function effective and business model profitable.

Labor scheduling is a crucial aspect of retail operations, and while new tools and processes bring about changes, the fundamentals remain the same. Retail companies that understand the basic building blocks of scheduling have an advantage in identifying and implementing innovative solutions into their store operations. Due to McKinsey research [1], retailers who have made strides in streamlining their processes and gaining efficiencies through lean-retailing initiatives reduced operating costs by 15%. However,

G. Antoniou et al. (Eds.): ICTERI 2023, CCIS 1980, pp. 288–299, 2023.
https://doi.org/10.1007/978-3-031-48325-7_22

with competition intensifying, retailers are seeking new ways to improve productivity further and enhance customer service.

One area of opportunity for improvement is workforce management, specifically labor scheduling and budgeting. The process of creating accurate staffing schedules and budgets for a large number of stores is inherently complex, and even sophisticated retailers can find room for improvement. Modern software solutions are helpful for monitoring employee attendance and managing payroll, but they often produce generic schedules that do not consider store-specific factors and workload fluctuations. This can lead to high labor costs and inconsistent customer service.

Leading retailers are taking a more data-driven approach to labor scheduling and budgeting by closely examining store activities to predict the number and skill set of employees needed every day, or even every hour, of the week. As a result, due to McKinsey research [2], some retail companies have achieved cost savings between 4% and 12%, improved customer service by shortening checkout queues or having more staff available to assist customers, and increased employee satisfaction.

There is no standard formula for resource modeling for grocery retail companies, as the specific approach may vary depending on the retailer's specific needs and goals. However, some common elements of resource modeling for revenue forecasting may include the following:

1. Determining the total revenue target for the period of interest (e.g., month, day).
2. Identifying the revenue contribution of each product category or department.
3. Analyzing historical sales data to identify seasonal patterns, trends, and outliers.
4. Developing a forecast based on historical data, market trends, and promotions.
5. Using the sales forecast to estimate the required staffing levels, inventory levels, and other resources needed to achieve the revenue target.
6. Monitoring actual sales performance and adjusting resource allocation.

In general, the calculation of the resource time allocation for the specific process ξ at the period t can be formalized in the following form:

$$T_t(\xi) = (\tau_t(\xi) \pm \Delta\tau_t) \cdot V_{demand} \tag{1}$$

where $\tau_t(\xi)$- target time for an activity (benchmark based on market research, empirical measurement, or historical metrics), $\Delta\tau_t$ - adjustment for specific store characteristics, V_{demand} - number of times the action is performed based on sales forecast (or a forecast of the particular process driver).

The process of adjusting target metrics for activities is flexible enough and can be adapted according to pre-specified business rules. However, resource planning remains quite sensitive to forecasting sales or other target metrics, which can have a high level of volatility, especially for companies in Ukraine. That is why accurate forecasting can play a significant role in decision-making. Some of the specific ways in which sales forecasting can help grocery retail business include:

1. Optimizing inventory: By forecasting sales, grocery stores can ensure they have the right amount of inventory on hand to meet customer demand without having too much excess stock. In addition, accurate time-series models can help minimize waste and reduce storage costs. Walmart uses a proprietary algorithm to analyze sales data and

forecast demand for each product at each store location. Such an approach allows Walmart to adjust inventory levels in real-time and minimize waste while ensuring that products are available for customers [3].

2. Managing cash flow: Accurate sales forecasting can help grocery stores manage their cash flow by ensuring they have enough inventory to meet demand without tying up too much capital in excess stock.
3. Planning promotions: By understanding customer demand patterns, grocery retailers can plan promotional activities and special offers more effectively, which can help boost sales and attract new customers. As an industrial application example, Tesco operates a Data-Driven approach to identify which products are likely to sell well during promotions and adjusts its inventory levels accordingly [4].
4. Pricing strategy: By understanding customer demand and the competitive landscape, stores can adjust prices to maximize revenue and profitability.
5. New product development: By analyzing customer demand and identifying gaps in the market, retailers can develop new products tailored to their customers' needs.

By leveraging data and analytics to make informed decisions, stores can optimize operations, enhance customer satisfaction, and drive growth and profitability. In our research, the main application of time series forecasting will focus on designing the resource model for planning operations in one modern retail function – online grocery delivery. The delivery function refers to the process of transporting groceries from the store or warehouse to the customer's location. In recent years, such service has become increasingly important as more customers opt for online grocery shopping and home delivery options.

Grocery retailers typically use different delivery models to fulfill customer orders, such as store pick-up, third-party delivery services, and in-house delivery fleets. Store pick-up allows customers to place orders online and then pick them up at a designated time from the store. Third-party delivery services involve partnering with delivery companies like DoorDash or Uber Eats to deliver groceries to customers' homes. Finally, in-house delivery fleets use company-owned vehicles and drivers to transport groceries from the store or warehouse to customers' homes.

2 Modeling the Resource Planning System

Effective delivery function in grocery retail requires careful planning and coordination to ensure that orders are fulfilled accurately, on time, and with minimal waste. Retailers need to consider factors like product availability, order volume, delivery times, and customer preferences when designing their delivery models. Efficient routing and scheduling of delivery vehicles can also help to reduce delivery costs and improve the overall customer experience.

For planning delivery resources, retailers need to assess the demand for delivery in each area they serve by analyzing customer data and delivery orders to identify peak times and locations of high demand. Based on this analysis, retailers can estimate the number of drivers, vehicles, and other resources needed to meet the demand. The number of drivers required may vary based on factors such as the size and weight of the orders, delivery distance, and expected delivery time.

The following architecture was designed to create a Decision Support System that helps plan resources (Fig. 1). It consists of three layers:

1. The Data and Business Inputs Layer serves as the foundation of the resource model and includes all relevant data inputs and business rules. This layer collects data from point-of-sale, labor management, and courier platform systems.
2. The Forecast and Optimization Layer utilizes machine learning algorithms to generate accurate sales forecasts, calculate the needed resources, and optimize staff schedules to meet customer demand while minimizing labor costs.
3. The Business Intelligence Layer provides insights and visualizations into the performance of grocery retail operations. This layer consolidates and presents data in user-friendly dashboards, enabling managers to monitor and track key performance indicators such as sales, labor costs, and productivity metrics.

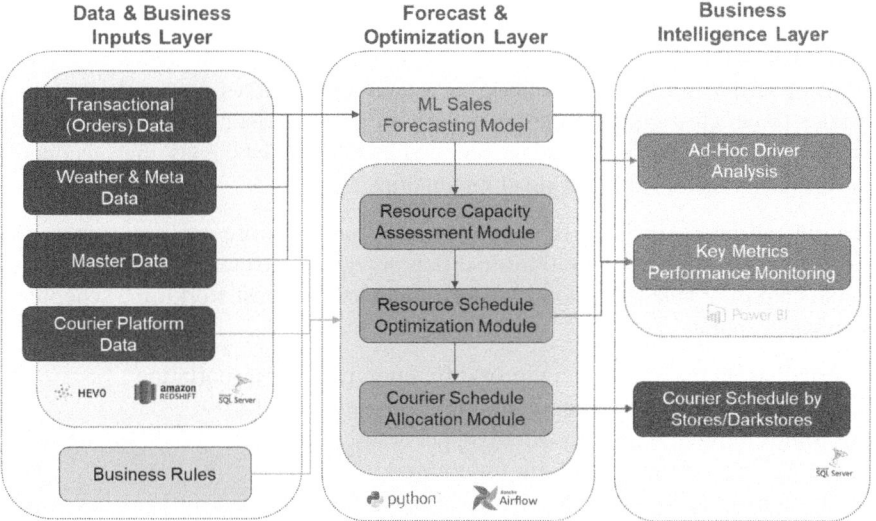

Fig. 1. Architecture of Resource Modeling System.

A comprehensive research project was undertaken to develop a resource planning system for a nationwide Ukrainian grocery company, which operates an in-house delivery service, and manages over a hundred locations across the country, including traditional chain stores and dark stores. The study's main objective was to devise a sophisticated system that would effectively optimize delivery resources, ensuring timely and efficient delivery service to customers.

2.1 The Data and Business Inputs Layer

The resource model for grocery retail utilizes several data sets within its data warehouse to generate accurate staffing schedules and budgets. These data sets include:

- Orders Data: The dataset includes the order amount, basket depth, address, order date, time, status, type, and items. To prepare the dataset for the ML model, this data is adjusted to polygon shifts based on client coordinates, allowing the model to accurately forecast demand and allocate resources accordingly.
- Weather Data: The resource model also incorporates weather data from Open-Meteo to adjust staffing levels during periods of extreme weather conditions that may affect customer demand [5].
- Air Siren Timestamps: Data on air siren alarms, including the duration of each alarm, is linked to each city-store level in the data warehouse, allowing the model to adjust staffing levels during periods of heightened security alerts.
- Holidays List: The model incorporates a list of holidays to adjust staffing levels and schedules, ensuring adequate coverage during peak customer demand periods.
- Courier Platform Data includes information on the delivery routes and logs for each courier, as well as data on the time taken for each delivery and any delays or issues encountered during the delivery process. This data can be used to optimize delivery routes, improve delivery times, and reduce the overall cost of delivery operations. Additionally, the platform may provide real-time visibility into the status of deliveries, allowing businesses to better manage their delivery resources.
- Master Data: This data includes store-specific characteristics and employee information, allowing the model to generate schedules and budgets that account for store-specific factors and workload fluctuations.

The overall dataset included three years of data from 118 stores and darkstores, covering grocery retail company operations in different regions. To create a resource scheduling model, specific business rules were also included to adjust workforce schedules to the company's policy:

- Restrictions on maximum weight for different types of transportation;
- Setting minimum and maximum durations for courier shifts;
- Ensuring a minimum number of orders per courier route;
- Imposing a maximum speed limit for couriers;
- Establishing a maximum radius between stores for combined shifts across different branches.

2.2 Forecast & Optimization Layer. Building Machine Learning Model

In order to develop an accurate time-series model for forecasting the number of orders, multiple data sets were utilized. The target metric was derived from orders data, while regressors were composed of weather data, air sirens duration dataset, and holidays list. The data were cleared of outliers to remove the effect of days when shops were closed or affected by unforeseen events, such as refitting periods, shopping malls or subways closures, and road closures for repairs. Such a result was achieved using anomaly identification methods based on clustering. An anomaly was defined as an extraordinary sales level on a specific day of the week, and the following algorithms were used to detect them: k-Nearest Neighbors Detector, Isolation Forest, Angle-Based Outlier Detection, Histogram-based Outlier Detection, and Local Correlation Integral. An ensemble of five models was used to calculate the weighted probability of each point being an outlier.

The data were cleaned once the anomaly observations were confirmed to be outside of holiday or lockdown periods. Time-series modeling was then used to fill in the missing data points with unbiased forecasts.

Orders dynamics in grocery retail often exhibit regular and predictable patterns that occur at fixed intervals over a year. These patterns may be related to seasonal events such as holidays, changes in weather, or other factors that affect consumer behavior. To effectively forecast time series data in grocery retail, it is important to capture and model these patterns. This can be done using various forecasting techniques, such as ARIMA or exponential smoothing. However, these methods may not always be suitable for capturing complex and irregular seasonal patterns. To address this challenge, Facebook developed a time series forecasting tool called Prophet. Prophet uses a decomposable time series model with three main components: a non-periodic trend component $g(t)$, a seasonality component $s(t)$ (weekly, monthly, yearly periodicity), holidays effect component $h(t)$, and additional regressors $r(t)$ [6]:

$$y(t) = g(t) + s(t) + h(t) + r(t) + e_t \qquad (2)$$

1. The non-periodic trend component represents the overall direction of the order dynamics over time using the logistic growth model in the form:

$$g(t) = \frac{c(t)}{1 + e^{-(\beta + \delta\alpha(t)^T)(t - (m + \gamma\alpha(t)^T))}} \qquad (3)$$

where $c(t)$ limiting value, which is driven by the district population growth, $\beta + \delta\alpha(t)^T$ represents a growth rate, that may vary over periods based on a binary operator for time events $\alpha(t)^T$ such as store refitting or competitor exit from a local market that lead to changepoints with offset parameter $m + \gamma\alpha(t)^T$:

$$\gamma_i = \left(s_i - m - \sum_{l<i}\gamma_i\right)\left(1 - \beta + \sum_{l<i}\delta_l \Big/ \beta + \sum_{l\leq i}\delta_l\right) \qquad (4)$$

2. The seasonality component captures daily, weekly, or monthly repeating patterns within the data by Fourier series decomposition:

$$s(t) = \sum_{n=1}^{N} (a_n \cos(2\pi nt) + b_n \sin(2\pi nt)) \qquad (5)$$

By composing a matrix of seasonalities $X(t) = [\cos(2\pi nt), ..., \sin(2\pi Nt)]$ the seasonal component can be fitted in the following way:

$$s(t) = X(t)\beta^T \qquad (6)$$

3. The model incorporates a holiday component to capture the impact of special events or holidays on the data. Additionally, the model includes regressor components such as capital projects and marketing campaigns, as well as external factors like COVID-19 and air sirens, which can be implemented using a binary operator.

$$h_i(t) = L_{holiday_i}(t)\kappa^T; \; r_j(t) = L_{regressor_j}(t)\kappa^T \qquad (7)$$

More recently, Facebook also introduced NeuralProphet [7], which is an extension of Prophet that uses a neural network architecture to model time series data. This allows for capturing more complex and non-linear patterns in the data and extends its capabilities by incorporating lagged regressors $L(t)$, future regressors $R(t)$, and customizable loss functions. Furthermore, in addition to including the effect of covariates, Neural Prophet can also take into account the past values of the target time series as inputs, which is useful in capturing the impact of autocorrelation in the time series $AR(t)$:

$$y(t) = g(t) + s(t) + h(t) + R(t) + \boxed{AR(t) + L(t)} + e_t \tag{8}$$

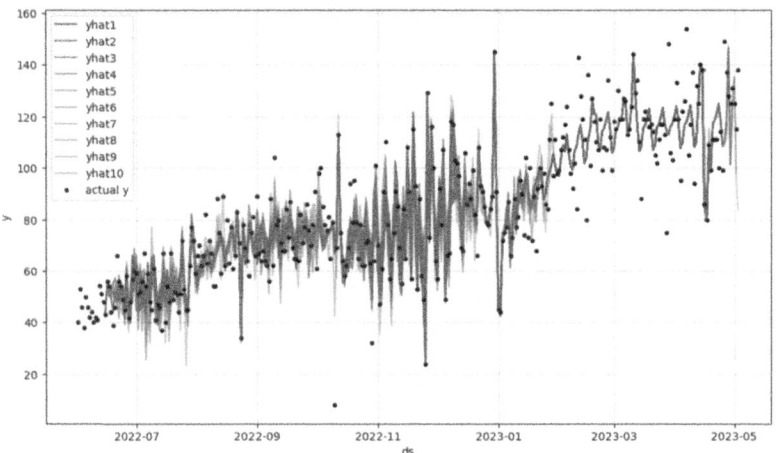

Fig. 2. Example of the output of NeuralProphet Forecasting model for a separate store.

The AR module in AR-Net can capture non-linear patterns in the data by using a fully connected Neural Network (NN) with hidden layers. By configuring the number and size of hidden layers, the model can improve the accuracy of the forecasts. However, this comes at the cost of interpretability since it becomes difficult to directly quantify the impact of a specific past observation on a particular prediction. Instead, we can only evaluate the relative importance of a past observation on all predictions by comparing the sums of the absolute weights of the first layer for each input position.

To adapt cross-validation for time series, a method involves dividing the test data into several subsets based on chronological order, each with a size that reflects the real-life scenario. In our case, we used a subset size of 14 days. Initially, the training data is utilized to forecast the first subset of 14 days. This subset is then included in the training data to forecast the subsequent subset of 14 days, and the model is retrained and updated. By implementing both methods, we can generate a forecast for the next two weeks that can be incorporated into the resource model (see Fig. 2).

By using the recommended approach, we can decompose the output into different components such as trend, seasonality, holidays, and regressors (see Fig. 3). The observed store exhibited significant changes in its trend during the summer of 2022, coinciding

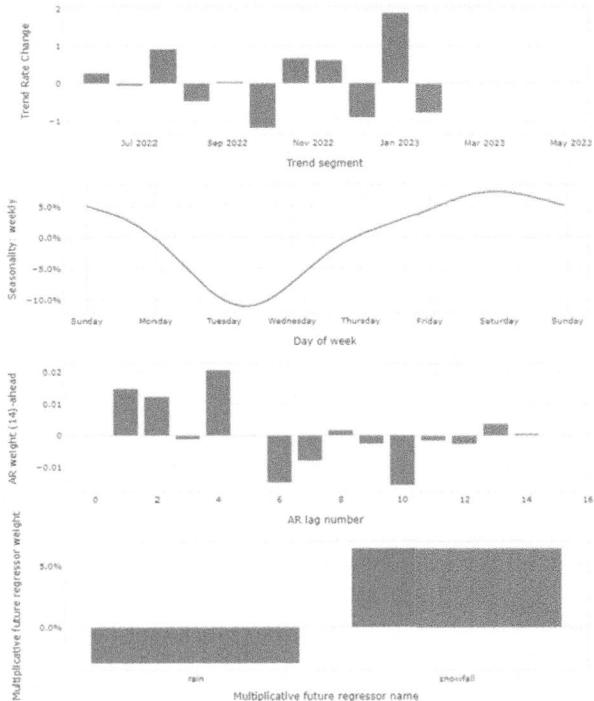

Fig. 3. Example of the time-series decomposition using NeuralProphet, representing the effect of trend rate change, weekly seasonality, AR weights, and weather impact (rain and snow).

with a large number of Ukrainians returning home, and during the fall when there were extensive attacks on the energy infrastructure. The branch displayed considerable weekly fluctuations, with traffic dropping to -10% of the average level on Tuesdays and rising to + 5% on weekends. The analysis revealed that the previous 1, 2, and 4 days had the most substantial positive impact on traffic dynamics, while the preceding 6, 7, and 10 days had a negative influence. Additionally, increased rainfall correlated with decreased traffic, while snowfall increased traffic by + 5%.

It is essential to emphasize that these observed behavior patterns are specific to the analyzed store, and variations are expected across different branches. The aforementioned example serves to illustrate the distinctive features of one store's traffic dynamics. The described decomposition model allows us to analyze the impact of each component on the overall forecasted values. This information can then be used to perform sensitivity analysis and scenario modeling to test different assumptions and optimize resource allocation strategies.

On a store level, we developed separate models using both fbprophet and Neural-Prophet methods, each with a hyperparameter grid search. For the fbprophet model, we experimented with varying the changepoint prior scale, seasonality prior scale, and seasonality mode. For the NeuralProphet model, we explored the number of hidden layers and the number of lags in autoregression. The RMSE metric was chosen as a target as it considers the magnitude of the forecasting errors, which is crucial for time series where outliers or significant deviations should be penalized appropriately. The model that resulted in the lowest RMSE metric was selected as the final model.

Upon analyzing the results of the cross-validation, we observed that the NeuralProphet method frequently outperformed the fbprophet method (Table 1), and we selected it as our primary forecasting framework for store-level predictions. This approach allowed us to train and fine-tune models for each store, considering its unique patterns and trends. The resulting models were able to generate more accurate forecasts, enabling better resource management and more informed decision-making.

Table 1. Model comparison by

Model	Weight	MAE	MSE	RMSE	RMSSE	MAPE	SMAPE
NeuralProphet	72,7%	5,303	56,339	6,981	0,629	16,224	7,424
Prophet	27,3%	5,579	52,622	7,118	0,704	16,998	7,920

2.3 Forecast & Optimization Layer. Building Machine Learning Model

After obtaining a two-week forecast from the trained time-series model, we further transform the output to an hourly capacity demand by leveraging product customer journey metrics and the historical hourly demand distribution. Such an approach allows us to more accurately predict the capacity required for each hour of the day.

To translate this demand into transportation requirements, we consider various route metrics specific to each store's delivery polygon. These metrics include the standard slot interval, maximum load capacity for each transport type per route, the distribution of order weights, the speed of each transport type, and the average route length. By combining these metrics with specific business rules, we are able to estimate the expected number of drivers required for each time slot. The resulting capacity demand is used to optimize the scheduling of transportation resources and ensure that each store receives the necessary deliveries in a timely and efficient manner.

In order to efficiently allocate delivery drivers to their respective shifts, an optimization model is developed. The model evaluates driver availability and preferences, considering factors such as work-hour restrictions and shift standards. The goal is to minimize overtime while ensuring that all deliveries are completed on time:

$$\min \sum_{d,p} |X_{ds} \cdot C_{sp} - RR_{dp}| \tag{9}$$

where d - day in the planning horizon, p- time slot in a day d, s- shift in a day d, X_{ds} - number of resources to schedule at day d for a shift s, C_{sp} - binary operator (shift s covers the period p), RR_{dp}- resources required at day d for the period p.

The restrictions of the model are:

$$X_{ds} \cdot C_{sp} \leq A \tag{10}$$

where A - maximum allowed capacity in a period.

$$X_{ds} \leq B \tag{11}$$

where B - maximum number of resources in a slot.

$$X_{ds}, RR_{dp}, A \in \mathbb{Z}^*, C_{sp} \in \{0, 1\} \tag{12}$$

The output of the optimization model is an efficient shift schedule for the drivers that meets the demand, route metrics, and driver availability while minimizing overtime.

An optimization model is utilized to efficiently create shifts for delivery drivers. The model takes into account various factors, such as productivity metrics for each driver, their personal rating, and previous schedules. Additionally, the algorithm considers driver preferences for free days and weekends when allocating shifts.

If the algorithm is unable to allocate all resources to the desired store, the driver may be assigned a shift in the closest stores based on the distance matrix. Such an approach ensures that the operation is optimized in terms of resource allocation and minimizes the driver route distance, reducing the overall delivery time and cost.

By utilizing these techniques, the optimization model ensures that each driver is allocated shifts in a fair and efficient manner, taking into account various factors such as productivity, personal preferences, and availability. The result is a well-optimized delivery schedule that can handle unexpected changes and variations.

2.4 Business Intelligence Layer

After the resource planning system was developed, a set of business intelligence (BI) components were designed as the output of the system. These BI components serve to provide key information and insights to the stakeholders and decision-makers of the organization.

The first BI component is the Courier Schedule by Stores/Darkstores. This component presents an overview of the schedules of the delivery drivers, including the stores or darkstores they are assigned to and the dates and times of their shifts. This component is important for ensuring that the delivery drivers are optimally allocated to the stores or darkstores that need them most (see Fig. 4).

The second BI component is Key Metrics Performance Monitoring. This component provides real-time monitoring of the resource planning system's key performance indicators (KPIs). The monitor board includes metrics such as order fulfillment rate, delivery time, and resource utilization rate. By monitoring these KPIs, the stakeholders can identify areas of improvement and make necessary adjustments.

The third BI component is Ad-Hoc Driver Analysis. This component allows stakeholders to perform ad-hoc analysis on the performance of individual delivery drivers. They can compare drivers' performance based on metrics such as productivity, delivery time, and customer ratings.

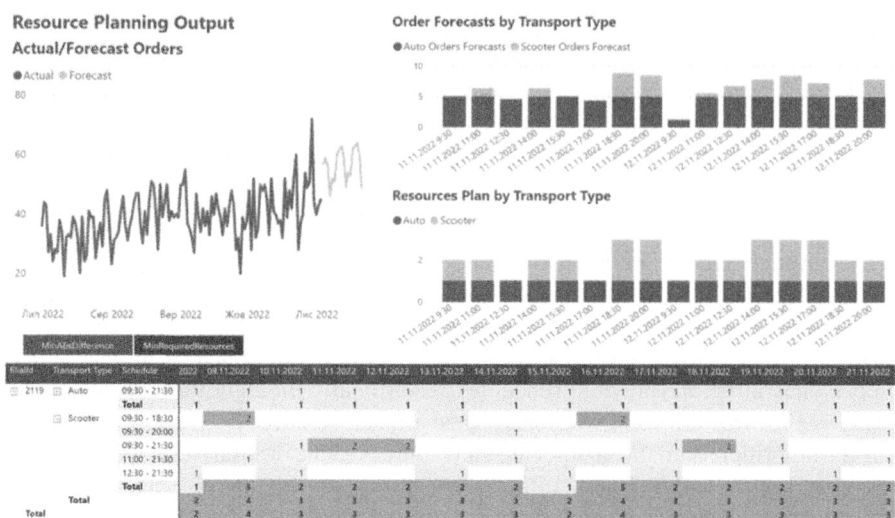

Fig. 4. Output of Business Intelligence module: Courier Schedule by Stores/Darkstores.

3   Conclusions

The model validation process extended over three months and involved piloting the model approach. Several challenges were encountered during this process. One significant difficulty was frequently updating work schedules to accommodate changes in polygon configurations. Additionally, accurately assessing speed and performance metrics for new cities without similar specifics to the stable network of stores proved challenging. The formation of standardized schedules for each courier also presented difficulties, especially during speculative traffic situations, and the inability to instantly fill courier vacancies further complicated the process.

Choosing appropriate benchmarks for performance evaluation proved to be a complex task due to the variations in operations, individual configurations for each store chain, and the diverse goals set by different companies. These objectives could range from improving service and cost reduction to maximizing profitability with potential risks of losing clients. Researchers focusing on the development of resource planning models in retail may adopt diverse target metrics. These metrics can encompass objectives like enhancing the net profit margin [8], improving resource utilization [9], and mitigating lost profits [10].

The resource planning model has shown promising results in forecasting demand and efficiently allocating resources to meet that demand. By incorporating multiple data sources and utilizing advanced modeling techniques, the system has been able to accurately forecast demand at the level of 16.3% MAPE during the pilot period. Furthermore, the optimization algorithm for shift scheduling can increase productivity by 10.1% and allow for better resource allocation, reducing costs by 8.6% in the simulated scenario described in the research.

The study was carried out to recognize the growing importance of effective resource planning in the grocery retail sector. With the increasing demand for online grocery shopping and delivery services, retailers are faced with the challenge of ensuring timely and efficient delivery of orders while optimizing the use of delivery resources. The research aimed to address these challenges and provide a solution that would enable retailers to meet their customers' evolving needs and expectations.

Overall, the recommended resource planning model is a valuable tool for grocery retail companies looking to optimize their delivery operations and improve customer satisfaction. However, it is important to carefully consider the benefits and limitations before investing in such a system and to continuously monitor and update it as needed to ensure its effectiveness.

References

1. Görgens, S., Greubel, S., Moosdorf, A.: How to mobilize 20,000 people. McKinsey & Company. https://www.mckinsey.com/industries/retail/our-insights/how-to-mobilize-20000-people. Accessed 07 May 2023
2. McKinsey & Company. Future of retail operations: Winning in a digital era. McKinsey & Company. https://www.mckinsey.com/~/media/McKinsey/Industries/Retail/Our%20Insights/Future%20of%20retail%20operations%20Winning%20in%20a%20digital%20era/McK_Retail-Ops-2020_FullIssue-RGB-hyperlinks-011620.pdf. Accessed 07 May 2023
3. Marr, B.: How Walmart Is Using AI, IoT And Big Data To Boost Retail Performance. Forbes, Forbes Magazine. https://www.forbes.com/sites/bernardmarr/2017/08/29/how-walmart-is-using-machine-learning-ai-iot-and-big-data-to-boost-retail-performance/?sh=72f852ac6cb1. Accessed 07 May 2023
4. Marr, B.: Big Data At Tesco: Real Time Analytics At The UK Grocery Retail Giant. Forbes, Forbes Magazine. https://www.forbes.com/sites/bernardmarr/2016/11/17/big-data-at-tesco-real-time-analytics-at-the-uk-grocery-retail-giant/?sh=499e361061cf. Accessed 07 May 2023
5. Open-Meteo Homepage. https://open-meteo.com. Accessed 07 May 2023
6. Taylor, S.J., Letham, B.: Forecasting at scale. Am. Stat. **72**(1), 37–45 (2018). https://doi.org/10.1080/00031305.2017.1380080
7. Triebe, O., Hewamalage, H., Pilyugina, P., Laptev, N., Bergmeir, C., Rajagopal, R.: NeuralProphet: explainable forecasting at scale (2021). https://doi.org/10.48550/arXiv.2111.15397
8. Chapados, N., Joliveau, M., L'Ecuyer, P., Rousseau, L.M.: Retail store scheduling for profit. Eur. J. Oper. Res. **239**, 609–624 (2014). https://doi.org/10.1016/j.ejor.2014.05.033
9. Salsingikar, S., Ganesan, V., Sengupta, S.: Labor scheduling in retail stores (2006)
10. Mac-Vicar, M., Ferrer, J.C., Muñoz, J., Henao Botero, C.: Real-time recovering strategies on personnel scheduling in the retail industry. Comput. Ind. Eng. **113**, 589–601 (2017). https://doi.org/10.1016/j.cie.2017.09.045

Poster Papers

Organization of Independent Work of Students in LMS Moodle Using a Metacognitive Approach (on the Example of Physical and Mathematical Disciplines)

Mariia Astafieva⬝, Dmytro Bodnenko(✉)⬝, Oksana Lytvyn⬝,
and Volodymyr Proshkin⬝

Borys Grinchenko Kyiv University, Kyiv, Ukraine
{m.astafieva,d.bodnenko,o.lytvyn,v.proshkin}@kubg.edu.ua

Abstract. Recent years of the covid pandemic and military threats have significantly exacerbated the need for young people to obtain university education remotely. This also increases the requirements for the effectiveness of using the didactic capabilities of digital means and tools of distance education, and the quality of relevant content for organizing and supporting students' independent educational activities. Able to effective independent cognitive activity can only be a person who has the appropriate cognitive skills, and is aware of his own cognitive processes. Therefore, filling students' independent work with metacognitive content should be considered as the subject and purpose of training, along with the traditional goal of the university, which is the formation of subject knowledge and skills. The subject of our research is the use of means and tools of LMS Moodle for the organization of independent work of students based on a metacognitive approach in the study of physical and mathematical disciplines. The article analyzes the problems of organizing students' independent work and offers some ways to solve them using the Moodle functionality. Examples of the use of individual Moodle activities by the article's authors for the organization and support of independent work of students are given.

Keywords: LMS Moodle · E-learning Course · Independent Work · Metacognitive Approach · Physical and Mathematical Disciplines

1 Introduction

The Reference Framework of competencies for democratic culture, developed by the Council of Europe's Department of Education, identifies a set of 20 competencies that educators should focus on to strengthen the capacity of learners to be competent and effective citizens of a democratic society. One of them is the ability to learn independently, which is necessary for a person "to pursue, organize and evaluate their own learning, following their own needs, in a self-directed and self-regulated manner, without being prompted by others" [1], because it is necessary to learn throughout life. Therefore, the effective organization of independent students' work is an important prerequisite for successful learning.

© The Author(s), under exclusive license to Springer Nature Switzerland AG 2023
G. Antoniou et al. (Eds.): ICTERI 2023, CCIS 1980, pp. 303–312, 2023.
https://doi.org/10.1007/978-3-031-48325-7_23

Thanks to the development of digital technology, Internet platforms have become an integral feature of the modern educational process. One such platform is LMS Moodle, which has a wide functionality for organizing independent work and is free, open-source and used by most universities. However, effective targeted (to form students' ability to self-learn) use of LMS Moodle requires not only knowledge of the platform's functionality and technical skills in using its tools but also a methodological approach to learning.

The proposed article considers the tools of LMS Moodle to organize and support the independent work of students in the study of physical and mathematical disciplines based on the metacognitive approach. The metacognitive approach is based on developing cognitive skills that allow students to learn and control their learning independently. It is the metacognitive approach to the organization of students' independent work that focuses on the formation of the ability to plan, control and evaluate their own learning.

The work aims to study the capabilities of LMS Moodle tools to support a metacognitive approach to the organization and guidance of students' independent work in studying physical and mathematical disciplines.

2   Literature Review

The use of digital technologies for the organization of independent work of students in the study of physical and mathematical disciplines is the subject of research by many scientists, for example [2, 3]. Here are some results of scientific research on the problem of organizing independent work of students in the digital age. The study [4] analysed the experience of Finland in organizing independent work of students, in particular, the influence of the level of digital competence of students on the quality of education is revealed.

The study [5] of emphasized the need to implement self-managed learning, that is, the participation of students themselves in determining the purpose of training, planning, presenting, and evaluating their work, etc. Various problems of training organisations through LMS Moodle became the subject of scientists' research [6]. The study [7] aims to specify, in a practical way, procedures and automation guidelines in Moodle based on the experience of instructional design; and the structuring of online activities through the programming of automated flows by configuring conditional rules and the integration of a virtual assistant to attend general queries with a Chatbot tool named Dialogflow from Google.

Opportunities and examples of using metacognitive learning strategies, and self-studying students in mathematics, are disclosed in [8, 9]. In particular, metacognitive activity is interpreted as a complex integrative formation aimed at controlling the processes of perception, storage, processing, and reproduction of information. So, the study [10] shows, how metacognitive strategies for teaching mathematics can improve students' metacognition, as well as how students and teachers of mathematics can use metacognitive strategies. In [11] the problems of increasing the motivation and self-awareness of students in the process of learning mathematics are investigated. To do this, it is proposed to use a metacognitive approach to learning. The tools of LMS Moodle were chosen to implement the research tasks.

Despite numerous scientific studies on the problems of organizing independent work of students, using means of LMS Moodle as one of the most common systems of distance learning, the organization of this work in the process of studying physical and mathematical disciplines based on a metacognitive approach has not yet been the subject of special scientific research. The capabilities of LMS Moodle to implement this task have not been sufficiently studied.

3   Research Methods

The research was carried out within the framework of the complex scientific theme of the Department of Mathematics and Physics of Borys Grinchenko Kyiv University "Mathematical methods and digital technologies in education, science, technology", DR No 0121U111924. The following research methods were used in the work: analysis, generalization of scientific literature to reveal the essence of the metacognitive approach; systematization of LMS Moodle tools for organizing independent work in physical and mathematical disciplines; empirical (questionnaires) to identify problems in the organization of independent work of students, observational (pedagogical observation, reflection of research activities) to implement independent work of students based on the metacognitive approach in LMS Moodle.

4   Research Results

4.1   Survey Results and Problem Statement

To identify the problems associated with the organization and implementation of independent work of students in the process of teaching physical and mathematical disciplines, as well as to clarify the attitude of teachers and students toward the role and place of independent work at the university, during March 2023, we conducted a survey. Respondents were 150 students who study or previously studied physical and mathematical disciplines. In addition, we interviewed 129 teachers from Borys Grinchenko Kyiv University and other universities in Ukraine.

The survey results showed the following.

Among the significant factors that negatively affect the implementation of independent work of students, teachers name the following: the reluctance of students to work independently and the teacher's lack of leverage in this situation – 59.1%; lack of methodological literature for teachers on the organization and management of students' independent work, control and evaluation of its results – 46.7%; lack of proper motivation of the teacher for this type of activity (management of most of the independent work of students is not taken into account in the teacher's educational load) – 46.7%. Students, on the other hand, see the problem mainly in their own insufficient metacognitive competencies: they do not know where to start solving a problem or working out a theory – 45.3%; they do not know how to overcome difficulties when they come across them when working out a theory or solving a problem – 39.3%; they have insufficient organizational skills (ability to plan and distribute time between different types of work) – 36.7%.

When asked about the use of the metacognitive approach in the organization of student's independent work, only 13.4% of teachers answered that they organize student's independent work based on this approach, and 19.5% (every fifth) chose the answer: "I do not know what it is."

This indicates the need to improve the methodological support of students' independent work, both on self-study issues (for students) and pedagogical guidance of students' independent work. It also confirms the relevance of unlocking the possibilities of a metacognitive approach to learning aimed at developing students' ability to learn and, in the conditions of distance learning - the importance of digital support for students' independent work and pedagogical guidance.

This also confirms the relevance of revealing the possibilities of a metacognitive approach to learning, aimed at developing students' ability to learn, and in the conditions of distance learning - the importance of digital support for students' independent work and pedagogical guidance.

4.2 LMS Moodle Tools for Organizing Students' Independent Work in Physical and Mathematical Disciplines

Analysis of the literature on the problems of organizing independent cognitive activity of students and our own experience give grounds to assert that the metacognitive approach in the educational process in general, in particular, and in the organization and management of independent work of subjects of learning, creates the conditions necessary for their effective self-development and the realization of individual creative potential, and the use of the capabilities of the LMS Moodle functionality helps to implement this approach more successfully.

The metacognitive approach to organizing students' independent work involves: teaching students metacognitive learning strategies to understand, control and regulate their cognitive activity; planning learning activities so that students can gradually take more responsibility for their learning; using active teaching methods and interactive platforms to engage students in the construction of their knowledge, rather than passively obtaining them; providing feedback to students on their use of metacognitive strategies; encouraging students to reflect on their learning experience; educating and developing students' ability to analyse the dynamics of their metacognitive skills, encouraging them to view errors as opportunities for learning.

The specificity of physical and mathematical disciplines (for example, the logic and interconnectedness and interdependence of various topics, a high level of abstraction - in mathematics, mathematical modelling of complex systems - in physics) dictates the need to have specific metacognitive strategies for successful learning. These are, in particular, conceptual understanding, not formal memorization, prediction of the result (hypothesis) and search for ways to prove or refute, not passive familiarization with the result, verification of the solution, and not a simple statement of the answer, search for alternatives and analogies, etc. Some metacognitive self-learning strategies (for the student) and learning (for the teacher) and possible Moodle digital tools to support these strategies are shown in Table 1.

Example 1. Assignments. Activities "Assignments" we use in project-based training. For example, to carry out research projects with small groups of students, you can build an

Table 1. Moodle LMS tools for organizing independent work in physical and mathematical disciplines

Competencies of students for self-study (*learning*)	Pedagogical metacognitive strategies	Moodle Tools
Ability to plan and distribute time between different types of work	Planning training	Checklist
Ability to determine (or anticipate) goals and ways to achieve them	Familiarization with the expected learning outcomes before studying the topic	Assignments (with "Know-Want to know-Learned" table), Choice
Analytical skills: analysis; ability to generate "correct" questions; awareness of difficulties and obstacles; monitoring the level of the material understanding; critical evaluation of the result obtained; search, consideration of alternative solutions, analysis of advantages, and disadvantages; logical thinking; ability to use mathematical tools	Active teaching methods that encourage the search and construction of knowledge. Formative assessment Ensuring the ability to build an individual educational trajectory, adaptive learning with the ability for everyone to receive assistance as needed	Lesson, Assignments, Quiz, Forum, Chat, Workshop, Wiki, Glossary, External tool, H5P
Communication skills: the ability to explain and argue their actions; the ability to seek (if necessary) help; the ability to use mathematical language	Joint training in groups, couples Execution of projects	Wiki, Glossary, Forum, Workshop, Chat, Database, External tool
Motivation	Implementation of practically oriented tasks, mathematical modelling	Assignments, Choice, Forum, External tool
Ability to read (study) mathematical literature, find the right sources	Formation of this ability through special tasks, relevant content, as well as "instructional memos" to support active reading	Lesson, Assignments, Quiz
Ability to self-assessment: the ability to assess their progress; the ability to adjust, if necessary, their educational trajectory	A questionnaire, survey, reflection; Formative assessment	Lesson, Quiz, Feedback, Choice, Workshop

external resource that allows participants to work together. The teacher creates an online service Miro Online Writeboard and embeds it into the activity of the Task (HTML tool "iframe").

Gives an example (see Fig. 1) of students performing research in the discipline of "Physics". After analysing the laws of horizontal motion (axis OX), the study's main task was to write down the laws of vertical motion (axis OY). It was important that this tool is needed to provide a common workspace for collective research work.

Example 2. **Lesson.** The activity "Lecture" can and should be used not just for placing certain theoretical material, but for organizing adaptive self-educational activity, with different trajectories, transactions between pages, various additional clusters, pages with

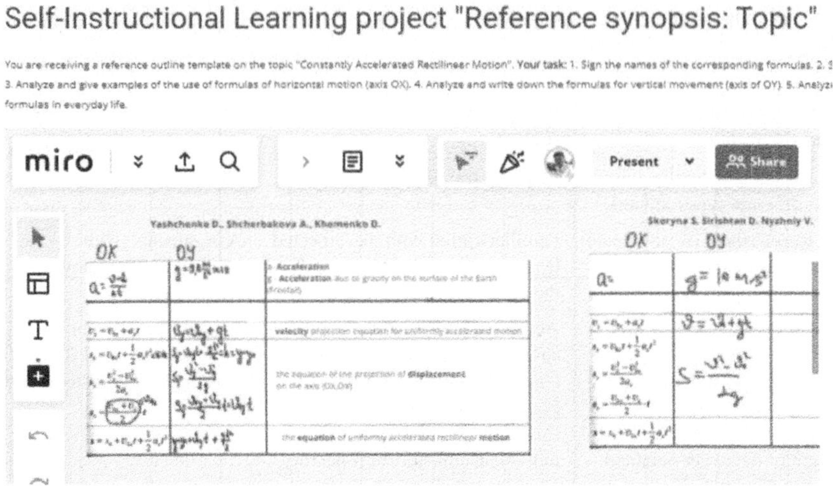

Self-Instructional Learning project "Reference synopsis: Topic"

You are receiving a reference outline template on the topic "Constantly Accelerated Rectilinear Motion". Your task: 1. Sign the names of the corresponding formulas. 2. $
3. Analyze and give examples of the use of formulas of horizontal motion (axis OX). 4. Analyze and write down the formulas for vertical movement (axis of OY). 5. Analyz
formulas in everyday life.

Fig. 1. Activities Assignments with built-in Miro Online Writeboard for small group projects

test questions or tasks (for automatic verification) after logically completed fragments of text, the answers to which would indicate an understanding of what was read and, accordingly, the possibility (or impossibility) to go further. In addition, for a virtual lecture to compensate for the lack of "live" communication between the teacher and students, there must be a special organization of content and a special style of presentation that models metacognitive behavior. All material should be divided into small logically complete blocks, and the presentation of the material (proof of the theorem, for example) should take place in the form of an imaginary dialogue (the author / lecturer asks a question and answers himself), so that the student has the feeling that he is thinking out loud. For more information about the virtual lecture, see [12].

Example 3. Forum. For an independent solution of the first-year students-mathematicians were offered 32 planimetric problems, which had a non-standard and unusual, yesterday's secondary school students, the formulation of the condition. This confused the students. Most, having read the conditions and not seeing the obvious (usual) algorithm of actions to obtain the result, simply gave up. And one wrote: "How can you solve such problems if almost nothing is known? I don't even know where to start."

An example of how a teacher took advantage of the situation to develop students' thinking and metacognition is given below.

At the Forum "Strong planimetric nuts" of the e-learning course at Moodle, the teacher fits the condition of one of the 32 tasks – "What is the square area?" (see Fig. 2.a) and invites students to: a) ask questions to "discover" more data (which are "hidden"); b) offer a solution.

The metacognitive strategy of the teacher was to educate observation, learn to ask the right questions, and use the Forum tool to motivate students to solve the problem, creating an atmosphere of excitement and competition (see Fig. 2 b).

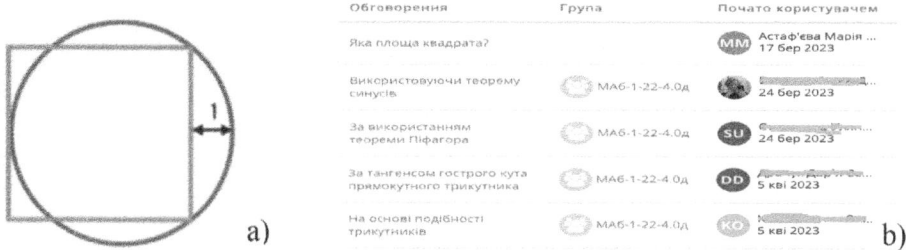

Fig. 2. a) images to the tasks; b) discussion of solution in the forum

As a result, the problem was solved in seven different ways. When posting his solution on the Forum, the student had to fill in the required field "Theme" (left column), which obliged him to briefly (in one phrase) indicate the method or key idea of the solution, for example, "Using the cosine theorem" (one of the important cognitive abilities for self-learning!). And, finally, the teacher initiates reflection (students discussed the proposed solutions, reflected on their thinking, described, and analysed the search for the idea of a solution, evaluated their own cognitive experience). Here is one of the posts, which shows that its author is aware of his metacognitive strategy (to perform additional constructions to see the idea) and evaluates it as effective:

> *"At first glance, it seemed that the problem could not be solved, because insufficient information was given. But, considering various variants of additions, a rectangular triangle immediately struck me, from which, according to the Pythagorean theorem, it is easy to find the side of the square."*

*Example 4. **Hidden text.*** To provide an individual approach in guiding the independent work of students, there is often a need to initially hide from students a part of the text on the page of the educational material, with the possibility of access to it if necessary. Moodle allows you to do this in HTML code editing mode. Hidden text can be the solution to the problem, hints (the idea of proof, the stages of solution), more detailed explanations of individual fragments, etc. Below (See Fig. 3) an example of Metacognitive Scaffolding to the task of working independently (according to the textbook) on the topic: "Double integral: definition, properties, geometric and physical content" (in the e-learning course of the academic discipline "Mathematical Analysis") is given.

*Example 5. **Quiz.*** We avoid test tasks of the closed type when you need to choose one (correct) answer from several of the above because such tasks are useless for forming of conceptual knowledge and metacognitions. In addition, in tests for self-checking, we usually use the interactive mode of multiple attempts. For this, it is advisable to program errors.

Forming such a task, it is provided, in addition to the correct answer, several wrong ones that the student will give under the condition of certain (quite specific!) gaps in knowledge. Therefore, for each such incorrect answer, a comment is provided, aimed at making the student realize his mistake and independently come to the right conclusion. An example of such a comment on an incorrect answer is given in (see Fig. 4).

Double integral: definition, properties, geometric and physical meaning ⊚

1. Formulate for yourself the questions to which you should receive answers after studying the topic. Write these questions down. After that, compare your questions with the proposed ones.

> ▼ **Suggested questions**
> *What is a double integral?*
> *What are the conditions of its existence?*
> *What geometric problems can be solved using the double integral?*
> *What is the physical meaning of the double integral?*
> *What properties does the double integral have?*

2. Before starting to read the material on the specified topic, answer the following questions:

- Does the term "integral" evoke any associations for you? Which exactly?
- Can you give a strict definition of the definite integral?

3. If you cannot answer the questions or are not sure that you correctly understand the concept of the definite integral, repeat the concept of the definite integral from the course of mathematical analysis 1, namely: what problems lead to this concept, strict definition, conditions of existence and properties of definite integral.

> ▶ Recommended sources

4. You should start studying the text about the double integral with a cursory reading in order to break it into logically complete parts. Compare your breakdown with the proposed one.

> ▶ The proposed division

5. Sequentially study each part, mastering new concepts, facts, methods, paying attention to the connections between them and with the previously studied material, apply the acquired theoretical knowledge in practice. If necessary, use more detailed methodical recommendations.

> ▶ Methodical recommendations

6. Make sure that you can answer the questions formulated by you in point 1.

7. Answer the questions for self-testing after the main text, solve the exercises and problems offered for independent solution.

8. If possible, talk to someone about the double integral, tell (explain) to someone the main things you learned.

Fig. 3. An example of "hidden text" in the guidelines for reading a book text of mathematics

> ✖
> You were wrong. I assume that you drew your conclusion from the fact that the function has two extremums. Therefore, I suggest you answer the following questions:
> 1) is the differentiability of a function at a point a necessary condition for the existence of an extremum at this point?
> 2) if f'(x)=0, then does it follow that x is the extremum point of the function?
> 3) and vice versa?
> After answering these questions, return to the test task.

Fig. 4. Corrective comment of the teacher in the test

After another attempt, there may again be a "corrective" comment on the wrong answer. The test in Moodle can be configured so that the student indicates how confident he is in the correctness of his answer. The use of this mode teaches students to an objective self-assessment and encourages them to solve the problem, analyse and check the solution, and not just send an immediate response.

*Example 6. **Wiki.*** The Wiki tool has wide opportunities for organizing students' independent work based on the metacognitive approach. Its functionality allows you to create collective pages that can be edited and commented on by users with access to these pages (the level of access is configured by the teacher). Moreover, the editing history is saved (unlike the Glossary tool). We use Wiki for students to perform collective projects (for example, create so-called conceptual tables on a certain topic [13]), as well as perform certain temporary roles: teacher, expert, and opponent. According to the history of editing, the teacher can "unobtrusively" track the evolution of each student's understanding of certain mathematical concepts, the dynamics of his metacognitive skills and make the necessary adjustments, as well as assess the knowledge and skills of students using the method of formative assessment. Here is an example of a task in the discipline "Mathematical Analysis 2", implemented by the Wiki tool.

Task: "Be in the role of a teacher".

The student gave the following definition of the limit of the sequence of points of the metric space: "Point a is the limit of the sequence of points of the metric space, if the members of this sequence approach point a as their ordinal number increases." Determine what the student misunderstands and offer your way to eliminate his gaps in understanding the specified concept.

5 Conclusion

1. As a result of a survey of teachers and students, it was established that the difficulties in organizing and implementing independent work of students are due, firstly, to the insufficient level of methodological support. In addition, the low efficiency of students' independent work is explained, among other things, by the inability to self-learn.

2. Based on the analysis of the scientific literature, the potential of the metacognitive approach for organizing students' independent work is revealed (using metacognitive learning strategies to understand, control and regulate cognitive activity; encouraging students to take responsibility for their learning; using active teaching methods and interactive platforms to involve students in the construction of their knowledge; providing feedback to students on their use of metacognitive strategies; encouraging students to reflect on their learning experience).

3. The most important competencies of students for self-study are highlighted, the corresponding pedagogical metacognitive strategies for the implementation of independent work of students are given. The Moodle LMS tools to support these strategies are offered. Examples illustrate the possibilities of LMS Moodle for organizing students' independent work using a metacognitive approach.

4. It is shown that using Moodle as a learning support tool allows creating an interactive and dynamic learning environment where students can actively learn, acquiring and improving metacognitive competencies. In particular, the fragmentary results of the study testify to the growth of students' self-regulatory competencies, namely, awareness of their metacognitive strategy and assessment of its effectiveness.

5. Analysis of the effectiveness of using Moodle tools for the organization of independent work of students in order to ensure a more active and aware of their learning activity, will be the subject of our further research.

References

1. Reference Framework of competences for democratic culture. https://www.coe.int/en/web/campaign-free-to-speak-safe-to-learn/reference-framework-of-competences-for-democratic-culture. Accessed 21 Mar 2023

2. Aranzo, R., Damicog, M., Macahito, C., Reyes, A.: Tancio, K., Luzano, J.: A case analysis of the strategies of students in learning mathematics amidst. Acad. Disruption Int. J. Multidisc. Appr. Stud. **2**(10), 1–15 (2023)
3. Batbaatar, N., Amin, G.: Students' time management during online class. In: International Conference Universitas Pekalongan, pp. 189–194. Universitas Pekalongan Publishing House, Indonesian (2021)
4. Ruhalahti, S., Lehto, T., Saarinen, S., Katto, L.: Identifying higher education first-year students' reported studying experiences studying during the pandemic. Eur. J. Educ. Stud. **8**(8), 1–21 (2021)
5. Vintere, A., Cernajeva, S., Gosteine, V.: Using E-study materials to promote mathematics self-learning at university. In: 19th International Conference on Cognition and Exploratory Learning in the Digital Age, pp. 285–290. Lisbon, Portugal (2022)
6. Indrawatiningsih, N.: Efektivitas learning management system (LMS) berbasis moodle sebagai sarana diskusi untuk meningkatkan kemampuan argumentasi matematika mahasiswa. Jurnal Pendidikan dan Pembelajaran Matematika **2**(7), 1–8 (2021)
7. Araya, F., Rebolledo Font de La Vall, R.: Rule-based automation in Moodle for self-instructional learning. Int. J. Soc. Sci. Educ. Res. Stud. **2**(10), 501–507 (2022)
8. Basokcu, O., Guzel, M.: Metacognitive monitoring and mathematical abilities: cognitive diagnostic model and signal detection theory approach. Egitim ve Bilim **46**(205), 221–238 (2021)
9. Amir, M.Z.Z., Nurdin, E.R., Azmi, M.P., Andrian, D.: The increasing of math adversity quotient in mathematics cooperative learning through metacognitive. Int. J. Instr. **4**(14), 841–856 (2021)
10. Du Toit, S., Kotze, G.: Metacognitive strategies in the teaching and learning of mathematics. Pythagoras **70**(209), 57–67 (2009)
11. Iannella, A., Morando, P., Spreafico, M.L.: Challenges in mathematics learning at the university: an activity to motivate students and promote self-awareness. In: Higher Education Learning Methodologies and Technologies Online: Third International Workshop. HELMeTO 2021 pp, pp. 321–332. Springer International Publishing, Italy (2022)
12. Astafieva, M., Zhyltsov, O., Proshkin, V., Lytvyn O.: E-learning as a mean of forming students' mathematical competence in a research-oriented educational process. In: Cloud Technologies in Education. Proceedings of the 7th Workshop CTE 2020, pp. 674–689. Kryvyi Rih, Ukraine (2020)
13. Astafieva, M., Boiko, M., Hlushak, O., Lytvyn, O., Morze, N.: Experience in implementing IBME at the Borys Grinchenko Kyiv university. The PLATINUM Project: monograph. Masaryk University Press, Brno (2021)

Cognitive Technologies and Competence Development: Bibliometric Analysis

Tetiana Ivanova[✉] ⓘ

The National Technical University of Ukraine
"Igor Sikorsky Kyiv Polytechnic Institute", Kiyv 03056, Ukraine
tetyana.v.ivanova@gmail.com

Abstract. The development of cognitive technologies and the formation of competences become an integral part of success in the conditions of a rapidly changing world. This document presents a bibliometric analysis of cognitive technologies and competencies. The dataset was retrieved from the Scopus database, analyzed and presented using VOSviewer. The search equation identified 281 studies. After analysis and application of filters, 60 studies were selected for further analysis. Among the journals that publish research on this topic, the most productive is Ceur Workshop Proceedings (5%). The most cited authors are González-González and Jiménez-Zarco. Authors from Indonesia, Spain and China published the most articles. The top 5 thematic categories include Computer Science; Social Sciences; Engineering; Business, Management and Accounting; Energy. The most cited article is Birjali, M., Beni-Hssane, A., Erritali, M. Research on cognitive technologies and the development of competencies open broad perspectives for improving learning, professional development, and personal growth.

Keywords: Cognitive technologies · Development of competences · Artificial Intelligence · E-learning

1 Introduction

One of the current areas of research that is attracting more and more attention is the use of cognitive technologies for the formation of competencies. In today's world, where digital innovations are changing all areas of life, these technologies have great potential to influence people's development and learning. Cognitive technologies are based on the use of intelligent algorithms and machine learning to analyze and process large amounts of information, which allows you to understand and use data more effectively. Their application in the field of training and development of competences opens up new opportunities for students, workers and everyone who seeks to improve their skills and acquire new knowledge.

Calışkan et al. [11] found that students should understand AI's role, formulate decisions, and consider AI results alongside their knowledge. Chao et al. [12] emphasize the importance of AI knowledge and competence and the need for comprehensive courses. Huang [22] supports AI education in fundamental stages to build students' key competencies.

© The Author(s), under exclusive license to Springer Nature Switzerland AG 2023
G. Antoniou et al. (Eds.): ICTERI 2023, CCIS 1980, pp. 313–324, 2023.
https://doi.org/10.1007/978-3-031-48325-7_24

Gutierrez et al. [19] study AI in e-learning, emphasizing the need to train AI programmers and define graduate competencies. López et al. [29] stress integrating cross-cutting competencies in higher education through e-learning tools.

Misthou and Paliouras [31] stress the need for government understanding and digital skill development for AI utilization. However, Gaol [15] explores AI-based performance management using the POAC method to enhance human resource competence, analyzing personnel competencies, educational-government interactions, budgets, and industry needs.

Permana and Pradnyana [35] developed an AI-based recommender system for student internships, enhancing the educational process by suggesting optimal places based on competencies. Sharma and Manchanda [38] used machine learning to predict and improve entrepreneurial competence in students by correlating personality traits with a lack of entrepreneurial skills.

Lestari et al. [27] discovered that knowledge of Robot, Artificial Intelligence, and Service Automation (RAISA) along with professional competence positively affect hospitality students' perceptions of future career opportunities. Margienė et al. [30] emphasize the necessity of e-learning systems and automated competency integration due to growing student mobility and diverse e-learning systems, which reduces workload and enhances resource sustainability.

Adinda [1] advocates for a competency-based approach to enhance students' self-regulation and self-direction in online learning, highlighting its positive impact on students' effectiveness in e-learning. Chirila [13] proposes an education system focused on competencies and incorporates dialog games to foster real-life skill development, addressing modern needs for practical skills and competitiveness. Garad et al. [16] explore the role of e-learning infrastructure and cognitive competence in distance learning effectiveness, confirming their positive influence on distance learning outcomes.

Birjali et al. [8] introduce an adaptive e-learning model based on big data and competencies, incorporating students' social activities. Hsu and Li [21] present the "competency-based guided learning algorithm" (CBGLA), which enhances e-learning adaptability.

Kassymova et al. [23] explored e-learning's role in enhancing cognitive competence, highlighting its advantages for self-improvement despite challenges. Keržič et al. [24] observed a strong link between computer skills competence and the utility of e-learning in a blended learning setup. Zulfiya et al. [42] developed an assessment model and methodology for student competencies in e-learning, spanning disciplines, modules, and program objectives.

Cao and Zhang [10] explored machine learning for competency models in HRM, enabling swift and precise assessment of employee competencies for optimal position placements. In turn, Burkhard et al. [9] introduced a learning structure embracing "human-intelligent machine" interaction as a norm for competency development in the AI era.

Lau [26] highlights the significance of information competencies for cognitive skill development in higher education, especially in the AI-driven context. Basantes-Andrade et al. [6] stress the importance of utilizing information and communication technology tools and aids. Khan et al. [25] emphasize the substantial influence of digital innovations on pedagogical digital competence and e-learning systems, mediated by computer self-efficacy.

Astafieva et al. [5] advocate effective e-learning approaches for enhancing students' mathematical competence. Heba et al. [20] also endorse information and communication technologies, including individualized learning systems and e-learning courses, to elevate students' mathematical competencies.

Albano and Ferrari [3] stress the significance of linguistic competence for academic achievement. Angelis et al. [4] recommend an individualized approach leveraging information technology and a dedicated platform to enhance language competence. Mujiono and Herawati [32] highlight a substantial gap in sociolinguistic competence between students who used e-learning with sociolinguistic instruction and those in traditional face-to-face settings.

D'Aniello et al. [14] emphasize the creation of an e-learning system to nurture intercultural competence, involving cognitive, behavioral, and affective skills for effective cross-cultural interaction. The study by Long and Lin [28] empirically explore the advancement of intercultural communicative competence in English-learning students with artificial intelligence, identifying a significant positive link with knowledge of foreign cultures, attitudes, and intercultural communication skills.

Al-Sharidah [2] asserts that e-learning platforms can enhance teachers' pedagogical competencies. Thomas et al. [39] identify the competency requirements for effective e-learning instruction by teachers. Wu [40] also introduces a model for enhancing teacher competence through an advanced machine learning algorithm. Zhao et al. [41] argue for the enhancement of teacher professional development systems in the age of artificial intelligence.

Batko and Szopa [7] emphasize the importance of developing computer technologies and new competencies in the era of robotics and artificial intelligence. Palacios-Marqués et al. [33] highlight how Web 2.0 tools in e-learning projects create new opportunities and modify traditional competencies based on knowledge management. Rivera-Kempis et al. [36] explore entrepreneurial competence and suggest that machine learning aids in entrepreneur classification. Sarangarajan et al. [37] show that artificial intelligence and e-learning help organizations remain adaptive and competitive, with a recommendation system optimizing skill development for business strategy execution.

The purpose of this article is a bibliometric analysis of cognitive technologies and competence formation. The study aims to answer the following questions regarding cognitive technologies and competence building: (1) Which journals are the most influential? (2) Which authors publish research on this topic? (3) In which countries are the authors interested in this topic? (4) What are the main thematic categories of research? (5) How heavily cited are works on cognitive technologies and competencies? (6) What terminology is used to conduct research on the relevant topic?

The manuscript is organized as follows: Sect. 2 presents the methodology of document search and their selection. Section 3 presents the results of the analysis. For this, journals, authors, countries, subject categories, types of publications, citations, terminology related to cognitive technologies and competencies were analyzed. Section 4 provides conclusions, limitations, and future research.

2   Materials and Methods

This study proposes a bibliometric analysis focusing on the study of publications related to the development of competencies in the context of the use of cognitive technologies. Within the framework of this article, an analysis will be conducted to identify cooperation between authors, organizations and countries. The main source of data for the study is the Scopus database, which is a recognized and authoritative resource for scientific research with significant global weight. This article proposes the use of bibliometric analysis for the identification of scientific publications corresponding to the topic of the implementation of cognitive technologies for the acquisition and development of competences.

The following equation was used for the search - TITLE ("Cognitive technolog?" OR "Artificial intelligenc?" OR "Artificial Intelligence Technolog?" OR "E-learning" OR "Innovation Education" OR "Machine Learning") AND TITLE ("Competenc? Development" OR "Competenc? Design" OR "Competenc?" OR "Competenc?-oriented education" OR "Competenc?-Based" OR "Modeling competence?" OR "Professional Competence") AND (EXCLUDE (AFFILCOUNTRY,"Russian Federation")) AND (LIMIT-TO (LANGUAGE,"English")) AND (LIMIT-TO (DOCTYPE,"ar") OR LIMIT-TO (DOCTYPE,"cp") OR LIMIT-TO (DOCTYPE,"ch") OR LIMIT-TO (DOCTYPE,"bk")) AND (LIMIT-TO (PUBYEAR,2022) OR LIMIT-TO (PUBYEAR,2021) OR LIMIT-TO (PUBYEAR,2020) OR LIMIT-TO (PUBYEAR,2019) OR LIMIT -TO (PUBYEAR,2018) OR LIMIT-TO (PUBYEAR,2017) OR LIMIT-TO (PUBYEAR,2016) OR LIMIT-TO (PUBYEAR,2015) OR LIMIT-TO (PUBYEAR,2014) OR LIMIT-TO (PUBYEAR, 2013) with the replacement of the last letters with the symbol "?" to search for all possible word endings.

Two hundred and eighty-one (281) documents were retrieved from the Scopus database search. Certain documents were selected for research and analysis using 4 stages.

At the first stage, studies related to the Russian Federation were excluded. The removal of these publications is due to the war and aggression that Russia has launched against Ukraine. As a result, 11 studies were excluded.

In the next step, all publications except articles, conference papers, book chapters and books are excluded. As a result, 33 studies were excluded and 237 documents were obtained.

As a result of the third stage, articles published in 2023 and before 2013 and not in English were removed (84 such documents were found). 153 documents were received.

At the fourth stage, articles were analyzed by title, abstract and main overview. As a result, the materials of 60 studies related to the specified direction will be used for the research.

The received articles are loaded into MS Excel for analysis and information visualization using the VOSviewer software.

3 Results

3.1 Journals

A total of 60 documents were published in 36 resources. This indicates a certain interest in publications related to the development of cognitive technologies, artificial intelligence, as well as the development of competencies as a result of their implementation. Out of 36 resources, 86.1% published only one document on this topic, the rest - 13.9% published from two to three documents.

In addition, the top journals were highlighted based on the number of published articles related to cognitive technology and competency development.

Table 1 presents the top 5 most productive journals, including publisher and indexing information such as number of publications, SJR 2021 indicator, CiteScore 2021 and SNIP 2021.

Table 1. Top 5 most productive magazines regarding cognitive technologies and competence development.

N	Journal	Publisher	№ of Publ	SJR-2021	Cite Score 2021	SNIP 2021
1	Ceur Workshop Proceedings	Conference Proceeding	3	0.228	1.1	0.317
2	Journal Of E Learning And Knowledge Society	Italian e-Learning Association	2	0.251	1.6	0.705
3	Journal Of Physics Conference Series	Conference Proceeding	2	0.251	1.6	0.705
4	Journal Of E Learning And Knowledge Society	Italian e-Learning Association	2	0.210	0.8	0.395
5	Sustainability Switzerland	Multidisciplinary Digital Publishing Institute	2	0.664	5	1.310

These top 5 journals published 18.3% of the total number of publications. The Ceur Workshop Proceedings conference published the most research, namely 3 articles (5.0%).

3.2 Authors

The five most productive authors in terms of number of publications, citations and publications as first author are presented in Table 2 . Inés González-González is in first place, with two publications, both of which are first author. Ana Jiménez-Zarco, who has 2 publications, but is not a co-author in any of them, ranks second in terms of citations. Next, the authors of one publication: as the first author - Adinda, not the first authors - Adeyanju, Adilbayeva.

Figure 1 presents the analysis of Co-authorship by Authors obtained using VOSviewer.

Table 2. Top 5 most productive authors regarding cognitive technologies and competence development.

No	Authors	Country of Author	Number of Publications	Number of Publications as the First Author
1	González-González, I. [18]	Spain	2	2
2	Jiménez-Zarco, A.I. [17]	Spain	2	0
3	Adinda, D. [1]	France	1	1
4	Adeyanju, J. [39]	Nigeria	1	0
5	Adilbayeva, U.B. [23]	Kazakhstan	1	0

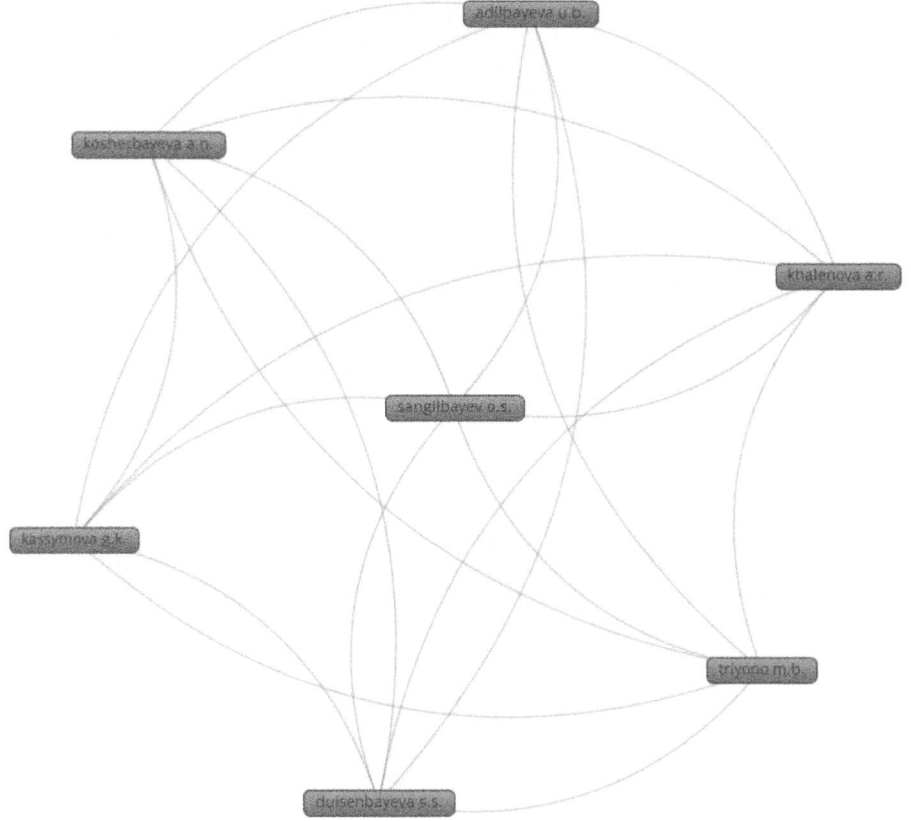

Fig. 1. Co-authorship by Authors (created using VOSviewer).

3.3   Countries

The number of publications on cognitive technologies and competencies, according to countries, was determined according to the Scopus database. Figure 2 shows the top 5 countries with the largest number of publications.

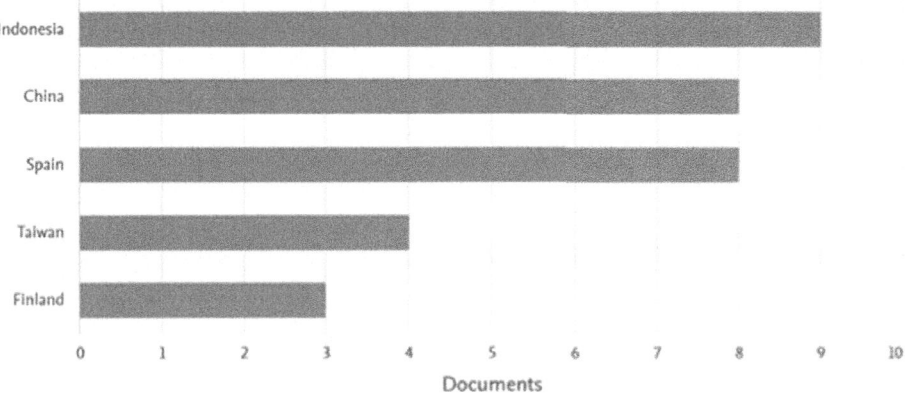

Fig. 2. Top 5 countries by the number of publications.

Overall, the country with the highest number of publications is Indonesia, which with 9 papers accounts for 15.0% of the total number of published articles.

3.4   Analysis of Citations

A corresponding study was conducted on the citation of authors and organizations among the selected articles. The results of the analysis of the top 5 cited authors are presented in Table 3 .

As a result of the analysis, it was found that the most cited article is Birjali, M., Beni-Hssane, A., Erritali, M. on the topic "A novel adaptive e-learning model based on Big Data by using competence-based knowledge and social learner activities". As a result, this study currently has 43 citations, which is 29.7% in the top 5.

Analysis of Citation by Authors is shown in Fig. 3.

3.5   Term Analysis

The analysis of all the terms used in the articles is shown in Fig. 4. In particular, 2 main clusters are highlighted, which are formed by all keywords.

The main cluster is formed by such keywords as Artificial intelligence, E-learning, Teaching, Machine learning.

Table 3. Top-5 publications with most citations.

No	Authors	Journal	Citations in Scopus	SciVal Topics
1	Birjali, M., Beni-Hssane, A., Erritali, M. [8]	Applied Soft Computing Journal	43	Computer-Aided Instruction; Adaptive Hypermedia; Intelligent Tutoring Systems
2	Palacios-Marqués, D., Cortés-Grao, R., Lobato Carral, C. [33]	International Journal of Project Management	32	Examiner; Education; Attendance
3	Garad, A., Al-Ansi, A.M., Qamari, I.N. [16]	Cakrawala Pendidikan	28	Blended Learning; Learning Management System; Distance Education
4	Huang, X. [22]	Education and Information Technologies	23	Computer Science; Education Computing; Computational Thinking
5	Parkes, M., Reading, C., Stein, S. [34]	Australasian Journal of Educational Technology	19	Internet Of Things; Transatlantic; Training and Development

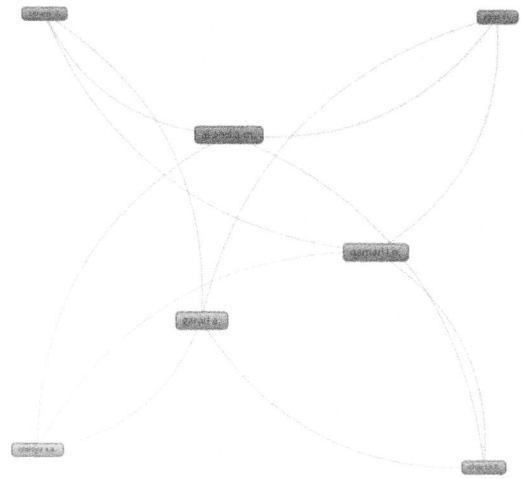

Fig. 3. Citation by Authors (created using VOSviewer).

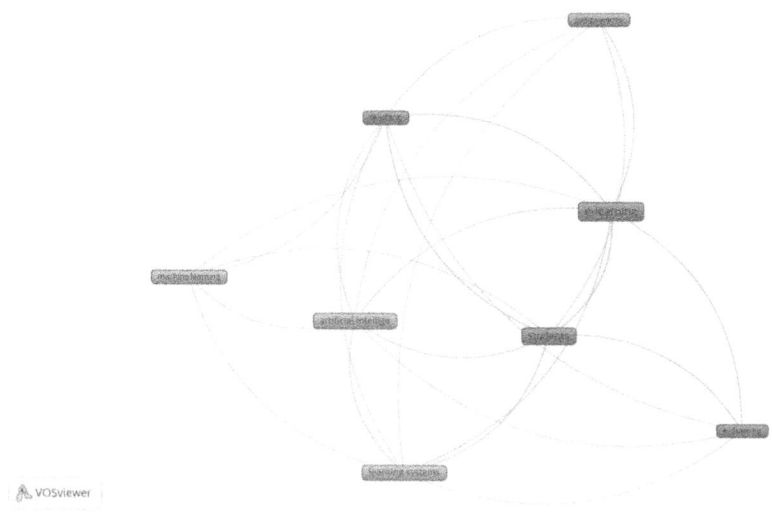

Fig. 4. Co-occurrence for All keywords (created using VOSviewer).

4   Conclusions

The study analyzed the Scopus publication database from 2013 to 2022 to analyze cognitive technologies and competencies. It was found that cognitive technologies show significant potential in improving learning processes and developing competencies, providing access to innovative methods and tools. They enable students and employees to develop critical thinking, problem-solving and creative thinking, communication skills and other important competencies.

A limitation of this study is the consideration of only publications from the Scopus database. Another limitation is the exclusion from the sample of publications from the Russian Federation, which is related to the war it started against Ukraine.

The study of ethical aspects of the use of cognitive technologies in the formation of competences, in particular the issues of confidentiality, security and risks associated with the use of data in the learning process, can become a topic for further research.

References

1. Adinda, D.: A competency-based approach to support e-learning during the COVID-19 situation. In: Proceedings of The European Conference On E-Learning, ECEL, pp. 1–9 (2021). https://doi.org/10.34190/EEL.21.043
2. Al-Sharidah, M.: The effectiveness of a training program via e-learning platforms in developing the technological competencies of pre-service teachers. J. Educ. Soc. Res. **12**, 131–142 (2022). https://doi.org/10.36941/jesr-2022-0128

3. Albano, G., Ferrari, P.: Linguistic competence and mathematics learning: the tools of e-learning. J. E-Learn. Knowl. Soc. **9**, 27–41 (2013). https://doi.org/10.20368/1971-8829/832

4. Angelis, L., Bohlouli, M., Hatzistavrou, K., Kakarontzas, G., Lopez, J., Zenkert, J.: The COMALAT approach to individualized e-learning in job-specific language competences. In: Linnhoff-Popien, C., Schneider, R., Zaddach, M. (eds.) Digital Marketplaces Unleashed, pp. 137–148. Springer, Heidelberg (2018). https://doi.org/10.1007/978-3-662-49275-8_16

5. Astafieva, M., Zhyltsov, O., Proshkin, V., Lytvyn, O.: E-learning as a mean of forming students- mathematical competence in a research-oriented educational process. In: CTE Workshop Proceedings, vol. 7, pp. 674–689 (2020). https://doi.org/10.55056/cte.421

6. Basantes-Andrade, A., Cabezas-González, M., Casillas-Martín, S.: Digital competences in e-learning. case study: Ecuador. In: Advances in Intelligent Systems and Computing AISC, vol. 1110, pp. 85–94 (2020). https://doi.org/10.1007/978-3-030-37221-78

7. Batko, R., Szopa, A.: Strategic Imperatives and Core Competencies in the Era of Robotics and Artificial Intelligence. Strategic Imperatives and Core Competencies In The Era Of Robotics and Artificial Intelligence, pp. 1–302 (2016). https://doi.org/10.4018/978-1-5225-1656-9

8. Birjali, M., Beni-Hssane, A., Erritali, M.: A novel adaptive e-learning model based on big data by using competence-based knowledge and social learner activities. Appl. Soft Comput. J. **69**, 14–32 (2018). https://doi.org/10.1016/j.asoc.2018.04.030

9. Burkhard, M., Seufert, S., Guggemos, J.: Paradigm shift in human-machine interaction: a new learning framework for required competencies in the age of artificial intelligence? In: International Conference On Computer Supported Education, CSEDU - Proceedings, vol. 2, pp. 294–302 (2021). https://doi.org/10.5220/0010473302940302

10. Cao, C., Zhang, Z.: Machine learning-assisted competency modeling for human resource management jobs. Mob. Inf. Syst **2022** (2022). https://doi.org/10.1155/2022

11. Çalışkan, S., Demir, K., Karaca, O.: Artificial intelligence in medical education curriculum: an e-Delphi study for competencies. PLoS ONE **17**, e0271872 (2022). https://doi.org/10.1371/journal.pone.0271872

12. Chao, P., Hsu, T., Liu, T., Cheng, Y.: Knowledge of and competence in artificial intelligence: perspectives of Vietnamese digital-native students. IEEE Access. **9**, 75751–75760 (2021). https://doi.org/10.1109/ACCESS.2021.3081749

13. Chirila, C.: A dialog based game component for a competencies based e-learning framework. In: SACI 2013–8th IEEE International Symposium on Applied Computational Intelligence and Informatics, Proceedings, pp. 55–60 (2013). https://doi.org/10.1109/SACI.2013.6609025

14. D'Aniello, G., Della Piana, B., Gaeta, M.: A situation-aware e-learning system to support intercultural competence development. In: CEUR Workshop Proceedings, vol. 3124, pp. 204–208 (2022)

15. Gaol, P.: Implementation of performance management in artificial intelligence system to improve Indonesian human resources competencies. In: IOP Conference Series: Earth And Environmental Science, vol. 717 (2021). https://doi.org/10.1088/1755-1315

16. Garad, A., Al-Ansi, A., Qamari, I.: The role of e-learning infrastructure and cognitive competence in distance learning effectiveness during the COVID-19 pandemic. Cakrawala Pendidikan. **40**, 81–91 (2021). https://doi.org/10.21831/cp.v40i1.33474
17. González-González, I., Gallardo-Gallardo, E., Jiménez-Zarco, A.: Using films to develop the critical thinking competence of the students at the Open University of Catalonia (UOC): testing an audiovisual case methodology in a distance e-learning environment. Comput. Hum. Behav. **30**, 739–744 (2014). https://doi.org/10.1016/j.chb.2013.09.013
18. González-González, I., Jiménez-Zarco, A.: Using an audiovisual case methodology to develop critical thinking competence in distance E-learning environment: The open university of catalonia (UOC) experience. E-Learning 2.0 Technologies And Web Applications In Higher Education. pp. 171–187 (2015). https://doi.org/10.4018/978-1-4666-4876-0.ch009
19. Gutierrez, S., Perez, S., Munguia, M.: Artificial intelligence in e-learning plausible scenarios in Latin America and new graduation competencies. Revista Iberoamericana De Tecnologias Del Aprendizaje. **17**, 31–40 (2022). https://doi.org/10.1109/RITA.2022.3149833
20. Heba, A., Kapounová, J., Smyrnova-Trybulska, E.: Theoretical conception and some practical results of the development of mathematical competences with use of e-learning. Int. J. Continuing Eng. Educ. Life-Long Learn.. **24**, 252–268 (2014). https://doi.org/10.1504/IJCEELL.2014.063098
21. Hsu, W., Li, C.: A competency-based guided-learning algorithm applied on adaptively guiding e-learning. Interact. Learn. Environ. **23**, 106–125 (2015). https://doi.org/10.1080/10494820.2012.745432
22. Huang, X.: Aims for cultivating students' key competencies based on artificial intelligence education in China. Educ. Inf. Technol. **26**(5), 5127–5147 (2021). https://doi.org/10.1007/s10639-021-10530-2
23. Kassymova, G., et al.: Cognitive competence based on the e-learning. Int. J. Adv. Sci. Technol. **28**, 167–177 (2019)
24. Keržič, D., Aristovnik, A., Tomaževič, N., Umek, L.: Evaluating the impact of e-learning on students' perception of acquired competencies in an university blended learning environment. J. E-Learn. Knowl. Soc. **14**, 65–76 (2018). https://doi.org/10.20368/1971-8829
25. Khan, W., Nisar, Q., Sohail, S., Shehzadi, S.: The role of digital innovation in e-learning system for higher education during COVID 19: a new insight from pedagogical digital competence. innovative education technologies for 21st century teaching and learning, pp. 75–100 (2021). https://doi.org/10.1201/9781003143796-6
26. Lau, J., Bonilla, J., Gárate, A.: Artificial intelligence and labor: media and information competencies opportunities for higher education. Commun. Comput. Inf. Sci. **989**, 619–628 (2019). https://doi.org/10.1007/978-3-030-13472-358
27. Lestari, N., Rosman, D., Putranto, T.: The relationship between robot, artificial intelligence, and service automation (RAISA) awareness, career competency, and perceived career opportunities: Hospitality student perspective. In: Proceedings of 2021 International Conference On Information Management And Technology, ICIMTech 2021, pp. 690–695 (2021). https://doi.org/10.1109/ICIMTech53080.2021.9535054
28. Long, J., Lin, J.: An empirical study on cultivating college students' cross-cultural communicative competence based on the artificial-intelligence English-teaching mode. Frontiers Psychol. **13**, 976310 (2022). https://doi.org/10.3389/fpsyg.2022.976310

29. López, J., Mar Eva Alemany Díaz, M., Vivas, D.: The integration of transversal competences in higher education in engineering through e-learning tools. The case of the ETSII at the UPV. In: CEUR Workshop Proceedings, vol. 3129 (2022)

30. Margienė, A., Ramanauskaitė, S., Nugaras, J., Stefanovič, P., Čenys, A.: Competency-based e-learning systems: automated integration of user competency portfolio. Sustainability (Switzerland), **14**, 16544 (2022). https://doi.org/10.3390/su142416544

31. Misthou, S., Paliouras, A.: Teaching artificial intelligence in K-12 education: competences and interventions. In: Auer, M.E., Tsiatsos, T. (eds.) IMCL 2021. LNNS, vol. 411, pp. 887–896. Springer, Cham (2022). https://doi.org/10.1007/978-3-030-96296-8_80

32. Mujiono, M., Herawati, S.: The effectiveness of e-learning-based sociolinguistic instruction on EFL university students' sociolinguistic competence. Int. J. Instr. **14**, 627–642 (2021). https://doi.org/10.29333/iji.2021.14436a

33. Palacios-Marqués, D., Cortés-Grao, R., Lobato Carral, C.: Outstanding knowledge competences and web 2.0 practices for developing successful e-learning project management. Int. J. Proj. Manag. **31**, 14–21 (2013). https://doi.org/10.1016/j.ijproman.2012.08.002

34. Parkes, M., Reading, C., Stein, S.: The competencies required for effective performance in a university e-learning environment. Australas. J. Educ. Technol. **29**, 771–791 (2013). https://doi.org/10.14742/ajet.38

35. Permana, A., Pradnyana, G.: Recommendation systems for internship place using artificial intelligence based on competence. J. Phys. Conf. Ser. **1165**, 012007 (2019). https://doi.org/10.1088/1742-6596

36. Rivera-Kempis, C., Valera, L., Sastre-Castillo, M.: Entrepreneurial competence: using machine learning to classify entrepreneurs. Sustainability (Switzerland). **13**, 8252 (2021). https://doi.org/10.3390/su13158252

37. Sarangarajan, S., Kavi Chitra, C., Shivakumar, S.: Automation of competency: training management using machine learning models. In: 2021 Grace Hopper Celebration India, GHCI 2021 (2021). https://doi.org/10.1109/GHCI50508.2021.9513995

38. Sharma, U., Manchanda, N.: Predicting and improving entrepreneurial competency in university students using machine learning algorithms. In: Proceedings of the Confluence 2020–10th International Conference on Cloud Computing, Data Science and Engineering, pp. 305–309 (2020). https://doi.org/10.1109/Confluence47617.2020.9058292

39. Thomas, O., Adeyanju, J., Popoola, B., Odewale, T.: Competency training needs of lecturers for effective e-learning instructional delivery in teacher education programs in south-west. Niger. J Negro Educ. **89**, 136–145 (2020)

40. Wu, J.: Construction of primary and secondary school teachers' competency model based on improved machine learning algorithm. Math. Probl. Eng. 2022 (2022). https://doi.org/10.1155/2022/6439092

41. Zhao, X., Guo, Z., Liu, S.: Exploring key competencies and professional development of music teachers in primary schools in the era of artificial intelligence. Sci. Program. **2021**, 1–9 (2021). https://doi.org/10.1155/2021/5097003

42. Zulfiya, K., Gulmira, B., Altynbek, S., Assel, O.: A model and a method for assessing students' competencies in e-learning system. In: ACM International Conference Proceeding Series (2019). https://doi.org/10.1145/3368691.3372391

Smart-Systems in STEM Education

Iryna Nikitina$^{(\boxtimes)}$ ⓘ and Tetyana Ishchenko ⓘ

Dniepropetrovsk State University of Internal Affairs, 26 Gagarina Ave, Dniepro 49005, Ukraine
N.I.P@i.ua

Abstract. The article "Smart-Systems in STEM Education" explores the significance of integrating "Smart-systems" technologies into STEM (Science, Technology, Engineering, and Mathematics) education. The article highlights the role of "Smart-systems" in preparing students for the future by providing practical experiences and fostering critical thinking and problem-solving skills. It discusses various technologies associated with "Smart-systems," such as robotics, IoT, AI, and data analytics, and their applications in STEM education. The article also presents a range of resources available for educators and students, including online courses, educational websites, maker spaces, and competitions. By leveraging these resources, educators can create engaging learning environments that inspire students to explore and pursue careers in emerging fields. The article emphasizes the benefits of incorporating "Smart-systems" in STEM education, including the development of technological literacy, interdisciplinary learning, and the cultivation of skills necessary for the digital age. Ultimately, embracing "Smart-systems" in STEM education empowers students to become the next generation of innovators and problem solvers who can contribute to a rapidly evolving technological landscape.

Keywords: STEM · Smart-Systems · Models · Technologies · Resources

1 Introduction

Smart systems refer to intelligent and interconnected technologies that leverage advanced computing, data analytics, and connectivity to enhance efficiency, convenience, and automation in various aspects of modern society. These systems incorporate a combination of sensors, actuators, algorithms, and communication networks to gather, analyze, and act upon data, enabling them to make informed decisions and perform tasks autonomously or with minimal human intervention. They play a significant role in transforming industries, improving quality of life, and addressing complex challenges in numerous domains. The important role they play in our life can be seen in the following cases:

- Efficiency and Automation: Smart systems enable automation and optimization of processes, leading to increased efficiency and productivity. They can streamline operations in manufacturing, transportation, energy management, and various other sectors, reducing costs, minimizing errors, and saving time.

G. Antoniou et al. (Eds.): ICTERI 2023, CCIS 1980, pp. 325–335, 2023.
https://doi.org/10.1007/978-3-031-48325-7_25

- Sustainable Resource Management: Smart systems facilitate intelligent monitoring and control of resources such as energy, water, and waste. They help identify patterns, predict demand, and optimize consumption, leading to improved sustainability, reduced environmental impact, and cost savings.
- Urban Development and Infrastructure: Smart systems contribute to the development of intelligent cities, often referred to as smart cities. These systems integrate technologies to manage transportation, energy grids, waste management, and public services efficiently. They enhance urban planning, optimize resource allocation, and improve the overall quality of life for residents.
- Connectivity and IoT: Smart systems heavily rely on IoT, which enables devices to connect and communicate with each other through the internet. IoT enables seamless data exchange and automation, allowing devices and applications to work in harmony and make informed decisions.

However, it's important to address potential challenges and concerns associated with smart systems, such as data privacy, security vulnerabilities, and the potential impact on employment due to automation. In the context of smart systems, it is STEM learning that encompasses various models, technologies, and resources that can facilitate hands-on and interactive education.

It's worth noting that the selection of models, technologies, and resources for STEM learning in smart systems should align with the age, grade level, and educational goals of the students. It's also important to provide guidance and mentorship to ensure students understand the underlying concepts and ethical considerations associated with smart systems.

Therefore, the main point of interest of the present paper is to consider smart-systems within STEM education. To achieve this purpose, we suggest to consider the following items:

- define main models of smart-systems
- define key features of STEM education
- various models for STEM education within smart-systems
- technologies for STEM education within smart-systems

2   Smart-Systems for Education in Modern Society

Smart systems in education refer to the integration of advanced technologies, intelligent tools, and data-driven approaches to enhance the learning process, improve educational outcomes and optimize administrative tasks. Smart systems, also known as smart technologies or smart devices, refer to a network of interconnected devices and applications that leverage advanced technologies such as artificial intelligence (AI), the Internet of Things (IoT), and data analytics to collect, process, and analyze data, automate tasks, and improve efficiency. These systems play a crucial role in modern society, revolutionizing various aspects of our lives. An overview of smart systems used in modern education is given in Table 1.

Table 1. Content of Main Smart Systems in Modern Education

Learning Management Systems (LMS)	LMS platforms serve as a central hub for organizing and delivering educational content. They enable educators to create, manage, and distribute course materials, assignments, assessments, and grades. LMS platforms also facilitate communication and collaboration between students and teachers
Adaptive Learning Platforms	Adaptive learning systems use algorithms and data analysis to customize the learning experience for individual students. These platforms assess each student's strengths, weaknesses, and learning styles, and then provide personalized content and recommendations to address their specific needs. Adaptive learning platforms can adjust the pace, difficulty level, and content of the curriculum to optimize learning outcomes
Virtual Reality (VR) and Augmented Reality (AR)	VR and AR technologies create immersive learning experiences that go beyond traditional classroom settings. Students can explore virtual environments, interact with digital objects, and simulate real-world scenarios, enhancing their understanding and engagement. VR and AR can be used in subjects like science, history, geography, and art to bring concepts to life
Artificial Intelligence (AI) Tutoring Systems	AI tutoring systems provide personalized guidance and support to students. These systems use machine learning algorithms to analyze student performance, identify areas of improvement, and offer targeted feedback. AI tutors can adapt their teaching strategies, provide additional resources, and track progress over time, assisting students in achieving their learning goals
Gamification	Gamification involves incorporating game elements, such as rewards, challenges, and leaderboards, into the learning process. It motivates students, enhances their engagement, and makes learning more enjoyable. Gamification can be applied in various educational contexts, from language learning apps to math puzzles and quizzes
Data Analytics and Learning Analytics	Data analytics tools enable educators to gather and analyze large amounts of data generated by students' interactions with digital platforms. Learning analytics provides insights into student performance, preferences, and behaviors. This information helps educators make data-driven decisions, identify areas of improvement, and personalize instruction
Mobile Learning	With the widespread use of smartphones and tablets, mobile learning has gained popularity. Mobile learning apps and platforms allow students to access educational content anytime, anywhere. These systems often offer bite-sized lessons, interactive exercises, and collaborative features, making learning more flexible and convenient
Cloud Computing	Cloud-based systems facilitate storage, sharing, and access to educational resources. Teachers and students can collaborate on projects, access materials from multiple devices, and seamlessly integrate various applications. Cloud computing ensures easy data backup, scalability, and centralized management of educational content

However, smart systems also come with challenges such as privacy and security concerns, data governance, and interoperability issues. Ensuring data privacy, security, and standardization are critical for the widespread adoption and success of smart systems.

3 STEM Application in Modern Education

The US National Science Foundation (an independent agency under the US government that provides basic research and an approach to education in all areas of science) introduced STEM as an acronym for Science, Technology, Engineering and Mathematics. This approach involves the integration of the mentioned areas into one educational process. Immersion of students in science and technology in their school years motivates students to choose professions related to them. [1].

Originating in the United States, STEM education has gradually spread throughout the country, finding roots in numerous public schools and universities that have established science and engineering programs. [2].

Over time, the scope of STEM has expanded to include new disciplines and letters added to its nomenclature. As a result, STEM transformed into STEAM, and then into STREAM. The inclusion of the "A" in STEAM and the "R" in STREAM recognizes that incorporating the arts and robotics into STEM education can enrich students' educational journeys, making them more capable of solving the challenges of tomorrow.

There is currently significant debate surrounding STEAM education. Sandy Buczynski underscores the interconnectedness of art and science in her article titled "Communicating Scientific Concepts Through Art: 21st Century Skills in Practice." She argues that incorporating the arts into STEAM education creates a tandem in which students are taught to think outside the box, as creativity and design thinking are equally important for innovation and solving technical problems. [3].

An attempt to intensify education only in the direction of science, engineering and mathematics without the parallel development of arts disciplines can lead to the fact that young Americans are deprived of creativity skills. [4].

Fundamentally, the introduction of these fresh acronyms represents a more inclusive strategy for STEM education. It seeks to amalgamate various fields of study and emphasize the importance of creativity, design-oriented thinking, and technical expertise in equipping students for the future's demands.

The acquired skills will allow the students in the future to effectively solve the most pressing global problems that they may encounter [5]. However, it is important to note that STEM should not be seen as a replacement for other academic disciplines. Rather, it complements them by instilling the ability to solve problems through engineering design while retaining the core teachings of science and mathematics.

4 Models for STEM Learning in the Context of "Smart-Systems"

When it comes to STEM learning in the context of smart systems, there are several models and approaches that can be employed to engage students and foster their understanding of these complex systems. Here are a few models commonly used in STEM education:

4.1 Project-Based Learning (PBL)

PBL involves students working on an extended project that requires them to apply their knowledge and skills to real-world problems and is an effective approach to engage students in STEM education. Project-based learning (PBL) emphasizes hands-on learning and real-world problem-solving. In this model, students work on projects that are designed to address complex problems or challenges, often with a focus on interdisciplinary and cross-curricular learning [6]. When it comes to the context of "smart-systems," PBL can be an excellent method to develop students' skills and knowledge in this field. Smart-systems refer to intelligent systems that utilize advanced technologies such as artificial intelligence, Internet of Things (IoT), robotics, and data analytics to enhance efficiency, connectivity, and automation in various domains.

Here's an outline of how you can design a project-based learning experience for STEM in the context of smart-systems:

- Select a relevant project topic: Choose a project topic that aligns with smart-systems, such as designing a smart home, developing an automated greenhouse, or creating a smart city infrastructure. Ensure the project integrates multiple STEM disciplines, allowing students to explore various aspects.
- Define project goals and objectives: Clearly define the goals and objectives of the project. These could include designing and prototyping a smart system, understanding the underlying technologies, exploring real-world applications, or addressing a specific problem using smart-system solutions.
- Form project teams: Divide students into teams to encourage collaboration and teamwork. Assign roles within the teams, such as project manager, researcher, programmer, engineer, and designer, to provide diverse opportunities for students to contribute their skills.
- Conduct background research: Encourage students to conduct research on smart-systems, including the technologies involved, existing applications, and potential challenges. They should gain a solid understanding of the concepts and principles before diving into the project.
- Develop project plans: Guide students in creating project plans, including timelines, milestones, and tasks. Help them break down the project into manageable components and establish a roadmap for the project's execution.
- Design and prototype: Based on their research and project plans, students can start designing and prototyping their smart-system solution. This might involve programming microcontrollers, utilizing sensors and actuators, integrating IoT devices, or developing algorithms for data analysis.

PBL typically involves several stages, including project planning, research, design, implementation, and evaluation. Teachers serve as facilitators, providing guidance and support as students work through each stage of the project.

Advocates of PBL argue that it can help students develop a range of skills and competencies, including critical thinking, problem-solving, collaboration, communication, and creativity. However, critics of PBL caution that it can be challenging to implement effectively, and that it may require significant time and resources to develop and implement high-quality projects.

4.2　Design Thinking

Design thinking is a problem-solving approach that focuses on empathy, creativity, and iteration. It is a valuable framework to incorporate into STEM education, particularly in the context of "smart-systems." Design thinking allows students to tackle complex problems related to smart-systems by considering user needs, exploring innovative solutions, and iteratively refining their designs. Here's how you can integrate design thinking into STEM education with a focus on smart-systems:

– Empathize with users: Start by fostering empathy among students by encouraging them to understand the needs and challenges of potential users of smart-systems. This could involve conducting interviews, observations, or surveys to gain insights into user preferences, behaviors, and pain points.
– Define the problem: Based on their empathetic understanding, guide students in defining a specific problem or challenge related to smart-systems. For example, it could be improving energy efficiency in buildings, enhancing transportation systems, or optimizing waste management.
– Ideate and brainstorm: Encourage students to generate a wide range of ideas and potential solutions to address the defined problem. Use brainstorming sessions, mind mapping, or other ideation techniques to stimulate creativity and innovative thinking. Emphasize the importance of quantity over quality during this stage.
– Prototype and experiment: Students should select the most promising ideas from the ideation phase and develop low-fidelity prototypes or mock-ups of their smart-system solutions. These prototypes can be physical or digital representations that allow for testing and gathering feedback.
– Test and gather feedback: Help students design experiments or simulations to test their prototypes and gather feedback from potential users or stakeholders. This feedback will provide valuable insights into the effectiveness of their solutions, usability concerns, and areas for improvement.

Design Thinking has several pros and cons when applied to STEM education. Advocates of Design thinking learning argue that it encourages students to think creatively and explore innovative solutions to real-world problems, cultivates critical thinking skills by challenging students to consider multiple perspectives and find solutions that are feasible, viable, and desirable, promotes communication, cooperation, and the exchange of ideas among students, etc.

As for Cons, Design Thinking in STEM can be time-consuming, requiring students to go through multiple iterations and stages. It involves dealing with ambiguous and ill-defined problems and focuses on the creative problem-solving process and user-centricity, which may not fully develop specialized technical skills in specific STEM domains. Assessing design thinking outcomes can be challenging since the traditional assessment methods may not capture the full range of skills and competencies developed through design thinking.

Despite these challenges, the benefits of incorporating design thinking in STEM education, including increased engagement, creativity, critical thinking, and collaboration, make it a valuable approach. Educators can address the cons by careful planning, providing guidance, and integrating design thinking with other instructional methods to ensure a well-rounded STEM learning experience.

4.3 Inquiry-Based Learning

Inquiry-based learning is a student-centered approach that fosters active exploration, questioning, and investigation. They can investigate how smart systems work, explore the impacts of different variables on system performance, and propose improvements or optimizations [7]. When incorporating "smart-systems" into inquiry-based learning for STEM education, students can explore and investigate the technologies, applications, and implications of intelligent systems. They can:

- Generate inquiry questions: Begin by prompting students to generate their own inquiry questions related to smart-systems. These questions should encourage investigation and exploration, such as "How do smart home systems improve energy efficiency?" or "What are the benefits and challenges of implementing smart transportation systems?"
- Design experiments or investigations: Encourage students to design experiments or investigations that allow them to explore specific aspects of smart-systems. For example, they might design experiments to measure the impact of different sensors on energy efficiency or investigate the data collection and analysis process in a smart city project.
- Collect and analyze data: Students should collect relevant data during their experiments or investigations. They can use sensors, data logging tools, or surveys to gather data and analyze it using appropriate quantitative or qualitative methods. This data analysis will help them draw meaningful conclusions and make connections to the concepts they are exploring.
- Make connections to real-world applications: Guide students in making connections between their findings and real-world applications of smart-systems. They should explore how their inquiry findings align with or inform existing smart-system solutions in various domains, such as healthcare, energy management, transportation, or agriculture, etc.

Incorporating "smart-systems" into inquiry-based learning for STEM education empowers students to explore the possibilities and challenges of intelligent systems and fosters a deep understanding of the applications and implications of technology in real-world contexts.

However, critics of Inquiry-based learning note that the open-ended nature of inquiry means that students may require more time to explore topics, conduct research, and reach conclusions. It relies on teachers to facilitate the process and provide guidance. Inadequate guidance can lead to confusion or incomplete understanding. Moreover, in collaborative inquiry projects, some students may dominate the group, while others may contribute less or disengage.

It's important to note that the cons of inquiry-based learning can be mitigated through effective instructional design, teacher support, and ongoing professional development. The benefits of active engagement, critical thinking, authentic learning experiences, and collaboration make inquiry-based learning a powerful approach for fostering deep understanding and preparing students for real-world challenges.

4.4 Gamification

Simply put, the definition of gamification is the use of game-design elements and game principals in non-game contexts. Gamification involves incorporating game elements and mechanics into the learning process. It can be used to make STEM learning in the context of smart systems more engaging and interactive. For example, students can participate in gamified simulations or virtual environments where they can design and manage smart systems, solve challenges, and compete with their peers [8]. Here's how smart systems can be incorporated into gamification:

– Smart systems play a crucial role in gamification. Especially in the context of STEM education. In STEM education, smart systems can be used to create immersive and interactive experiences that make learning more enjoyable and effective.
– Virtual Reality (VR): VR technology provides a fully immersive experience by simulating real or fictional environments. In STEM education, VR can be used to create virtual laboratories, engineering simulations, or scientific explorations. Students can interact with virtual objects, perform experiments, and visualize complex concepts, making learning more interactive and engaging.
– Augmented Reality (AR): AR overlays digital content onto the real-world environment, enhancing it with additional information or interactive elements. In STEM gamification AR can be used to provide real-time data overlays during field trips or science experiments, making the learning process more dynamic.
– Robotics and Programming Kits: Smart systems in the form of robotics and programming kits allow students to engage in hands-on learning experiences. These kits often include programmable robots or microcontrollers that students can assemble, code, and control.

Gamification in the context of STEM education has both pros and cons.

Supporters of Gamification claim that Gamification makes learning more enjoyable and engaging for students. Opponents insist that there is a risk of students focusing more on the game mechanics rather than the underlying educational content.

Overall, by combining game mechanics, immersive technologies, and data-driven approaches, educators can create engaging learning experiences that promote curiosity, critical thinking, and a deeper understanding of STEM subjects.

4.5 Collaborative Learning

Collaborative learning involves students working together in groups to solve problems, discuss ideas, and share knowledge. In the context of smart systems, students can collaborate on projects or challenges that require them to analyze, design, and implement solutions. This model encourages communication, teamwork, and the exchange of diverse perspectives.

Here are some examples of smart systems in collaborative learning for STEM:

– Online Collaboration Platforms: Smart systems provide online collaboration platforms that allow students to work together on STEM projects, regardless of their physical location. These platforms offer features such as real-time document editing, video conferencing, chat, and file sharing.

- Cloud-based Storage and Sharing: Smart systems enable cloud-based storage and sharing solutions, allowing students to access and collaborate on STEM-related documents, code, datasets, and other resources. This ensures that all team members have equal access to relevant materials and can contribute to the project, even outside of physical classrooms or lab environments.
- Virtual Lab Environments: Smart systems can create virtual lab environments that simulate real-world laboratory settings. These environments allow students to conduct experiments, manipulate variables, work together on experiments, share data and observations, and collectively analyze and interpret results.
- Remote Sensing and Robotics: Smart systems can integrate remote sensing technologies and robotics into collaborative STEM learning. Students can control remote sensors or robots from different locations and collectively collect data, perform experiments, and explore environments. This promotes teamwork,

Collaborative learning in the context of STEM education has both advantages and challenges. On the hand, collaborative learning promotes active engagement and participation, leading to improved learning outcomes and cultivates important social skills, such as communication, teamwork, and leadership. Students can rely on their peers for clarification, guidance, and feedback.

On the other hand, in collaborative learning, there is a risk of some students contributing more than others. Unequal participation can lead to one or a few individuals dominating the group, while others may become passive or disengaged. Differing opinions, communication issues, or conflicting work styles can hinder progress and create tension within groups. Assessing individual contributions in collaborative learning can be challenging.

Overall, collaborative learning in STEM education offers numerous benefits, including improved learning outcomes, social skill development, and exposure to diverse perspectives. However, challenges such as unequal contribution, conflicts, and individual learning needs need to be addressed to maximize the effectiveness of collaborative learning experiences. With careful planning, guidance, and support, collaborative learning can be a powerful approach to enhance STEM education.

5   Technologies for Teaching STEM in the Context of "Smart-Systems"

"Smart-systems" technologies play a crucial role in STEM education. Introducing these technologies in STEM education helps students understand their practical applications, preparing them for future careers in fields such as robotics, AI, IoT, and data science [9]. Besides, in an increasingly digital world, it is crucial for students to develop technological literacy. This familiarity with technology equips them with the necessary skills to adapt and thrive in a technology-driven society.

Here are some specific "Smart-systems" technologies suitable for STEM education:

- Internet of Things (IoT): IoT technologies enable the connection of physical objects and devices to the internet, allowing data collection, analysis, and control. Students can learn about sensors, communication protocols, and cloud-based platforms to create IoT projects that demonstrate the integration of hardware, software, and data.

- Robotics: Robotics combines hardware, software, and control systems to create intelligent machines that interact with the physical world. STEM education can involve building and programming robots, understanding kinematics and dynamics, and exploring areas like computer vision and autonomous navigation.
- Artificial Intelligence (AI): AI technologies, such as machine learning and deep learning, are at the core of many "Smart-systems." Students can learn about AI algorithms, training models, and data analysis techniques to develop intelligent systems.
- Sensor Technologies: Understanding various sensors and their applications is essential for "Smart-systems." Students can learn about different types of sensors, such as temperature sensors, accelerometers, or proximity sensors, and explore their use in environmental monitoring, robotics, or home automation projects.
- Virtual and Augmented Reality: Virtual reality (VR) and augmented reality (AR) technologies provide immersive and interactive experiences. Students can explore these technologies to simulate and visualize "Smart-systems" in virtual environments, enhancing their understanding and enabling experimentation without physical constraints.

Here are some examples of how technologies based on "Smart-systems" can be applied in STEM education:

Robotics Workshops: Organizing robotics workshops where students learn to build and program robots using kits like LEGO Mindstorms or Arduino. They can explore concepts of kinematics, sensors, and programming while working on challenges and competitions.

IoT Data Analysis: Students can collect data from various IoT sensors, such as temperature, humidity, or light sensors, and analyze the data using programming languages like Python or data analytics tools like Jupyter Notebook. They can gain insights from the data and understand the impact of environmental factors on "Smart-systems."

Environmental Monitoring Systems: Students can design and develop "Smart-systems" for monitoring environmental parameters like air quality, water quality, or noise levels. They can utilize IoT sensors, data analysis techniques, and visualization tools to create interactive dashboards or reports to study and understand environmental changes [10].

These examples demonstrate how technologies based on "Smart-systems" can be effectively applied in STEM education to provide hands-on experiences, foster interdisciplinary learning, and cultivate essential skills needed for future careers in science, technology, engineering, and mathematics.

6   Conclusion

In conclusion, integrating "Smart-systems" technologies into STEM education opens up a world of possibilities for students. By incorporating robotics, IoT, AI, data analytics, and other related technologies, students can engage in hands-on, experiential learning that bridges the gap between theory and real-world applications.

Through "Smart-systems," students gain practical skills, develop critical thinking and problem-solving abilities, and cultivate a deep understanding of STEM concepts.

They learn to work collaboratively, leveraging interdisciplinary knowledge and applying it to design and implement innovative solutions.

By embracing "Smart-systems" in STEM education, students are prepared for the evolving demands of the digital era. They become technologically literate, ready to tackle the challenges and opportunities presented by automation, intelligent systems, and the Internet of Things.

In this ever-changing technological landscape, "Smart-systems" in STEM education paves the way for a future where innovation, collaboration, and transformative thinking drive progress and shape a better world.

References

1. Why is STEM Hard to Define? https://www.invent.org/blog/trends-stem/stem-define
2. U.S. Department of Education (2022). https://www.ed.gov/news/press-releases/us-depart ment-education-launches-new-initiative-enhance-stem-education-all-students
3. Buczynski, S.: Communicating science concepts through art: 21st-century skills in practice. Sci. Scope **35**(9), 29–35 (2012)
4. Nikitina, I., Ishchenko, T.: Transforming STEM into STEAM. The role of psychology and pedagogy in the spiritual development of modern society. Publishing House "Baltija Publishing", pp. 173–176 (2022). http://www.baltijapublishing.lv/omp/index.php/bp/catalog/boo k/245, https://doi.org/10.30525/2592-8813-2022-4-18
5. National inventor Hall of Fame. https://www.invent.org/blog/trends-stem/stem-define
6. PBLWorks. WhatisPBL? https://www.pblworks.org/what-is-pbl
7. Bauld, A.: What Is Inquiry-Based Learning? (IBL). https://xqsuperschool.org/rethinktoget her/what-is-inquiry-based-learning-ibl
8. Blankman, R.: Gamification in Education: The Fun of Learning. https://www.hmhco.com/ blog/what-is-gamification-in-education
9. Smart education systems: what they are and how they're designed. https://profuturo.educat ion/en/observatory/trends/sistemas-educativos-inteligentes-que-son-y-como-se-disenan/
10. Internet of Things (IoT) in Education for Sustainability: Using STEMTech Model to Design Smart Home System for Experiential Learning at Secondary Schools. https://www.researchg ate.net/publication/358671924_Internet_of_Things_IoT_in_Education_for_Sustainability_ Using_STEMTech_Model_to_Design_Smart_Home_System_for_Experiential_Learning_ at_Secondary_Schools

Digital Technologies as a Means of Forming Subject-Methodical Competence Future Primary School Teachers

Svitlana Palamar[✉] ⓘ, Liudmyla Nezhyva ⓘ, Kateryna Brovko ⓘ,
and Dmytro Bodnenko ⓘ

Borys Grinchenko Kyiv University, Bulvarno-Kudriavska St., 18/2, Kyiv 04053, Ukraine
{s.palamar,l.nezhyva,k.brovko,d.bodnenko}@kubg.edu.ua

Abstract. The article updates the problem of the application of digital technologies in various professional methods and their integration in the educational process of institutions of higher pedagogical education. According to the authors, the formation of subject-methodical competence of future primary school teachers is not possible without the formation of abilities: navigate in the information space, use existing and, if necessary, create new electronic resources; use modern digital technologies in the educational process. For the formation of professional skills and abilities in students of pedagogical specialties, the quality of educational content of a methodological nature, created by means of digital technologies, is essential.

In accordance with the purpose of digital technologies and their application in professional teaching methods, the most effective sets of programs and applications were determined for students to create educational and methodological support for the educational process in primary school. Such as: mind maps (summary diagram of interactive lessons), presentations (demonstration of educational materials), comics (visualization of educational information in frames-pictures), word clouds (visualization of educational content in keywords), infographics (visualization and structuring of a large amount of educational information), virtual boards (coordination of work in groups in class, during the web quests, organization of student communication), interactive tasks and online tests (creation of exercises to acquire skills, development of tasks and tests to reveal knowledge and skills).

Keywords: Digital Technologies · Subject-methodical Competence · Information-Digital Competence · Professional Training · Future Primary School Teachers

1 Introduction

At the current stage of development, humanity is experiencing complex processes of globalization and integration, which lead to radical changes in all spheres of life. The methods of distribution and the volume of information have changed, the virtual E-environment is rapidly modernizing, which affects the reformatting of the educational

G. Antoniou et al. (Eds.): ICTERI 2023, CCIS 1980, pp. 336–347, 2023.
https://doi.org/10.1007/978-3-031-48325-7_26

space. The introduction of innovative technologies in education, the development of students' critical thinking and emotional intelligence and at the same time avoiding their overload, orientation towards a happy child who actively learns about the world and prepares for successful socialization - conceptualize both the education of the European standard and the new Ukrainian school. The recommendations of the European Parliament and the European Council on competences for lifelong learning identify a number of important aspects within the Reference Framework, such as: critical thinking, creativity, initiative, problem solving, risk assessment, decisiveness and constructive management of feelings [1].

Teacher training of the New Ukrainian School must meet the requirements of the times and be based on the principles of the competency approach emphasized in the paradigm of modern education [2]. The professional standard for the professions «Teacher of primary classes of a general secondary education institution», «Teacher of a general secondary education institution», «Primary school teacher (with a diploma of a junior specialist)» embodies a modern approach to the definition and description of professional competencies of a teacher, in the list of which one subject-methodical competence is among the most important. Subject-methodical competence involves: «The ability to model the content of learning in accordance with the mandatory learning outcomes of students, in particular, to apply modern methods and technologies for modeling the content of learning students' subjects (integrated courses)». Also, «The ability to form and develop in students key competencies and skills common to all competencies; the ability to select and use modern and effective methods and technologies of teaching, education and development of students, in particular, to apply innovative learning technologies; the ability to develop students' critical thinking; the ability to evaluate and monitor the results of student learning based on the competence approach» [3].

The formation of subject-methodical competence in future primary school teachers is not possible without the formation of the ability to navigate in the information space, use existing and, if necessary, create new electronic resources; use modern digital technologies in the educational process. Modern youth prefer using of electronic textbooks, and interactive information technologies that combine text, sound, images, video, graphics, and augmented reality. The harmonious combination of traditional teaching tools with the use of digital technologies in the form of presentations, videos, mind maps, interactive exercises, etc. significantly increases the effectiveness of forming students' readiness for future pedagogical activities in new life conditions.

Therefore, the problem of activation of digital technologies in industry methods and their integration in the educational process of institutions of higher pedagogical education is relevant in modern pedagogical science.

2 Literature Review

Research by R. Gurevich, L. Konoshevskyi, and N. Opushko [4] is devoted to the problems of digitalization of modern education and its impact on the educational environment, which contributes to the improvement of the quality of training of future teachers and the convergence of education with science. The study of T. Vasilieva, Yu. Petrushenko,

O. Kryklii [5] is devoted to theoretical, methodological and applied approaches to the organization of distance learning in higher educational institutions. The authors in the collective monograph also present the developed practices of using digital technologies and the results of innovative educational activities.

Also, the work of Yu. Zhuk, O. Sokolyuk, N. Dementievska, O. Pinchuk [6] is dedicated to the organization of educational activities in a computer-oriented educational environment. In the research of L. Sushchenko, O. Andryushchenko, P. Sushchenko [7], the theoretical foundations of the process of digital transformation of higher education institutions in the conditions of digitalization of society were revealed, the organizational and pedagogical conditions for the formation of digital competence of future teachers were defined and scientifically substantiated.

The study is about the reorientation of the modern teacher to his deep awareness of competitively oriented requirements for his professional activity: readiness for the maximum use of digital tools that increase the efficiency of the educational process; introduction of distance educational innovations based on new possibilities of digital technologies; learning new teaching methods; creation of digital space – an environment with a powerful potential for ensuring the educational activity of an individual. The authors claim that information technologies are aimed at developing the competencies of future teachers, giving them competitive advantages: dynamism of cognitive activity; motivation (encouragement of students of higher pedagogical education to independently learn new things); the availability of information that simplifies the learning process; interdisciplinary content.

Of particular interest in our research are studies related to ways of introducing effective visualization in the creation and use of immersive storytelling [8–10]. A number of works are devoted to the analysis of gamification of artificial intelligence activities to improve cognitive skills among primary school students [11]. Separate studios [12, 13] consider the use of Pixton Bitstrips, Graphix Comic Builder, Comic Life, Cartoon Maker software for creating thematic electronic comics and consider tools for analyzing their use in the educational process.

The problems of using web technologies in the process of psychological and pedagogical training of future primary school teachers were considered by Yu. Kulimova [14], L. Nezhyva [15], Ogier, K.Ghosh [16], J.Catalá, J. Scorer [17], the peculiarities of using comics in the pedagogical practice of the New Ukrainian School were determined by N. Rudenko [18]. The ideas of the analyzed studies indicate that digital technologies influence the improvement of teaching and assessment methods, and their use positively changes the process of acquiring knowledge, skills, and abilities, and contributes to the development of the competencies of education seekers. According to scientists, digital transformations are based on such trends as efficiency, competitiveness, and the creation of new values.

3   The Aim of the Research

The development of digital technologies affects the reformatting of all areas of the educational system. In the process of reforming Ukrainian education, digital trends are gaining credibility. The latest information technologies are being actively implemented in educational institutions, in particular, in the educational process of training future primary school teachers. At the Faculty of Pedagogical Education at the Borys Grinchenko Kyiv University, the educational process is carried out taking into account modern digital trends. The content of the Educational and Professional Program 013.00.01 Primary Education at the first (bachelor) level of higher education, work programs of educational disciplines in the specialty 013 Primary Education, and the development of educational and methodological support provide for the active use of modern digital technologies. The information technology content of the disciplines is constantly being improved.

For example, in the content of the educational discipline «Native language education», which consists of two blocks «Ukrainian language with teaching methods» and «Children's literature with teaching methods», the use of the following digital tools are provided: a virtual online board Padlet, on which methodological problems are discussed; mind maps Bubbl.us for creating interactive lesson plans; various infographic platforms for making didactic material; LearningApps for developing interactive exercises and Kahoot! For the development of online tests, etc.

The purpose of this article is to substantiate the prospects of using digital technologies in the process of forming the subject-methodical competence of future primary school teachers.

4   Research Methodology

To implement the study, the following methods were used: theoretical (analysis and synthesis of pedagogical and methodological sources, handbooks, programs, systematization and generalization of theoretical material; study of the experience with the research problems; clarification of the basic knowledge of the studied problem), empirical (pedagogical observation, interviews with students and questionnaires, formulating conclusions).

The study was conducted based on Borys Grinchenko Kyiv University, during October – to December 2022. The study was attended by 64 students of the specialty «Primary Education» of 2–3 years of study. The research was performed within the framework of a complex scientific topic of the Faculty of Pedagogical Education of Borys Grinchenko Kyiv University «System of training of primary school teachers for professional activity in the context of the reform «New Ukrainian School»» state registration No. 0121U113726 and the framework of a complex scientific topic the Department of Mathematics and Physics of Borys Grinchenko Kyiv University "Mathematical methods and digital technologies in education, science, technology", DR No 0121U111924.

5   Discussion and Results

Subject-methodical competence is mostly formed in classes on multi-disciplinary professional methods, which provide for the formation of special (professional) competence (SK-2 under the Educational and Professional Program 013.00.01 Primary Education at the first (bachelor) level of higher education). The ability of the future teacher to model the content of learning in accordance with the mandatory learning outcomes of students, to apply modern methods and technologies for modeling the content of learning students of subjects (integrated courses) requires integration with another important professional competence, namely: information and digital. This competence, according to the Professional Standard, involves the formation of «the ability to navigate in the information space, to search and critically evaluate information, to operate with it in professional activities; the ability to effectively use existing and create (as needed) new electronic (digital) educational resources; to use digital technologies in the educational process» [3].

A modern graduate who graduated from an institution of higher pedagogical education becomes competitive in the labor market only under the condition of education, mobility, mastery of new information tools, and readiness to act in a programmed environment. Undoubtedly, it is necessary to provide methodological tools and at the same time computer literacy, media literacy, and media culture of future primary school teachers.

Undoubtedly, it is necessary to provide methodological tools and at the same time computer literacy, media literacy, and media culture of future primary school teachers. Under this condition, the digital component of subject-methodical competence contributes to the formation and development of future teachers' knowledge about the possibilities and advantages of digital technologies in primary education. The ability to work with digital devices and educational resources, the skills of digital communication and interaction, the ability to present educational material in an interesting way, to create educational and didactic tools in a digital educational environment, to use innovative technologies to evaluate the results of student learning. Thanks to this knowledge, skills and abilities, future primary school teachers have the opportunity to organize the educational activities of primary school children with the greatest efficiency. The skillful use of digital technologies diversifies the possibilities of modeling lessons and conducting them, creating digital content, educational and methodological support, and developing formative assessments.

For the formation of subject-methodical competence, the quality of educational content of a methodological nature, created using digital technologies, is essential. This is reproduced in Table 1.

On the basis of the Faculty of Pedagogical Education of Borys Grinchenko Kyiv University, we conducted a study of the content of regulatory documents (professional standard; work programs of educational disciplines integrated with professional methods; programs of pedagogical practices) on the subject of digital literacy development and the impoving of skills in the use of e-resources during the practice. We also conducted a survey of intern students regarding the frequency and ease of using digital platforms for solving educational tasks at the New Ukrainian School. Investigating the possibilities of using the digital resources described above by future primary school teachers, a questionnaire was conducted among students of the specialty 013 Primary education of the first (bachelor) educational level of the Faculty of Pedagogical Education of Borys Grinchenko Kyiv University. The survey was conducted in order to find out the priority of using digital technologies during the performance of practical tasks on professional methods for the preparation and use of educational and methodological support in the educational process in the New Ukrainian School (primary). 64 2nd and 3rd year students participated in the survey (see Fig. 1).

Table 1. Digital educational technologies and their application in professional teaching methods in primary school

The purpose of digital technology	E-resources	Application in professional methods
Mind maps (Mind Map, Mind Mapping)	MindMeister FreeMind Bubbl.us	Structuring of interactive lessons; creation of educational schemes to activate the creative thinking of primary school children
Presentations	PowerPoint Google презентації Prezi	Demonstration of educational materials
Comics	Comic Master Pixton Storyboardthat MakeBeliefsComix Write Comics Witty Comics	Visualization of educational topics, concepts, rules, experiments, plots of works, actions of heroes, communication situations, etc. in frames-pictures
Word clouds (visualization of thematic content in keywords)	WordArt	Word clouds: according to the topic of the lesson; on the basis of the artistic work; according to the characteristics of the image; according to the topic of written student work; to the concept and so on
Infographics (visual elements and minimal text, providing an easy understanding overview of the topic; structuring complex, large-scale information)	Canva Piktochart ThingLink	Didactic media products for primary school in the form of schemes
Virtual boards (coordination of work in groups during the lesson; organization of students' communication)	Padlet Trello Whiteboard	Interactive lesson plan; the plan of the web quest; discussing the school project
Interactive tasks and online tests (creation of interactive tasks, exercises, online tests)	LearningApps Kahoot!	Interactive training exercises; online tests; crossword puzzles, quizzes

In the process of carrying out practical tasks on professional methods, students learn to select digital resources: for presentation and visualization of educational material; for the organization of students' performance of various exercises and game activities during the lessons; to provide formative and summative assessment. Also, students of higher pedagogical education learn to evaluate the functionality and effectiveness of digital resources for the implementation tasks of practical teaching methods in primary school. In accordance with the purpose of digital technologies and their application in professional teaching methods, the most effective sets of programs and applications were determined for students to create educational and methodological support for the educational process in primary school. Mind maps (creation of mind maps of interactive lessons), presentations (demonstration of educational materials), and comics (visualization of educational information in frames-pictures). We also used word clouds (visualization of educational content in keywords), infographics (visualization and structuring of information), virtual boards (coordination of work in groups in class, on web quests, organization students communication), interactive tasks and online tests (creation of exercises for acquiring skills, tasks and tests development to identify knowledge and skills).

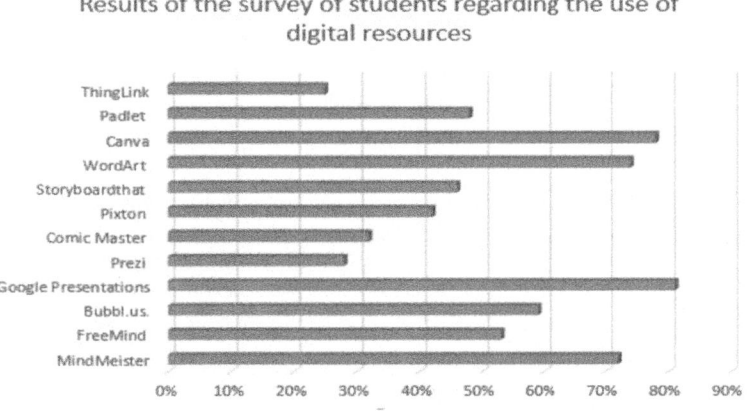

Fig. 1. Results of the survey of students regarding the use of digital resources in the professional methods classes.

Judging by the analysis of students' answers, students of higher pedagogical education are aware that using of mental maps of interactive lessons in primary school improves the perception of educational material. The most popular platforms for creating mental maps among surveyed students are MindMeister (72%), FreeMind (53%), Bubbl.us. (59%). Among the advantages of MindMeister for the organization of educational activities of junior high school students, the respondents mentioned: the ability to organize teamwork, organize online brainstorming, plan projects, create strategies, as well as presentations. With the help of a mental map, it is convenient for the teacher to reproduce the content of the lesson, display it graphically and present it in a clearly structured form. Intelligence maps of interactive lessons contain the subject of the lesson, from which the stages of the lesson are displayed by triggers. At each stage, textbook pages with the necessary material, tasks, exercises, calls to audio fragments, videos from various information resources can be placed. Mind maps of interactive lessons during distance learning has gained particular popularity, as their use facilitates the perception of information thanks to visualization, allows you to quickly process a large amount of material, improves the understanding of concepts, processes thanks to the activation of associative thinking, etc. Also, in the survey, students noted that during pedagogical practice, mind maps were used as a way of organizing the creative thinking of primary school children with the help of schemes. In particular, the mental map of a simple plot problem contributes learning the structure of the problem (condition, question), facilitates students' understanding of numerical data, the strategy of solving and finding the answer. The mind map of the story helps to remember the plot; it can display the plan of the work, main events, characteristics of the characters, etc. According to this map, primary school children successfully retell the work, analyze the images, determine the main idea of the work, etc.

To create presentations, students use traditional methods Power Point from Microsoft. However, it should be noted that Google Presentations (81%) and partially Prezi (28%) are gaining more and more popularity among the respondents. In the student practice of preparing future primary school teachers for professional activities, comics are gaining relevance. Students use e-resources to create comics (Comic Master (32%), Pixton (42%), Storyboardthat (46%)). During the classes on professional methods, students are offered to create comics in order to present any topic in the form of an interesting story in pictures. Students learn to use comics in a variety of subjects. Thanks to them, you can visualize: basic concepts, rules; show the course of experiments and experiments, depicting their algorithm on several frames; the plot of the work of art. Moreover, it is possible to create scenes with the participation of the main characters in order to observe their characters and actions; dialogues between students and outstanding scientists, writers, travelers, artists; life situations of communication; speech constructions, etc.

Fig. 2. Infographics created by students in the program Canva.

The WordArt program becomes popular among students (74%). Future primary school teachers skillfully create and use word clouds in professional techniques as a visualization of thematic content in key words. A separate study by the authors of the article [15] is devoted to this problem. According to the results of the survey, it was found that respondents mainly use the Canva platform (78%) to create infographics addressed to primary school children. It is a graphic design platform that allows users to create graphics, presentations, posters and other visual content for social networks. The service offers a library of templates, a large bank of images, fonts and illustrations. Among Canva's advantages, students noted ease of use; the possibility of saving your works in jpg and pdf formats; convenient interface, attractive design. Students consider the Canva platform to be the most relevant for visualizing and structuring complex, extensive educational information in primary school practice. Respondents name various topics of creating infographics and their purpose. This is a visualization, a graphic representation of interesting facts, a comparison (of subjects, objects, facts), an explanation of a concept (of educational material), a scheme of work, a structure of educational material. Infographics can be designed to help primary school children understand educational material, remember information, and complete educational tasks. In classes on professional methods, students check the created infographics in an academic group. During the presentation, group mates find out whether the presented information is clear, what questions arise when viewing it, whether additions or explanations of the existing content are needed. Didactic infographics are being improved according to the comments of group mates and the lecturer. Examples of student infographics created for primary school children of the New Ukrainian School are presented (see Fig. 2).

Answering about using platforms to create virtual whiteboards, students preferred Padlet (48%) due to its wide range of features. Students of higher pedagogical education are aware that it is appropriate to use a virtual interactive whiteboard during distance learning. This network resource as a learning tool is used to coordinate group work with various content with the possibility of joint editing, communication in the lesson in real time. Thanks to the interactive online board, it is possible to combine text, images, video, audio in an interactive format. Students noted the advantages of the Padlet interactive board in the work of a primary school teacher. This is convenient for the location of educational information, which the teacher can prepare in advance according to the topic of the lesson. Thanks to this property of the virtual board, future teachers can organize the educational process using interactive methods. When students solve educational tasks, it is enough to activate the corresponding block and the algorithm of actions will appear on the Padlet wall. Also, it is convenient to place completed tasks of students on the virtual board for joint discussion (see Fig. 3).

The described digital technologies are used less for the organization of quests. Padlet, ThingLink (25%) are used to develop and conduct web quests. Creating a web quest is a rather complex and difficult task, as it requires the use of several digital resources at once: a virtual whiteboard, infographics, programs for creating tasks, etc. But creating and conducting an educational web quest is an interesting and creative process that covers all the subjects of the educational process and provides casual anchoring during gaming activities. Webquest is a powerful educational technology. This is a mental and dynamic game, which consists in teams of participants going through certain stages of the «route» (developed with the help of mental maps or virtual boards) with the performance of special tasks (using QR codes to perform them). The tasks at each stage correspond to a specific learning goal. Texts, images, photos, video files and calls to e-resources with tasks are placed on the interactive whiteboard. It is possible to move elements, adjust the background image and organize joint activities with other users. A ready-to-use board can be published on social networks, embedded in a website, exported in various formats.

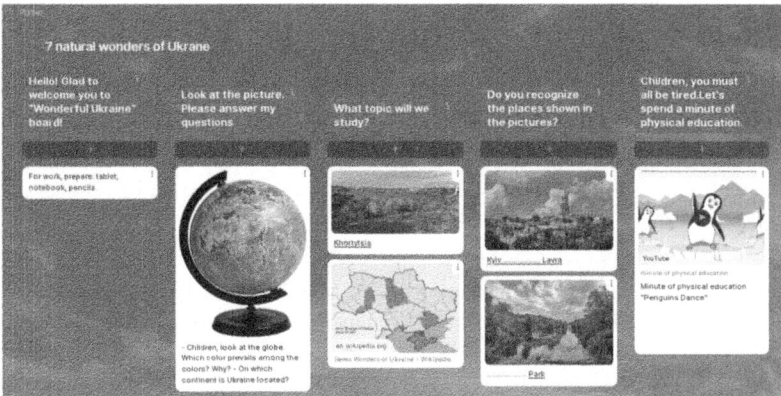

Fig. 3. An example of using a virtual interactive whiteboard in a primary school.

To the question of which of the programs is optimal for creating interactive tasks and online tests for students, students chose the following applications: LearningApps – 89%. Kahoot! – 71%, Liveworksheets – 43%. Students named the convenient free LearningApps platform as the undisputed leader in use for creating a database of interactive tasks for primary school children. Among the advantages of using this service for professional activities, students named: a variety of templates for creating tasks, including multiple choice, filling in the blanks, assigning a pair, determining the sequence, filling in the form, quiz, crossword, filling in the table, etc. They also mentioned the ability to add text, images, sound, and video to the author's exercises; the ability to get a web link for placing interactive exercises on an educational platform, website, or teachers' blog; availability of ready-made exercises that you can apply in your activity.

6 Conclusions and Prospects for Further Research

In the course of the conducted research, it was found that the work programs of educational disciplines integrated with the methodology contain tasks and recommendations for working with digital platforms. However, teachers should pay more attention to the development of technological skills to use digital resources for the educational and methodological support of the education of primary school children.

The results of the study confirm students' understanding according to the importance of using digital tools for the implementation of practical tasks in professional methods. Graduates of higher pedagogical education showed readiness to choose and use convenient free digital resources, taking into account their multi-functionality.

Based on the results of the survey, the most popular digital technologies among future primary school teachers were determined, which are effective for the implementation of methodical tasks: creation of intelligence maps of interactive lessons – MindMeister; presentations of educational material – Google Presentation, Power Point. Visualization of the educational topic, concepts, rules, experiments, plots of works, actions of heroes, communication situations, etc. in frames-pictures of comics – Comic Master,

Storyboardthat; visualization of thematic content using a word cloud – WordArt. Creating didactic tools for teaching primary school children using infographics – Canva; organization of educational activities, communication between students, preparation of web quests using a virtual board – Padlet, preparation of interactive tasks – Learning Apps, online tests - Kahoot!.

References

1. SocioEco.org, Council Recommendation of 22 May 2018 on key competences for lifelong learning. https://base.socioeco.org/docs/council_recommendation.pdf. Accessed 22 Mar 2023
2. Mon.Gov.Ua, New Ukrainian School: Conceptual principles of reforming the secondary school. https://mon.gov.ua/storage/app/media/zagalna%20serednya/Book-ENG.pdf. Accessed 22 Mar 2023
3. MON.Gov.Ua, The standard of higher education of Ukraine in the specialty 013 "Primary education" in the field of knowledge 12 «Education/Pedagogics» for the first (bachelor's) level of higher education. https://mon.gov.ua/storage/app/media/vishcha-osvita/zatverdzeni%20s tandarty/2021/07/28/013-Pochatk.osvita-bakalavr.28.07.pdf. Accessed 22 Mar 2023
4. Gurevich, R., Konoshevsky, L., Opushko, N.: Digitalization of education in modern society: problems, experience, prospects. Educ. Discour. **38–39**, 22–46 (2022)
5. Petrushenko, Yu., Kryklii, O.: Digital technologies in education: modern experience, problems and prospects. Sumy State University, Sumy (2022)
6. Zhuk, Yu., Sokolyuk, O., Dementievska, N., Pinchuk, O.: Organization of educational activities in a computer-oriented educational environment: manual. Pedagogical Opinion, Kyiv (2012)
7. Sushchenko, L., Andrushchenko, O., Sushchenko, P.: Digital transformation of higher education institutions in the conditions of digitalization of society: challenges and prospects, Scientific Bulletin of Uzhhorod University. Ser. «Pedag. Soc. Work» **51**, 157–162 (2022)
8. Pilgrim, J., Pilgrim, J.M.: Immersive storytelling: virtual reality as a cross-disciplinary digital storytelling tool. In: Connecting Disciplinary Literacy and Digital Storytelling in K-12 Education on Proceedings, pp. 192–215. IGI Global (2021)
9. VanFossen, L., Gibson-Hylands, K.: Interactive storytelling through immersive design. In: MacDowell, P., Lock, J. (eds.) Immersive Education, pp. 221–247. Springer, Cham (2022). https://doi.org/10.1007/978-3-031-18138-2_14
10. Lytvynova, S., Semerikov, S., Striuk, A., Striuk, M., Kolgatina, L.: AREdu 2021-Immersive technology today. In: S. Lytvynova, S. Semerikov (eds.) Proceedings of the 4th International Workshop on Augmented Reality in Education, vol. 2898, pp. 1–40. CEUR Workshop Proceedings (2021)
11. Utem.Edu.My, Kamarudin, N., Ikram, R., Azman, F., Salahuddin, L.: A design of gamification artificial intelligence coding activities to improve cognitive skills among primary students. https://www3.utem.edu.my/care/proceedings/merd22/pdf/06%20Enginee ring%20Education/091_p191_192.pdf. Accessed 25 Feb 2023
12. Zaibon, N., Shiratuddin, N.: Pedagogical analysis of comic authoring systems for educational digital storytelling. J. Theor. Appl. Inf. Technol. **89**, 461–469 (2016)
13. Prihandini, S.: Development of thematic e-comic based on augmented reality. J. Digit. Educ. Technol. **2**, 1–5 (2022)
14. Kulimova, Y.: Using the web-technologies in the process of psychological and pedagogical training of Future Primary School Teachers. Open Educ. e-Environ. Mod. Univ. **8**, 34–41 (2020). https://doi.org/10.28925/2414-0325.2020.8.5

15. Nezhyva, L., Palamar, S., Marienko, M.: Clouds of words as a didactic tool in literary education of primary school children. In: Proceedings of the Cloud Technologies in Education, vol. 3085 pp. 381–393. CEUR Workshop Proceedings (2022)
16. Ogier, S., Ghosh, K.: Exploring student teachers' capacity for creativity through the inter-disciplinary use of comics in the primary classroom. J. Graph. Novels Comics **9**, 293–309 (2017). https://doi.org/10.1080/21504857.2017.1319871
17. Catalá, J., Scorer, J.: Comics, Cartoons, Graphic Novels. Latin American Studies. Oxford Bibliographies, Oxford (2023). https://doi.org/10.1093/obo/9780199766581-0279
18. Rudenko, N., Shirokov, D.: Application of e-resources in the process of creating comics in mathematics lessons in a primary school. Innov. Pedag. **41**, 138–143 (2021)

Approaches and Economic Benefits of Property Registers Digitalization: Evidence from Ukraine

Yevhenii B. Shapovalov[1,2]([✉]) [ID], Oleksander P. Zakusilo[1] [ID],
Viktor B. Shapovalov[2] [ID], Oleg I. Burba[3], Oleg L. Pilat[4],
and Andrii G. Martyn[4,5] [ID]

[1] Ministry of Digital Transformation of Ukraine, Kyiv 03150, Ukraine
sjb@man.gov.ua
[2] National Center "Junior Academy of Science of Ukraine", Kyiv 04119, Ukraine
[3] e-Governance Academy Foundation, Rotermanni 8, 10111 Tallinn, Estonia
[4] Individual enterpriser, Kyiv, Ukraine
[5] National University of Life and Environmental Sciences of Ukraine, Kyiv 03041,
Ukraine

Abstract. Modernization of the system of the property registers and
cadastres is an essential component for the development of society,
automation of services providing, optimization of register's operating
costs, efficient operation of subsidy and tax systems, and reduction of
property fraud. An analysis of the current state of the property public
registers system has been given. It has been established that considering
Ukraine's historical development, it is expedient to modernize the existing information systems and ensure information interaction between the
existing registers and cadastres, but not develop the new system (property cadastre). The interoperability of the data in property information
systems has been described. The functions of the main stakeholders on
property registration (land surveyors, a state employee of the state architectural service, local government architects and private engineers, and
state registrars of property rights) have been described in the form of a
case UML diagram.

Keywords: public register · property · interoperability · data
exchange · property cadastre

1 Introduction

Quality data, public register completeness, and data interoperability are vital
for a democratic, citizen-friendly, developed country. Efficient public registers
simplify service provision.

Different countries have varying register conditions due to history and poli-
tics. Ukraine is favorably positioned for register development and public service
simplification. The "Diia" web portal and mobile app utilize public registers,
requiring high-quality data.

© The Author(s), under exclusive license to Springer Nature Switzerland AG 2023
G. Antoniou et al. (Eds.): ICTERI 2023, CCIS 1980, pp. 348–359, 2023.
https://doi.org/10.1007/978-3-031-48325-7_27

Operating public registers needs funding. Ukraine has 135+ state registers, each needing 620,000 EUR/year [1], possibly exceeding 350. Thus, total digital register costs range from 83.7 to 150 million EUR/year.

Property-related government services rely on various state registers and cadastres. Data quality needs improvement due to duplication or missing data. Manual data input delays services and introduces errors. Inter-register data exchange is limited in construction and property domains.

Hence, analyzing Ukraine's register system development is pertinent. The article assesses property-related public register status, proposes development avenues, and estimates potential economic impacts.

2   Methods of the Research

The article examines the current state of the Ukrainian public property system and explores different development approaches. It also considers global experiences and analyzes the advantages and disadvantages of existing approaches, taking into account the unique characteristics of Ukrainian registers. Additionally, it discusses the risks and economic and social effects within the framework of sustainable development.

Entities related to the domain of property are identified and described to analyze the interaction and assess the feasibility of the localization of entities in information systems. To substantiate the feasibility of developing a system of property registers, we analyzed the feasibility of implementing measures from both economic and organizational legal points of view.

The exchange schemes of entities in the registers and the UML use case diagram for the primary users of the proposed way of property registers system development have been developed and shown. The economic effect was evaluated using data from the government budget of Ukraine [2].

3   Main Definitions and Entities of Property Registers

Entities on Property. The effective functioning of property registers systems requires the exchange of data on entities such as owners (legal entities and individuals), property, and property rights between registers or services. These entities include land plots, buildings, apartments, and various property rights [3,4].

Registers and Cadastres. Registers and cadastres are the main types of information and telecommunication systems related to properties. A public digital register is an information and communication system owned by the state, municipality, or self-regulatory organization. It is responsible for accumulating, processing, protecting, accounting, and registering property-related information. On the other hand, cadastres are more specific to property systems. The term "cadastre" has evolved over time and does not have a single agreed definition. It can refer to a register with a cartographic basis or a system for collecting taxes or integrating property data with the property rights of owners [3,4].

In Ukraine, there are various cadastres, including cadastres of natural resources (land, water, forest, plant, animal, etc.), urban planning cadastre, and environmental cadastres (waste and hazardous waste cadastres, avalanche cadastre, cadastre of anthropogenic emissions, etc.) [3,4].

Data Interoperability in the Public Sector. One of the challenges in the public sector is the duplication of data in different information systems. Different government systems may contain data on the same entities but in different formats and versions. To address this issue, it is recommended to use a unique identifier for entities related to property, such as buildings, land plots, and property rights. This unique identifier can be used to request data from different state information systems. For example, the State Register of Real Property in Ukraine contains master data on property rights, while other registers may contain data on buildings or owners. The use of a unique identifier and data exchange through APIs or systems like "Trembita" can facilitate access to data from different registers [5,6].

Trembita and Data Exchange. "Trembita" is a digital exchange system for public registers in Ukraine. It is based on the service-oriented architecture of the X-Road software and hardware solution. X-Road has been successfully implemented in Estonia, Finland, and Azerbaijan and aligns with the conceptual model of European public services. "Trembita" aims to establish easy access to information from public registers while ensuring data security and minimal technical changes to existing systems. It provides a distributed and secure platform for the exchange of digital data within the information space of public authorities. The infrastructure solutions of "Trembita" enable fast and efficient implementation, scalability, and interoperability through web services with a single interaction protocol. It also incorporates Ukrainian cybersecurity requirements, ensuring the protection of data during electronic message exchange through encryption and TLS tunnels [6–12].

In conclusion, the exchange of data on entities related to property is crucial for the effective functioning of property registers systems. Registers and cadastres play a vital role in managing property-related information. To ensure data interoperability in the public sector, the use of unique identifiers and systems like "Trembita" can facilitate data exchange between different registers and improve access to information.

4   Primary Information and Telecommunication Systems in Ukraine

In Ukraine today, there are the following cadastral systems (cartographic): urban planning cadastre, the single digital public system in the field of construction (SDPSiFC; as part of the urban planning cadastre), land cadastre, and national geospatial data infrastructure [4]. Also, there are property registers in Ukraine, such as the State Register of Real Property Rights and the State Register of Geographical Names. A list and description of property information and telecommunication systems are presented in Table 1.

Table 1. List and description of property information and telecommunication systems.

	National geospatial data infrastructure	Urban planning cadastre	SDPSiFC	Land cadastre	State Register of Real Property Rights	State Register of Geographical Names
Holder / administrator	State Geocadastre* / Institute of Cartography	Ministry of Regional Development	Ministry of Regional Development / State Enterprise "Diia"	State Geocadastre* / State Enterprise "Land cadastre Center"	Ministry of Justice / State Enterprise NAIS	State Geocadastre*
Contains master data on an entity	–	Vast data range, including environment and society.	Buildings (built and under construction)	Land plots	Property rights, building	Names of geographical objects, administrative-territorial structure
Receives and displays data about	All types of property and geodata, etc.	–	Land plots and register rights	–	–	–
Cartographic basis	Various satellite images, including an open street map	General plan and cities orthophotos	Map of Land cadastre and open street map	Orthophoto (made in 2009–2012)	Not carto-graphically oriented	Contains just coordinates
Compliance with international geographical standards	Yes, including INSPIRE	Not corre-sponding to international standards	Yes, including INSPIRE	Not corresponding to international standards	Not comparable with geodata standards	Not comparable with geodata standards

Analyzing the existing systems (Table 1) for interaction is advisable. The most focused on data exchange is the National geospatial data infrastructure. It is supposed that different geodata will be displayed on a single cartographic basis. To do this, it is necessary to set up an API with various state registers and cadastres. However, using "Trembita" in this case may not be relevant because some exchanges contain available information, and providing a highly secured exchange is unnecessary. However, as of today, such interactions are mostly absent.

The urban planning cadastre exists only in the legislative field but is still not implemented. SDPSiFC is one of the newest and most advanced systems, and it has widely implemented "Trembita" with another public informational system. State Register of Real Property Rights is also entirely actively implementing "Trembita" and other types of data exchange. In particular, the State Register of Real Property Rights and SDPSiFC are exchanging data using API, but it is supposed to provide "Trembita"-based exchange in short terms. The State Register of Real Property planned to introduce "Trembita" data exchange services even more widely [13].

The state land cadastre is a rather conservative system, and using the "Trembita"-based exchange system with other registers is not used enough. However, it is highly essential because many services on land plots based on usage of state land cadastre services are often used by citizens (for example, land plots registration or declaration rights on the land plots). Also, the interaction of geosystems in Ukraine is complicated because different information systems can use different coordinate systems. This problem is especially acute for local government geosystems, where even their coordinate systems can be used.

5   Analysis of Approaches to Public Property Registers Development

Today, the issue of the development of the system of property registers and cadastres is essential and has been previously considered [14]. Historically, different property registry systems in different countries are provided. Some countries have started right away building their registers in digital. On the other hand, some countries have a long history of using paper-based registers that have evolved with changes in society, such as in the case of Ukraine. Previous proposals have categorized the development of property systems into four main groups based on the characteristics of cadastral systems. [15]:

1. Napoleonic systems;
2. German systems;
3. Scandinavian systems;
4. English-speaking systems.

The main difference in these groups is the purpose (focus). The focus may be on the property itself or the right of ownership [15]. Also, the registers in different groups may be designed as separate systems, single systems, or single systems based on a microservice architecture (in the latter case, different registers located in the single informational system but using client-server architecture interact with each other).

Each system has its advantages and disadvantages if we consider them both from a registration point of view and a technological point of view. Therefore, it is advisable to analyze such systems in terms of feasibility.

The most promising approaches for Ukraine are the Napoleonic and German models, which have separate registers or cadastres that can exchange data. Alternatively, the Scandinavian approach, which utilizes a single property cadastre system, is also considered favorable. It is worth noting that the Scandinavian approach is similar to the integrated property cadastre systems, as in the United States (Houston Cadastre) and Belarus, where they are pretty efficient. Therefore, it is appropriate to consider the main advantages, disadvantages, and analysis of creating a new system (cadastre of property) and providing interoperability of a few registers and cadastres, which is presented in Table 2.

As can be seen from Table 2, both approaches involve the implementation of the same functionality. Both approaches will be based on a microservices architecture or may be characterized by a certain number of failures in the case of a new single system. Of course, in the case of the development of a new system (cadastre), implementing modern BIM and GIS technologies [16, 17]

Table 2. Main advantages, disadvantages, and analysis of the creation of a new system (cadastre of property) and providing of interoperability of few registers and cadastres

	Creating a new system, "Property Cadastre"	Use of a few existing systems and implementation of the "Trembita" exchange system
The main advantages	Data centralized, no need for state information resource exchange.	Fast implementation, no additional costs for new system creation, no ownership issues.
System functions	Accounting for buildings, land, property rights, and their relationship with owner identifiers; data used for subsidy and tax calculations.	
Holder and administrator issue	Need to be defined	All holders are defined
Costs for software upgrades and debugging processes	High. It is necessary to develop actually from scratch	Existing systems and processes in place; need development and digital information exchange
Architecture	May be slow and prone to failures without microservice architecture; separate services for buildings, land, ownership, and taxes.	Separate services that interact on a client-server architecture
Dependence on the cartographic basis	Dependent on cartographic basis; need to develop basis before detailed implementation	Slight. It is possible to start implementation without a cartographic basis
Development time	Approximately three years; potential delays due to map update and implementation process	It can be implemented throughout the year. The realistic implementation period is two years

will be possible. However, providing such technology may cause a delay in the implementation of a system of a property registration system for several years. The organizational issues (for example, data relocation and reengineering) may delay the process even further. Also, in the case of providing modern technologies without relevant data, it may not afford the necessary effectiveness. The creation of a new system will depend entirely on the availability of a cartographic basis (satellite map updating) because it will be needed to obtain buildings' footprints using neural networks based on satellite maps [18–20].

So, it seems more important to ensure the completeness of property data in existing registers, and then it will be possible to apply modern technologies. In this case, if there are updated maps and building footprints, it will be possible to merge data from the registers with the coordinates of the footprints.

The proposed property register system foresees data exchanging on all necessary entities using "Trembita": data on the owner-citizens (Demography register; Ministry of Interior; unique ID - unique registry entry number; in Ukrainian - unikalnyi nomer zapysu reiestru (UNZR)) and legal entities (Legal entities register, Ministry of Justice; unique ID - Unified State Register of Enterprises and Organizations of Ukraine; in Ukrainian; Edynyi derzhavnyi reiestr pidpryiemstv ta orhanizatsii Ukrainy (EDRPOU)); property rights (State Register of Real Property Rights Ministry of justice; unique ID - property right number); and property itself - land plots (Land cadastre, State Service of cartography and geodesy; unique ID - cadastre number) and building (SDPSiFC, Ministry of regional development; unique ID - building and buildings part identificators). Entities exchange in the proposed property system is presented in Fig. 1.

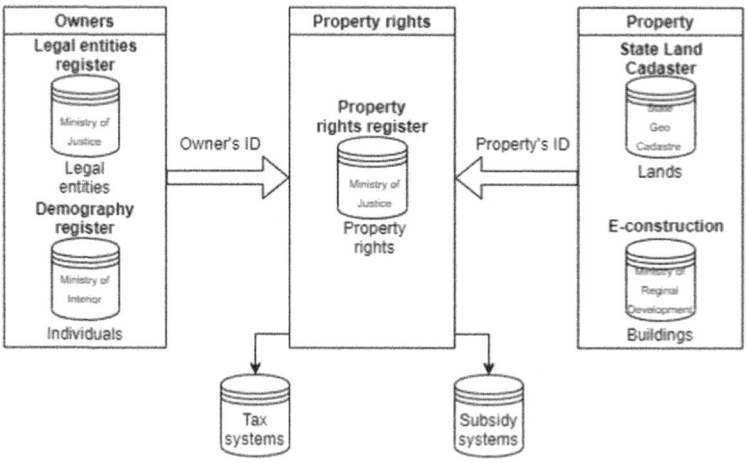

Fig. 1. Entities exchange in the proposed property system

6 The Functionality of the Proposed Public Property System

The main advantage of the proposed approach is that the main actors and actions are the same as before, but some processes are simplified and automatized. Land surveyors register data on land plots. The function of land managers and their functions in the system are new land plots registration (creation of a record), changing the purpose of land plots, and determination of boundary land plots. After providing data interaction, it will be possible to automatically generate a request to register rights in the State Register of Real Property Rights in case of relevant data in the State Land Cadastre.

E-constructii is filled by various construction systems, including a state employee of the state architectural service, local government architects, and private engineers. The employee of the state architectural service has the function of certification of buildings by the state standards (as a result, it will be possible to submit applications for property rights registration automatically) and uploading engineering documents. The architect of local governments also has the function of uploading engineering documents and urban construction restrictions (for new constructions). In addition, E-construction is used by city developers (private companies) to upload construction designs. Registration (of property rights) can declare the status of ownership that is removing, adding, and changing property rights. Use a case diagram in the registry system is shown in Fig. 2.

In addition, the proper functioning system of property registers system is one way to conduct an effective modern census. Thus, when the system on the property is filled, given the documentation on the number of building apartments, it will be pretty simple to conduct a census [21] and faster than using the demography register's data (but according [21] somewhat less effective).

7   The Economic Effect of the Register

Ukraine has lost more than 1.52 Bn EUR/year due to the lack of data on buildings and structures. About 1.37 Bn EUR/year of subsidies were paid out due to the state budget, and about 150 M EUR/year may be paid for the local budget as property taxes. Considering the gradual implementation, it is recommended to calculate the effects based on a 1% change. The impact will be most significant in the short term and gradually decrease over time. Table 3 provides calculations of housing subsidy and property tax changes with the proposed system.

Table 3. Calculations of housing subsidy and property tax changes with the proposed system

	Subsidy	Property tax
Legal basis	Resolution of the Cabinet of Ministers № 848	Tax Code
Subjects	Recipients (owners): The apartment not over 120 m2 or houses not over 200 m2	Payers (owners): The apartment over 60 m2 or houses over 120 m2
Type of budget	State budget	Local budgets
The annual amount of funds [2]*	1.37 bn. EUR/year	150 mil. EUR/year
Annual budget effect in case of change of 1%	13,7 mil. EUR/year	1,5 mil. EUR/year

* data related to 2019 [2]

Thus, it is advisable to consider the economic effect of the proposed approach based on a different implementation hypothesis. With the most optimistic calculation, the reduction in unpaid taxes can be up to 100% of the amount of 0.060 Bn EUR/year additional revenues. The decrease in subsidies paid with a

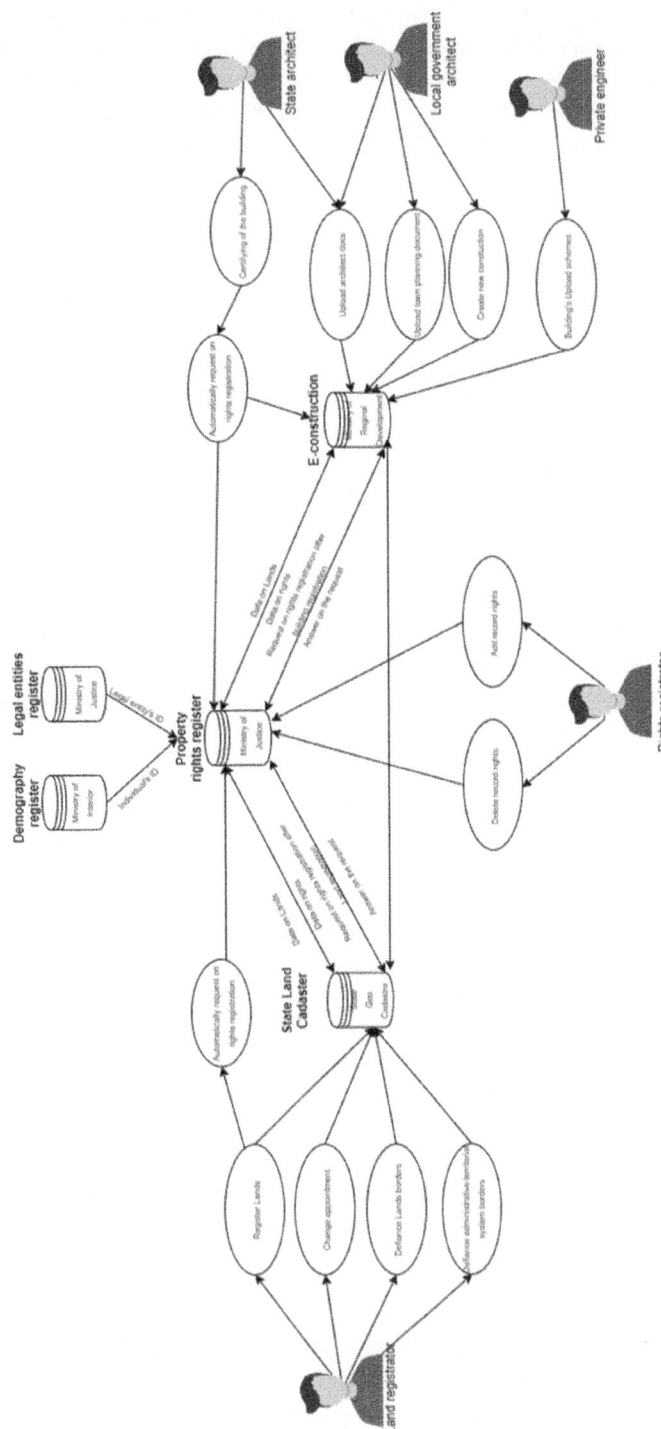

Fig. 2. Use a case diagram in the registry system

law violation, even in the optimistic version, can reach 40%, amounting to 0.548 Bn EUR/year. Therefore, the maximum economic effect of the creation of the registers of buildings and can be 0.608 Bn EUR/year Assessment of effectiveness is presented in Table 4.

Table 4. Assessment of the economic effect of the implementation of the effective functioning of the system property registers

	State budget keeping by optimizing subsidies payment		Additional property taxes to local budgets		Total effect
	% of total	bn. EUR/year	% of total	bn. EUR/year	bn. EUR/year
Minimum	10	0.137	25	0.015	0.152
Realistic	20	0.274	50	0.030	0.304
Maximum	40	0.548	100	0.060	0.608

8   Conclusions

In conclusion, modernizing public information registers is essential for county and public sector development, streamlining administrative services. Key elements in property register systems include owners (legal entities and individuals), properties (land or buildings), and ownership rights.

The "Trembita" exchange system, modeled after Estonia's X-road, is favored for data exchange between registers and cadastres over creating a new property cadastre in Ukraine.

The proposed property register mechanism involves data exchange on owners from the Demography Register of the Ministry of Interior and Legal Entities Register of the Ministry of Justice. Property rights data originates from the State Register of Property Rights under the Ministry of Justice. Property-specific information comes from the Land Cadastre of the State Service of Cartography and Geodesy and the E-construction system of the Ministry of Regional Development. This approach capitalizes on existing processes, potentially expediting implementation compared to a new property cadastre.

This approach could yield annual savings of up to 0.608 billion EUR-via subsidy payment optimization (0.548 billion EUR/year) and additional property tax revenue for local budgets (0.060 billion EUR/year).

Future steps involve a feasibility study to adapt "Trembita" to Ukraine, a cost-benefit analysis for financial validation, and researching successful register modernization cases from other countries to enhance the approach.

These steps will fortify the research, benefiting the public sector and citizens.

References

1. TAPAS. «Stan ta perspektyvy rozvytku derzhavnykh elektronnykh informatsiinykh resursiv» (2018)
2. Zakon Ukrainy. Pro Derzhavnyi biudzhet Ukrainy na 2020 rik (2020)
3. Sevlisian, H.F. Obzor kadastrov, kotorue funktsyonyruiut v Ukrayne. 3222.ua (2014)
4. BRDO. Zelena knyha. Systemnyi perehliad yakosti derzhavnoho rehuliuvannia. Formuvannia ta otrymannia zemelnoi dilianky pry novomu budivnytstvi (2018)
5. Postanova Kabinetu Ministriv Ukrainy. Deiaki pytannia orhanizatsii elektronnoi vzaiemodii derzhavnykh elektronnykh informatsiinykh resursiv (2018)
6. EGov4Ukraine. EGOV4UKRAINE - empowering administrative service delivery (2016)
7. Willemson, J.: Pseudonymization service for X-road eGovernment data exchange layer. In: Andersen, K.N., Francesconi, E., Grönlund, Å., van Engers, T.M. (eds.) EGOVIS 2011. LNCS, vol. 6866, pp. 135–145. Springer, Heidelberg (2011). https://doi.org/10.1007/978-3-642-22961-9_11
8. Vallner, U.: Secure data exchange platform. Principles and implementation. X-Road (2017)
9. Paide, K., Pappel, I., Vainsalu, H., Draheim, D.: On the systematic exploitation of the Estonian data exchange layer X-road for strengthening public-private partnerships. In: Proceedings of the 11th International Conference on Theory and Practice of Electronic Governance (ICEGOV 18), pp. 34–41 (2018). https://doi.org/10.1145/3209415.3209441
10. Secure and efficient electronic information exchange in the public sector (2019)
11. Nyadimo, E.: Republic of Kenya Report on Electronic Land Transactions (2019)
12. Saar, A., Rull, A.: Technology transfer in the EU: exporting strategically important ICT solutions to other EU member states. Baltic J. Eur. Stud. **5**(2), 5–29 (2015). https://doi.org/10.1515/bjes-2015-0011
13. Technical assessment of registers supervised by the MoJ (Administrator - NAIS) (2019)
14. Sabaliauskas, K., Grinevičiūtė, K., Žaliukas, M., Mensevičius, K.: Support to Justice Sector Reforms in Ukraine (2005)
15. Saufi, M. : Kadastrovye i registracionnye sistemy v Rossii i za rubezhom. Mezhdunarodnyj zhurnal prikladnyh nauk i tehnologij «Integral», vol. 4, pp. 120–128 (2018)
16. Shojaei, D., Olfat, H., Rajabifard, A., Briffa, M.: Design and development of a 3D digital Cadastre visualization prototype. ISPRS Int. J. Geo Inf. **7**(10), 384–397 (2018). https://doi.org/10.3390/ijgi7100384
17. Rašković, V., Muchová, Z., Petrovič, F.: A new approach to the registration of buildings towards 3D land and property management in Slovakia. Sustainability (Switzerland) **11**(17), 4652 (2019). https://doi.org/10.3390/su11174652
18. Rastogi, K., Bodani, P., Sharma, S.A.: Automatic building footprint extraction from very high-resolution imagery using deep learning techniques. Geocarto Int. **37**, 1501–1513 (2020). https://doi.org/10.1080/10106049.2020.1778100
19. Schuegraf, P., Bittner, K.: Automatic building footprint extraction from multi-resolution remote sensing images using a hybrid FCN. ISPRS Int. J. Geo Inf. **8**(4), 1–16 (2019). https://doi.org/10.3390/ijgi8040191

20. Chawda, C., Aghav, J., Udar, S.: Extracting building footprints from satellite images using convolutional neural networks. In: 2018 International Conference on Advances in Computing, Communications and Informatics, ICACCI 2018, pp. 572–577 (2018). https://doi.org/10.1109/ICACCI.2018.8554893
21. UN and Department of Economic and Social Affairs - Statistics Division. Principles and Recommendations for a Vital Statistics System - Revision 3. Statistical Papers, Series M No. 19/Rev. 3 (2014). https://doi.org/10.1515/9783110459335-016

Author Index

G. Antoniou et al. (Eds.): ICTERI 2023, CCIS 1980, pp. 361–362, 2023.
https://doi.org/10.1007/978-3-031-48325-7

GPSR Compliance

The European Union's (EU) General Product Safety Regulation (GPSR) is a set of rules that requires consumer products to be safe and our obligations to ensure this.

If you have any concerns about our products, you can contact us on ProductSafety@springernature.com

In case Publisher is established outside the EU, the EU authorized representative is:

Springer Nature Customer Service Center GmbH
Europaplatz 3
69115 Heidelberg, Germany

The manufacturer's authorised representative in the EU is Springer
Nature Customer Service Centre GmbH, Europaplatz 3, 69115 Heidelberg,
Germany. If you have any concerns regarding our products, please
contact ProductSafety@springernature.com

Printed and bound by CPI Group (UK) Ltd, Croydon, CR0 4YY
05/05/2026
02102981-0009